气候变化全球治理丛书

Series on

Global Governance for Climate Change

论全球气候治理

——构建人类发展路径创新的国际体制

Global Governance for Climate:

Shaping an International Architecture to Boost Innovation of Human Development Paths

邹　骥　傅　莎　陈　济　祁　悦
王　克　崔学勤　刘俊伶　张小丽　等著

中国计划出版社

图书在版编目（CIP）数据

论全球气候治理：构建人类发展路径创新的国际体制 / 邹骥等著．—北京：中国计划出版社，2015.11
（2016.8 重印）
（气候变化全球治理丛书）
ISBN 978-7-5182-0287-4

Ⅰ．①论… Ⅱ．①邹… Ⅲ．①气候变化—治理—国际合作—研究 Ⅳ．① P467

中国版本图书馆 CIP 数据核字 (2015) 第 259867 号

论全球气候治理
——构建人类发展路径创新的国际体制
邹　骥　傅　莎　陈　济　祁　悦
王　克　崔学勤　刘俊伶　张小丽　等著

中国计划出版社出版
网址：www.jhpress.com
地址：北京市西城区木樨地北里甲 11 号国宏大厦 C 座 3 层
邮政编码：100038　电话：(010) 6390433（发行部）
北京京华虎彩印刷有限公司印刷

787mm × 1092mm　1/16　22.5 印张　304 千字
2015 年 11 月 第 1 版　2016 年 8 月第 3 次印刷

ISBN 978-7-5182-0287-4
定价：68.00 元

目录

Contents

Contents

序言

近来，党中央就参与全球治理提出了明确的号召。2015 年 10 月召开的中国共产党第十八届五中全会提出：积极参与全球经济治理，促进国际经济秩序朝着平等公正、合作共赢的方向发展，加快实施自由贸易区战略。积极承担国际责任和义务，积极参与应对全球气候变化谈判，主动参与 2030 年可持续发展议程。

中共中央总书记习近平同志在 2015 年 10 月 12 日主持中央政治局第二十七次集体学习时，就全球治理格局和体制问题强调，我们参与全球治理的根本目的，就是服从服务于实现“两个一百年”奋斗目标、实现中华民族伟大复兴的中国梦。要审时度势，努力抓住机遇，妥善应对挑战，统筹国内国际两个大局，推动全球治理体制向着更加公正合理方向发展，为我国发展和世界和平创造更加有利的条件。

李克强总理在 2015 年 6 月 12 日国家应对气候变化和节能减排工作领导小组会议上指出：积极应对气候变化，不仅是我国保障经济、能源、生态、粮食安全以及人民生命财产安全，促进可持续发展的重要方面，也是深度参与全球治理、打造人类命运共同体、推动共同发展的责任担当。中国作为负责任的大国，将坚持共同但有区别的责任原则、公平原则和各自能力原则，承担与自身国情、发展阶段和实际能力相符的国际义务，中国将不断提高减缓和适应气候变化能力，为促进全球绿色低碳转型与发展路径创新尽自身最大努力。

我国党和国家领导人的相关论述，是基于对全球绿色低碳发展趋势的判断，明确了我国参与全球气候治理顶层设计的指导思想，是对促进全球气候治理、推动中国绿色低碳转型、创新发展理念的贡献。

我们正处在一个重要的历史时代。伴随着过去20多年全球化进程的推进，在新能源、信息、新材料等领域科学技术和产业的革命与发展，世界范围内绿色低碳发展的趋势日益明显，其间又经历了全球性金融危机之后长达7年的艰难缓慢复苏过程，世界经济发展正处在一个识别新的调整方向、寻找新的增长引擎的深刻调整时期。各国发展与世界经济和全球环境的联系日益密切，使得有关全球秩序与全球公共财富管理的全球治理问题日益凸显。与全球政治、经济、资源能源、安全等治理内容具有广泛关联的全球气候治理也一直是国际政治中元首级别的热点话题。气候变化领域在2012—2015年德班平台特设工作组的平台上，世界各国围绕着2015年巴黎气候大会的成果展开了紧张的谈判，同时也都在国家元首或政府首脑级别活跃地开展了气候外交活动，以求通过对全球气候治理的影响提高自身在整个全球治理中的话语权和影响力。积极参与全球气候治理对参与全球治理具有综合的外溢效应，意义重大。

与此同时，中国社会经济发展也进入了新常态。全国上下形成了一个广泛的共识，就是：加快转变经济发展方式、调整经济结构、深化体制改革，努力克服“中等收入陷阱”，已经成为中国实现两个“百年目标”、实现中华民族伟大复兴中国梦的必由之路。而这些具体体现为国家更加民主富强、人民生活更加富裕美好、生态环境更加健康优美。这与积极应对全球气候变化的目标高度一致。对中国而言，积极应对气候变化，不是因为别人要我们做，而是我们自己要做，中华民族的利益和全人类的利益在应对气候变化方面是高度一致的，是打造人类利益共同体的重要切入点。

自2009年哥本哈根气候大会中国政府宣布2020年前应对气候变化目标以来，中国在减缓和适应气候变化方面采取了一系列重大政策，取得了世人瞩目的成绩。2015年6月，中国政府又向联合国提交了2020年后国家自主贡献文件，展示了一个负责任发展中大国应对气候变化的坚强决心和务实态度。近几年，在多边和双边国际气候合作进程中，特别是在一系列国际气候变化谈判中，中国日益发挥着主持公道、敢于担当、积极促进合作的全球气候治理建设者的积极作用。中国正以积极自信的战略姿态投入到全球气候治理的构建之中。

当前，构建全球气候治理的一件大事，就是按照2011年《联合国气候变化框架公约》缔约方大会德班决议的要求，在德班平台上开展2015年巴黎协议的谈判，以期在《公约》下，以《公约》的原则为指导，形成一份旨在加强履行《公约》行动的国际法律协议，将巴黎气候大会开成具有里程碑意义的气候治理大会。《公约》是全球气候治理在道义、政治和法律方面的基础，是在当今世界秩序和全球治理构架下开展国际气候合作的主渠道。《公约》所确立的“共同但有区别的责任”原则、公平原则和各自能力原则，是发达国家和发展中国家参与全球气候治理应当遵循的根本原则。尽管《公约》缔结已20多年，世界战略格局发生了深刻的变化，但全球范围内发达国家与发展中国家的历史责任、发展阶段、各自能力没有发生根本改变，在发展水平上的差距甚至有扩大之势；在联合国体系内处理国际政治经济事务时依然划分为发达国家与发展中国家两大基本阵营。作为气候变化历史责任基础的累积排放格局而言，发达国家依然是主要排放贡献者。即使到2012年，就1750年工业革命以来与能源和水泥生产相关的二氧化碳等主要温室气体累积排放而言，《公约》附件一所载发达国家以不到20%的世界人口达到近70%的全球累积排放。美欧自1750年来的历史累积排放已达到中国的5倍，据有关情景研究预估，即使到2030年美欧的累积排放也将高达中国的2.6倍。这些都表明《公约》所确定的基本原则的基础仍然存在，《公约》的原则依然适用于指导巴黎协议谈判。

建设全球气候治理结构，必须站在对人类前途负责的高度，充分反映不同发展阶段的国家人民可持续发展的诉求，主持正义，坚持公平。在界定不同国家减缓与适应气候变化责任时，根据它们的历史责任、发展阶段、各自能力和国情，有所区分。发达国家要率先在全经济范围内实现充分的绝对量化减排，并向发展中国家提供充分、及时的资金、技术和能力建设支持，以帮助发展中国家实现低碳转型。发展中国家要在发达国家的资金技术支持下，采取积极应对气候变化行动并努力探索适合本国国情的低碳转型的道路。通过南北合作与南南合作最终形成合作共赢的全球气候治理体系。

中国作为发展中国家在参与全球气候治理的过程中，一方面正确判断自身的发展阶段、能力和国情，把可持续发展作为第一要务，与广大发展中国家一道维护自身的发展权益，主动承担与自身发展阶段、能力相适应的责任，主动与发展中国家开展南南合作；另一方面，也要以对历史和全人类负责任的态度，变应对气候变化挑战的自身可持续发展为机遇，积极探索发展路径的创新，将保护全球气候与解决自身发展过程中的发展方式转变、提质升级、结构调整、深化改革等重大战略任务紧密结合，最大限度地实现应对气候变化与经济社会发展的多赢。

由邹骥教授领导的国家应对气候变化战略研究和国际合作中心与中国人民大学“能源与气候经济学项目”联合团队，在过去几年里积极投身于全球气候治理的研究和政策咨询工作，其中很多成员亲自参与、经历了《联合国气候变化框架公约》下多边进程的谈判和多边、双边气候外交的技术支持实践活动，以理论联系实际的学风进行了大量有意义的探索，为政府政策决策提出了许多有益的建议。特别是他们自 2013 年以来在与国际上谈判圈和智库圈的交流和辩论过程中，提出了以“人类发展路径创新”概念框架统一解释和论述国际气候政策，对于在形成气候谈判立场中反映公平原则、“共区”原则和各自能力原则，把握应对全球气候变化长期发展趋势和目标，准确给全球气候治理各主要利益相关方定位，提出气候谈判的解决方案、落实统筹国内国际两个大局的战略思想等，都取得了较大进展。

尽管本书的论述由于时间的限制还略显仓促和不足，但它在系统论述全球气候治理问题上所进行的探索具有重要意义。希望中国和国际上有更多的专家学者和智库能够持续深入地对此问题进行长期研究，为全球气候治理的发展提供更加充分的智力支持。

特别值得一提的是：在这个作者队伍中有很多是刚刚毕业而走上工作岗位不久的、甚至是在读的博士生。他们受过很好的学术训练，又有机会亲身参与重要的实践活动，充满热情和活力，经过实践锻炼迅速成长，给中国气候变化领域智库的发展带来了希望，令我十分欣慰。

是为序。

中国气候变化事务特别代表

解振华

前言

自 1992 年签署《联合国气候变化框架公约》（以下简称《公约》）以来，中国作为《公约》缔约方一直参与这一重要国际进程。20 多年来，中国经济社会发展发生了翻天覆地的变化，世界战略格局也发生了深刻变化。中国一跃成为全球第二大经济体，在国际事务中的影响力和话语权越来越大，但同时温室气体排放量也伴随着经济发展快速增长，已成为年度瞬时排放最大的国家。不论从国际影响力、温室气体年度排放总量还是从加快推进中国绿色低碳发展转型的力度和行动上看，中国在以《公约》为基础的全球气候治理体系中都扮演着史无前例的重要角色。与上世纪 90 年代参与《公约》多边进程谈判时的中国相比，今天的中国在参与建设全球气候治理体系的过程中，越来越多地主导谈判进程、提出详细方案，越来越多地扮演制度设计者和领导者的角色。习近平总书记在中共中央政治局第二十七次集体学习时提出推动全球治理机制更加公正更加合理，中共中央十八届五中全会指出中国将积极参与应对全球气候变化谈判并积极承担国际责任。中央的这些战略决策为深度参与全球气候治理奠定了坚实的政治基础，为将全球气候治理打造成中国全面参与全球治理的抓手创造了条件。

国家应对气候变化战略和国际合作中心是国家发展和改革委员会直属政府智库。中心的一项重要职能就是为中国参与国际气候变化谈判提供技术支撑，这项职能具体由中心国际合作部负责落实。中心国际合作部在中心分管副主任邹骥教授的带领下，积极参与气候变化多边谈判及气候国际合作等其他相关多边双边进程，配合国家发展改革委以及外交部、财政部等其他相关部门组织开展谈判相关重大战略性问题研究，为中国代表团参加气候变化国际谈判提供了直接的技术支撑。

中国人民大学"能源与气候经济学项目组"长期以来从事应对气候变化政策研究工作，在国际气候体制、低碳政策分析、减排路径模拟和低碳技术发展战略等方面积累了大量的研究经验，取得了丰富的成果。

本书的作者大多是来自中心国际合作部和中国人民大学的青年研究人员。他们有幸成为中国参与全球气候治理的亲历者和执行者。他们直接参与了《公约》下"长期合作行动特设工作组"、"京都议定书特设工作组"和"德班平台特设工作组"以及《公约》两个附属机构的谈判，在共同愿景、2013—2015审评、国家信息通报、京都议定书方法学、多边评估、德班平台第二方面工作以及巴黎协议谈判长期目标、减缓、全球盘点以及技术等具体议题的谈判中发挥了重要作用。他们参与了中国国家自主贡献文件的研究和起草工作，开展中国国家自主贡献的分析和评估，开展各缔约方国家自主贡献的分析评估，参与国际组织和研究机构对中国和全球的排放情景分析，为中国政府制定谈判长期战略和近期对策以及形势研判提供研究支持。他们还为"政府间气候变化专门委员会"（IPCC）第五次评估第三工作组报告政府审评工作提供支撑，开展IPCC评估报告解读分析工作，协助中国政府组织相关研究机构和专家参与IPCC进程，向中国政府提供关于IPCC未来工作安排的建议。他们还积极参与气候变化相关的双边交流与合作，先后参与中美、中加、中俄、中欧、中英、中法、中德、中日、中韩等政府间对话的技术支持工作，并亲自主持相关专家对话。组织与世界资源研究所、美国进步中心、哈佛大学、法国可持续发展与国际关系研究所、瑞典斯德哥尔摩环境研究所、南方中心等智库间对话交流活动，组织基础四国、立场相近发展中国家专家对话。他们在智库专家层面上直接参与了具有历史意义的中美气候变化联合声明谈判的技术支持工作。

另外，中心国际合作部还积极开展气候资金和技术相关战略和政策研究。先后开展了金砖银行气候变化战略、G20 框架下气候资金问题研究；参加政府间可持续发展融资专家委员会谈判，协助中国政府参与绿色气候基金董事会会议谈判；组织开展中国气候技术需求评估，承担公约下气候技术中心与网络（CTCN）中国政府联络员工作，参与公约下技术执行委员会（TEC）谈判；组织开展短寿命气候污染物（含 HFCs）相关技术的政策研究。他们还开展了气候变化政策国别研究，对美国、欧盟、日本、俄罗斯、澳大利亚等主要发达国家以及巴西、南非、印度、马来西亚、印度尼西亚等主要发展中国家气候变化战略和政策的分析评估；开展气候变化南南合作机制研究，为中国政府建立气候变化南南合作基金提供政策咨询。

这些工作虽然繁重且琐碎，但本书的主要作者凭着高度的国家荣誉感和历史使命感以最大的工作热情全身心的投入到每一个工作细节。同时，他们也不断思考研讨，充分利用国家平台的信息和智力资源优势，在工作细节中不断分析、提炼和总结全球气候治理理论基础、历史沿革、数据信息、分析框架、主要治理主体以及与全球治理的关系等重大问题。本书主要由这些在工作中的思考以及信息和知识积累编撰形成。

本书从气候变化科学事实入手，分析了应对气候变化问题的特殊性及其给全球治理和人类发展带来的挑战，回顾了世界工业化历史进程，在可持续发展和公平的框架下，结合《联合国气候变化框架公约》下的国际气候合作进程和应对气候变化在全球治理体系中的演进和发展，论述了以人类发展路径创新为核心概念的全球气候治理理念。全书共分九章：第一章主要由王克、邹骥、祁悦和陈济等完成，介绍了全球气候治理的背景，包括气候变化的科学认识、应对气候变化的特殊性、全球气候管理体制的总体要求以及不同国家面临的发展约束；第二章主要由傅莎、崔学勤、刘俊伶、张小丽、邹骥等完成，通过回顾发达国家的历史排放轨迹，在讨论环境库兹涅茨曲线的基础上分析了不同排放轨迹的差异、变动和驱动因素；第三章由崔学勤、傅莎和邹骥完成，着重阐述了全球发展与环境公平问题，主要包括公平的伦理学基础、内涵、基于公平的碳空间分配方案等；第四章主要由傅莎和邹骥撰写，论述了人类发展路径创新概念框架的内涵及其在设计应对气候变化国际体制中的意义和应用；第五章由祁悦、陈济、傅莎等完成，论述了《联合国气候变化框架公约》在应对全球气候变化国际合作进程中主渠道的作用，回顾了《公约》下国际气候合作的历史演进，综述了《公约》下应对气候变化国际合作的现有机制；第六章由陈济、王克、刘俊伶、傅莎和邹骥等完成，考察了应对气候变化与全球可持续发展进程、全球贸易进程、全球金融合作进程等的联系和影响，论述了气候变化作为非传统安全问题在全球治理中的特殊性和外溢效应；第七章由陈济和祁悦等完成，分析了美国、欧盟和主要发展中国家在全球气候治理中的责任、作用和诉求；第八章由傅莎和邹骥等完成，论述了中国在全球气候治理中的定位；第九章由邹骥主笔，总结提炼了以人类发展路径创新为核心的全球气候治理理念及政策含义。

除了上述主要作者的贡献外，本书的完成离不开很多机构和同事的帮助。在此特别感谢前任国家发展和改革委员会副主任、中国气候事务特别代表解振华先生，他的智慧、胸襟、勇气、无数具体指导和远见卓识对作者的观点产生了根本性的深刻影响。作者还要感谢国家发展和改革委员会应对气候变化司为我们提供的研究平台和业务指导，感谢中国首席谈判代表苏伟先生及其团队的宝贵支持，还要特别感谢国家应对气候变化战略研究和国际合作中心的领导对我们研究工作的支持，特别感谢张晓华、胡晓、阿柔娜和刘林蔚在过去的研究工作中的重要贡献，特别感谢徐丽丽、陈奕颖、钟洋、马红岩和孙倩以及中国人民大学“能源与气候经济学项目组”的同事们在书稿写作和出版过程中的辛勤工作。本书的编写和出版得到了中国清洁发展机制基金赠款项目的有力支持，特别感谢以下课题的资助：巴厘路线图谈判进展及可能前景研究、关于德班平台重大战略问题和总体战略研究、基础四国专家机制及发展中国家智库交流合作平台技术支撑项目、2020后“公平、有效、多赢”的国际合作强化行动安排设计和中国对策研究、中美气候变化务实合作技术支撑项目、公约下2013—2015年全球长期目标审评技术支撑项目。

由于时间仓促和作者水平有限，书中仍有不少遗憾，缺点错误在所难免，敬请读者予以指正。

邹骥及所有作者

2015年深秋于北京玉渊潭

摘要

1. 急需发展全球气候治理体系，对日益稀缺的全球大气碳容量进行管理

政府间气候变化专门委员会（Intergovernmental Panel on Climate Change, IPCC）第五次评估基于最新科学进展，进一步建立了温度、温室气体浓度和累积排放量之间的对应关系，提出如果要在本世纪末将自工业革命以来的温升控制在2℃以内从而最大限度控制温升带来的全球气候风险，需要从现在起迅速采取行动，通过控制全球温室气体瞬时排放流量而控制彼时的累积排放存量，全球2050年温室气体排放量相比2010年需要减少40%—70%，到2100年实现近零排放。留存给人类向大气层中继续排放温室气体的数量空间已经变得日益有限，按照2010年全球排放水平，2050年前剩余的温室气体排放空间仅够排放17—31年。

因此，虽然存在科学上的不确定性，但气候变化科学评估结论已经对全球温室气体排放提出了上限要求，世界各国未来排放空间将受到极为现实的紧约束，这将对各国的社会经济长期发展带来巨大而深远的影响。因此，日益稀缺的全球大气碳容量已经成为全球公共财富，急需发展全球气候治理结构对其进行分配和管理。

2. 气候变化超大时空尺度外部性特征给全球气候公共财富管理带来巨大挑战

二氧化碳等温室气体在大气层中留存的寿命周期长且影响范围涉及全球，同时具有全球跨国外部性和跨代外部性的特征，并存在很大的不确定性。这不仅使得气候变化问题超越了国家及国家以下水平现有决策主体的常规决策时空视野，并且由于目前还不存在一个凌驾于各国主权之上的超主权治理主体，只能通过跨国政府间谈判和协商进程，通过各国共享的价值观和发展理

念等形成不断演进和强化的全球气候治理结构。这一进程是缓慢的，与应对气候变化的紧迫要求形成矛盾，给全球气候治理体系带来巨大的挑战。此外，在全球现有政治体系下，大时空尺度的外部性管理依然要求各国家主体发挥关键作用，因此国家决策者的决策思维和视野也需要扩展至全球并跨越时间长河。

3. 全球气候治理体系正逐渐形成并不断强化对温室气体排放的机制性约束

为了对温室气体排放权益进行管理并构造减缓和适应责任体系，国际社会于 1992 年达成了《联合国气候变化框架公约》（以下简称《公约》）。作为全球气候治理体系政治和法律基石的《公约》，已经对全球气候治理目标、重点行动和合作领域、各主体定位与作用、各方责任、全球合作模式和机制等关键性问题进行了框架性的原则规定，并得到 190 多个缔约方在政治和国际法意义上的接受。以《公约》为基础，在公平原则、共同但有区别的责任和各自能力原则指导下，全球气候治理机制逐渐形成并不断完善，对温室气体排放的机制性约束不断强化。此后签署的《京都议定书》和历届缔约方大会所作出的决定，以及计划于 2015 年年底签署的巴黎协议，都是旨在操作层面上、在新的历史时期不断加强履行《公约》的行动。

4. “共同但有区别的责任”和公平原则依然是全球气候治理的基石

主权国家是全球气候治理体系的主体。在气候治理体系中，主权国家的地位和责任主要是由其基于历史累积排放的历史责任、发展阶段、能力以及国情所决定的。从战后、特别是《公约》生效以来的世界格局变化看，尽管发达国家和发展中国家在经济、人口、能源消耗和年度排放等的全球占比上发生了巨大的变化，但从总体上看发展中国家与发达国家在经济发展水平、发展阶段和能力上的差距依然显著存在，发达国家以较少人口占比依然在历史累积排放总量和人均量上占有支配地位。因此，保持发达国家和发展中国家的区分并在此基础上坚持《公约》的公平原则、共同但有区别的原则（以下简称“共区”原则）和各自能力的原则，对于构建全球气候治理体系并不过时，依然重要。《公约》20 多年的进程还表明：“共区”原则并没有阻碍像中国这样的发展中新兴经济体向低碳发展方向转型，也不意味着消极行动，而是恰恰相反，促进了新兴经济体国家根据自己的历史责任、发展阶段、能力和国情，积极探索发展路径的低碳创新。

“公平、有效、共赢”是评价全球气候治理体系是否成功的三个基本标准。公平作为全球气候治理的根本性原则，由缔约方的历史责任、发展阶段、能力和国情四个核心要素所决定。自工业革命以来，发达国家以较少的人口占比占有了近 70% 的二氧化碳累积排放量，美欧迄今的累积排放量约是中国的 5 倍。发达国家对今天及今后一个很长时期发生的全球气候变化负有主要的历史责任。发达国家今天所享有的发达基础设施和丰富物质基础以及在全球所处的有利竞争地位和国际进程主导地位，是在其大量占用全球公共碳容量资源的基础上形成的。而且发达国家还在全球化过程中通过投资、贸易和价值观、知识、技术的扩散在世界范围内向后起的发展中国家传播其所开创的传统高碳发展模式，实现了碳排放的转移。

对发展中国家而言，广泛的就业、脱贫、基础设施建设、社会保障体系的建设、努力提高中等收入人口的占比，依然是社会经济发展的首要议程，工业化和城镇化在一个可预见的未来依然是经济发展的主题。但是人口众多的后起发展中国家在实现工业化、城镇化的发展过程中，受到了前所未有的全球范围碳排放制约。沿袭发达国家近现代传统高碳发展路径实现社会经济现代化与碳排放制约形成了趋势性的时代冲突。

正是由于历史责任以及当前发展阶段、能力和国情的差异，发达国家和发展中国家存在共同但有区别的责任。发达国家一方面需要通过自身迅速和深度的减排腾出更大的碳排放空间用于支持发展中国家的转型，另一方面则需要基于其自身资金、技术的优势和在治理上的能力，积极对发展中国家实现低碳转型和跨越提供资金、技术和能力建设的支持。发展中国家则需要转变发展观，在保证实现人类发展指数持续上升的前提下，以新的要素组合、新的技术、低碳的能源、新的基础设施设计和生产力布局，实现低碳发展转型。

因此，“共区”原则是全球气候治理中落实公平的具体体现，是确定《公约》各缔约方减缓与适应目标、出资和技术援助责任和受资与接受技术权力、履约透明度要求等的根本依据。如果全球气候治理失去公平的基础，进而从政治上失去广大发展中国家对国际气候进程合法性的认同和参与的动力，将会严重削弱多边国际合作的政治信任基础，也不能充分反映不同发展背景下排放源的特性和驱动因素，进而难以形成反映历史与现实情况的有效的应对气候变化战略和政策。在2015年巴黎协议谈判中，能够在多大程度上保证“共同但有区别的责任”这一根本性原则在协议中得以具体体现，事关巴黎协议在未来将把全球气候治理引向何方，也是判断巴黎协议是否成功的重要标志。

5. 通过创新发展路径管控全球排放并打破全球气候治理发展僵局

通过对主要先行工业化国家和发展中国家自工业革命以来的历史排放轨迹进行实证考察后发现，主要发达国家在完成工业化、城市化进程后，其能源及水泥生产相关 CO_2 排放总量普遍已达到峰值，但表现出不同的峰值特征。美国、加拿大和澳大利亚的人均峰值水平较高，接近或超过 20 吨 CO_2/ 人左右。欧盟和日本等国家的人均峰值普遍在 10 吨 CO_2/ 人上下。这说明在历史实践中大体存在着人均排放峰值水平相差两倍的两类排放轨迹（“美、加、澳”模式和“欧、日”模式），即有可能以相对较低的人均排放水平达到相近的人均收入水平。

从各国人均排放峰值与国内人均 GDP 关系来看，纵观工业革命以来的全球 CO_2 排放历史，目前还没有一个经济体能够摆脱随人均 GDP 或人类发展水平提高，人均 CO_2 排放水平“先增长后下降”的倒 U 型库兹涅茨曲线现象，但在不同的国家会有不同的表现。发达国家实现人均排放峰值时对应的人均 GDP 水平，大体分布在 2 万—2.5 万美元（2010 年价）左右。这一现象体现了一个经济体在工业化、城镇化进程中，在从不发达到发达的过渡过程中，经济质量不断提升，对化石燃料依赖下降，而对技术、服务依赖上升的过程。存在这一现象并不意味着发展中国家无所作为。发展中国家目前仍普遍处于曲线“左侧”的排放快速增长期，占世界排放比重不断上升，且这一态势在一定时期内还将持续，但从历史累积排放和人均排放看，发展中国家仍远低于发达国家。与此同时，考虑到国内外的资源、环境、排放空间约束，发展中国家已不具备延续发达国家老路的条件，未来是沿着“高发展水平高排放的美、澳、加模式”还是“相似发展水平对应较低排放水平的欧、日模式”的轨迹，抑或是走出一条创新的不同于发达国家传统路径的更为低碳的发展路径，对其自身的社会经济可持续发展和全球应对气候变化的努力，都将具有十分重要的意义。

改变和管控排放路径，实质上是改变和管理发展路径的问题。发展路径的内涵十分丰富，总体而言涉及经济增长的发动机何在，增长率主要来源于要素投入的增长还是来源于要素效率的提高；收入的构成主要来自投资还是消费和净出口；人口的数量变化及收入分配结构与城乡和地理分布结构的变化；能源结构中高碳密度能源或低（零）碳密度能源的占比；受经济结构和部门产品结构变化和技术变动影响的单位 GDP 所消耗的能源量（能源强度），这些都会体现到产业部门和基础设施建设中去。随着收入水平的提高，生产和生活服务业在国民经济中的占比提高，高能耗或高排放密度的重化制造业的国内市场渐趋饱和，占比甚至绝对数量下降；制造业内部技术进步，产业链延长，产品结构变化，附加值上升，也会导致能耗和排放强度下降。

发展中国家需要创新一条不同于美欧等发达国家的、更早达到更低峰值的发展道路。创新发展路径并不是空洞的学理概念，而是从对发达国家工业化历史进程的反思中提炼而成的。发展中国家有望通过适当的政策和体制管控措施并充分发挥后发优势，有望影响排放轨迹并实现发展路径创新，促成峰值更早更低到来。而要实现上述峰值，降低单位 GDP 能源强度和使能源结构低碳化将是发展中国家未来排放下降的主要抓手。

发展路径创新的出发点是处于不同发展阶段的国家都要以积极的姿态行动起来，探寻与本国发展阶段相符的发展路径创新；其着眼点是影响排放轨迹和路径的发展驱动因素，以求在更实质、更实际、更具体的层面上打破全球气候治理谈判的僵局。这一概念框架着眼于实质的行动和措施，在可持续发展的总体框架下，围绕发展路径的创新和变迁，识别出发达国家与发展中国家各自行动的方向和重点，在政策含义上具有积极且重要的建设性意义。

发展路径创新的概念，还进一步确立了坚持发达国家和发展中国家区分的必要性与合理性。从发展路径的角度看，发达国家除了对累积排放有历史责任，还有开创、主导扩展和强化高碳发展路径的责任。发展中国家大多在全球供应链中处于高能耗、高排放但是低附加值的不利地位。尽管发达国家大多进入后工业化社会的排放下行阶段，但在全球化背景下与发展中国家的排放增长趋势和路径依赖仍然脱不了干系。因此，改变发展中国家未来的发展路径，是全球系统调整的问题，发达国家对此也负有不可推卸的责任。

6.《联合国气候变化框架公约》是全球气候治理的主渠道，反映了全球气候治理的道义、政治和法律基础

发达国家与发展中国家的存在与划分既是世界政治经济的客观现实，也是几十年来联合国政治运作的基本谈判集团，有着深刻的历史与现实渊源，这种划分还会长期存在。在一个很长的时间尺度内，《公约》的目标、原则、条款依然适用。《公约》的目标体现了通过保护大气环境实现可持续发展的指导思想；《公约》建立了公平原则、发达国家和发展中国家对减缓与适应气候变化担负“共同但有区别的责任”原则和各自能力原则；据此要求各缔约方都要承担制定国家应对气候变化战略和规划、向联合国报告排放清单和国家信息通报等共同责任，但要求发达国家率先实现全经济范围内的绝对量化减排，向发展中国家提供技术、资金和能力建设方面的援助支持；发展中国家履行《公约》责任的程度取决于发达国家提供资金支持的程度；减缓与适应气候变化都是应对气候变化战略的重要组成部分。

7. 正确设置全球气候治理的目标并处理好“自下而上”与“自上而下”管理模式的关系

全球气候治理的目标是一系列目标的集合。就全球长期目标而言，应当在政治上确认气候变化问题的迫切性和重要性，在战略上确认全球和地区、国家走低碳发展道路的方向，确认国家的相关责任和体制政策发展方向，给社会经济的决策者和投资者发出明晰的信号。但不宜简单机械地将具有不确

定性的科学模拟数字直接变为全球目标的数量依据，从而增大全球目标决策的“数字游戏”和“谴责游戏”的风险，给多边谈判造成困难。以全球低碳发展、路径创新的共同愿景为基础形成政治的共识，加之发达国家确定全经济范围的绝对量化减排目标、发展中国家确定降低 GDP 碳强度等相对控排指标，附之以非化石能源占比、森林碳汇等驱动因子指标，再加上密切关注和反馈技术、市场等信息，促进利益相关者不断更新其决策目标，就足以达到促使各方把握低碳发展大方向并实质性推进低碳转型的效果。

选择“自上而下”还是“自下而上”的管理模式，是《公约》20 多年进程中一直在争论的问题。经历了 2001 年美国布什政府宣布不向国会提交申请批准《京都议定书》的决定之后，美国国内政治对多边气候进程的影响凸显，“自上而下”的模式在没有超主权国际强制力的情况下陷入困境。在新一轮的多边气候协议谈判过程中，作为“自下而上”管理模式标志的“国家自主贡献”目标在 2013 年华沙决定中应运而生，且为大多数缔约方所接受。但围绕着旨在对“国家自主贡献”目标进行周期性全球盘点、审评乃至影响进一步更新的“自上而下”的安排，依然是一个争论的焦点，有待在巴黎协议中得到安排。透明度问题也被视为适用于“自上而下”管理模式的领域。此外，法律强制约束力适用的内容、法律形式等，也是落实全球气候治理中需要解决的问题。如何处理好 “国家自主”确定其目标的“自下而上”与通过“全球盘点”和提高透明度并促进各方更新今后的目标的“自上而下”之间的平衡，将“自上而下”与“自下而上”的管理模式恰当地运用于合适的领域，以在保证全球气候治理有效的前提下最大限度地让美国参与并鼓励广大发展中国家参与，是全球气候治理中要解决的一个关键问题。

8. 平衡处理好大国影响与广泛参与之间的辩证关系至关重要，广大发展中国家的权益必须得到充分的尊重和保障

基于战后建立的“联合国”这一处理多边国际事务的基本国际政治秩序，《公约》决策规则的核心在于“协商一致”原则。它要求无论大小国家和穷国富国，在决策中都具有平等的地位，要求充分考虑包括公民社会在内的各

方诉求，要求广泛参与的国际民主和旨在加强政治互信的透明。在国际气候进程的实际推进中，大国发挥着带头作用和示范作用，具有更大的实际影响力。平衡处理好大国基于带头示范作用的较大影响力和广大中小缔约方广泛参与之间的辩证关系是保证多边国际气候进程公正有效的关键。由于历史的原因，发达国家在国际规则制定过程中占有明显的优越位置，这就提出了如何使广大发展中国家（特别是那些位处亚、非、拉的最不发达国家、脆弱的小岛屿发展中国家等）的权益、诉求和根本利益得到尊重和保障的问题。要切实加强发展中国家参与全球气候治理的能力，在议题设置、协议内容要素等问题上满足在资金、技术和能力建设方面对发展中国家提供援助的要求。

9. 全球气候治理是全球治理的重要有机组成部分，事关打造人类命运共同体和全球公共财富，涉及全球政治、经济和安全治理

加强全球气候治理是实现维护全球气候公共财富、实现人类共同利益、打造人类命运共同体的需要，应以合作共赢为核心。它涉及人类可持续发展总体战略目标，也涉及全球政治、经济、安全治理的诸多重要方面。在 2015 年 9 月联合国可持续发展首脑峰会上刚刚确立了 17 项全球可持续发展目标，气候变化是独立的一项。在政治上涉及全球事务的主导权和话语权、影响力；在经济上，涉及未来世界经济的总体趋势和面貌、贸易投资与金融的标准和竞争力；在安全上由于气候负面影响的巨大损失以至于成为非传统安全的重要内容。气候变化是横跨全球治理各主要领域的综合性议题。

10. 发达国家与发展中国家要在不同历史责任、发展阶段、能力和国情下探索各自发展路径的创新并处理好竞争与合作的关系

全球气候治理已经成为国际战略的新领域，被一些学者视为国家的“新边疆”，具有重新定义各国长期竞争力的深刻含义，进而关系到国际竞争的战略格局。其历史沿革以及创新改革诉求，正深刻地反映了 21 世纪以来全球政治和力量格局的变迁。全球气候治理中的各主体都已经不同程度地行动

起来，力求通过参与全球气候治理保持或提升国际影响力、竞争力，掌握在全球治理中的主导权。美国、欧盟等发达国家或集团一直在争夺全球气候治理的领导权，影响治理模式、要素、发展节奏、议题设置、规则制定、道义和法律基础等重要决策。

发达国家在先行进入后工业化发展阶段以后，能源结构普遍得到优化，产业结构通过全球化进行了调整，大量高排放产业转移到发展中国家，自身占据了全球产业链的高端，加之技术进步，大大降低了排放强度，人口保持稳定，收入长期以低增速保持在高位水平，这些都使得发达国家在年度排放量方面开始得到显著改善，其政策、技术、做法有值得借鉴之处。这些对形成全球低碳发展大势起到了一定的引领作用。

但发达国家在直接造成全球气候变化的累积排放指标上依然以不到 20% 的人口占比达到近 70% 的累积排放贡献率，预期即使到 2030 年发达国家的累积排放量依然多于发展中国家。美国、加拿大、澳大利亚等国的人均排放量依然处于 15—20 吨 / 人的高位，远远高于世界平均水平。欧盟在交通部门的排放量依然在增长。发达国家在整体上减排的力度与 IPCC 第四次评估报告所要求的到 2020 年比 1990 年减排 25%—40% 的力度依然有较大差距。

发达国家在对发展中国家提供技术和资金援助方面的步伐依然缓慢，不能满足应对气候变化的要求，与广大发展中国家的期望还有不小的差距。在谈判2020年之后的全球气候治理安排过程中，一些主要发达国家意欲背离《公约》原则和条款重写治理规则的倾向明显。

主要新兴经济体发展中国家随着 20 多年来在世界政治经济战略格局中地位的提升，在《公约》原则和条款的指导下，不断实现战略调整，将应对气候变化纳入社会经济发展战略，通过《坎昆协议》和正在谈判中的巴黎协议，根据各自能力和国情建立了相对的控排目标和适应气候变化的国家战略，带动发展中国家探索发展路径的低碳创新，取得了显著的成绩，但依然面临艰巨的挑战。

总体而言，具有不同历史责任、处于不同发展阶段、拥有不同能力和国情的发达国家和发展中国家，尽管进展各异，但在走低碳发展道路、需要国际合作等重大问题上已经形成普遍战略共识。在进一步发展全球气候治理过程中，需要进一步凝聚发展路径创新的共识，识别发达国家和发展中国家在不同发展阶段和起点、基于不同能力和国情的更加有针对性的路径创新战略、目标、行动和政策措施与支持体系。

11. 中国所处的发展历史阶段和新的全球战略定位决定了创新发展路径是中国的内在需求

中国作为最大的、也是正在经历快速工业化和城镇化并向更高发展阶段迈进的发展中国家，经过 30 多年的改革开放，取得了巨大的发展成就，经济总量居世界第二位，人均 GDP 国内生产总值增至 7800 美元左右，已经基本解决了温饱问题，正进入 2020 年全面建成小康社会的决胜阶段。

从增长经济学的角度看，劳动力、资本及相应的能源、土地、水、矿产品和环境容量资源等要素投入的增加为过去 30 多年的高速增长作出了主要的贡献。经济发展转入“新常态”意味着中国经济长期依赖要素投入实现增长的发展模式已经走到尽头。在经济起飞的原始资本积累阶段，通过超量蚀本占用消耗包括自然资源与环境容量资源完成资本原始积累，这个过程的社会代价已经十分突出。随着以煤为主的能源资源禀赋在日益趋紧的全球碳约束下成为比较劣势，如果不及时转向创新驱动的经济发展方式，推进以低碳为重要特征的科技革命和产业变革，中国可能陷入以资源、环境为主要制约的“中等收入陷阱”，这将严重影响“两个一百年”奋斗目标的实现。另一方面，随着社会财富存量提高和收入达到中等水平，中国也应有条件转换到依靠知识、技术、治理而提高效率支撑增长的发展阶段，从而缓解透支自然资本付出的社会代价。

向更高收入水平和更发达的发展阶段迈进，依然是中国面临的时代主题和今后几十年的奋斗目标。但显然传统的发展模式非但不能支撑中国向更高发展阶段迈进，而且已经转变为中国面临的中等收入陷阱的一个特征。

在全面深化改革和推进生态文明建设的背景下，探索和开创新的发展模式和发展路径已经成为中国发展战略和政策的紧迫理论和实践课题。低碳发展作为一种高效的集约、内涵经济模式，通过应用与新的资源环境约束条件相匹配的技术和制度，催生新的经济增长点和推动传统行业新的增长模式，从而促进经济发展质量和效益持续提升，将成为倒逼推进中国经济转型发展的强大动力。

中国需要努力探索一条优于发达国家传统发展道路的创新的低碳发展路径，促成经济增长模式转向“新常态”，将低碳发展作为提高要素生产率和驱动经济增长与转型的重要推动力，实现能源效率提高、能源结构转变、产业结构升级、收入结构优化、人力资本提升，从而全面实现发展路径的创新。要将创新发展路径，实施低碳发展战略上升到中国实现“两个百年”目标、确立在未来国际低碳发展大趋势下和低碳科技革命浪潮中国际竞争地位的认识高度。

12. 中国创新发展路径对全球绿色低碳转型具有重要意义

经历了 30 多年的高速增长和近 20 多年的全球化进程，中国在世界经济中的地位已经举足轻重并与世界经济融为一体。因此，中国的新发展阶段和中国的转型，都必须置于整个世界经济体系中去看。中国创新发展路径对全球绿色低碳转型具有重要意义。

当前在全球温室气体自工业革命以来的累积排放依然以发达国家为主的格局下，年度瞬时排放的格局已发生显著变化，从长期看最终会改变累积排放的格局。中国碳排放总量和人均碳排放量的快速增长在全球范围内日益凸显，经济体量大、能源消费多、碳排放总量高的特征明显，虽然历史累积排放相比美、欧仍有较大差距，但是随着经济发展和中国工业化和城市化进程的进一步推进，中国碳排放的增长趋势不可避免，未来历史累积排放份额也将显著上升。因此，中国未来的碳排放水平对于全球的排放格局和路径至关重要，中国创新发展路径实现碳排放管控也是全球实现碳排放控制目标的重要组成部分。

中国经济向更加清洁高效的低碳发展方向转型还将创造一个巨大的清洁能源技术、产品和服务的国内市场，并形成对发达国家先进的制度、技术、标准和管理经验的大规模需求。中国基于所拥有的全球领先的制造能力和优质产能、相对充裕的资本能力以及庞大的国内市场潜力，通过多边和双边合作，与美欧等国家在低碳技术产业链中加强协作，共同深化全球供应链布局和专业化分工，共同加强低碳技术研发和商业模式方面的创新能力，将有力推动低碳技术应用成本的降低，扩大全球范围内低碳技术和产品的市场空间，从而减少温室气体排放，并促进经济繁荣和创造就业机会，实现全球经济的再平衡。同时低成本的低碳解决方案，也将促进发达国家国内基础设施的翻新，降低发达国家实现低碳转型的难度。

作为最大的发展中国家，中国可通过创新发展路径，开辟一条低碳可持续发展路径，为后发的发展中国家提供可资借鉴的示范模型和经验，通过南南合作支持这些发展中国家规避传统高碳路径依赖和锁定效应，尽早走上“低污染低排放”的高效创新发展道路。同时，中国也能通过自身转型促进全球转型，为重塑全球发展路径作出贡献。

13. 中国参与全球气候治理应在坚持发展中国家定位基础上实现从参与者到建设者的角色转变

在当今世界经济、能源和气候变化体系中，中国的基本定位是一个具有独特地位的发展中大国。尽管中国已成为全球第一大年度排放国和第二大经济体，但其主要发展指标和长期存在的显著“二元结构”特征均显示了其发展中国家的定位，和发达国家在所处的发展阶段、面临的发展需求、承担的历史责任和具备的综合能力等方面仍存在巨大差别。此外，中国还面临能源资源禀赋以煤为主，地域辽阔，人均资源存量、质量、有效人均活动空间均十分紧缺等特殊国情。中国在参与全球气候治理时需充分考虑其发展阶段、能力和国情。

气候变化已成为事关国际道义、国际责任、国家形象、国家发展权利的重要议题。中国作为一个对世界负责任的发展中大国，在全球气候治理中一方面要倡导尊重历史事实，冷静而理直气壮地坚持发展中国家定位，坚持“共区”原则，主持公道，坚持要求发达国家按照《公约》原则和条款履行其率先和示范减排的责任及对发展中国家提供足额技术、资金支持的责任；另一方面也要充分认识到新时期自身发展阶段和面临形势的变化，认识到应对气候变化、创新发展路径不是因为别人要我们做，而是我们自己需要做。在中国经济迈向更加发达阶段的努力中，积极参与全球气候治理，是中国转方式、调结构、升档次、克服中等收入陷阱、顺应国际低碳发展和科技革命潮流以及提高在国际领域低碳竞争力的内在要求。

中国应立足于本国国情和客观定位，在可持续发展框架和《公约》原则下，运用大国思维以积极务实姿态参与国际气候进程，发挥负责任大国的建设性作用，主动承担与自身发展阶段、能力、定位和国情相符的应对气候变化全球责任，实现从参与者到建设者的角色转变和从“顺势而为”到“主动出击”的战略转身，推动“公平、有效、共赢”的全球气候体制建设。

总体而言，中国参与全球气候治理时应统筹考虑如下目标：一是为完成工业化、城镇化和发展路径转型的平稳过渡，避免硬着陆，需要争取峰值到来之前合理的必要排放空间；二是推动形成“公平、有效、共赢”的国际气候体制，支持应对气候变化，最大限度降低气候变化带来的不利影响；三是争取实现更低更早峰值的发展机会和发展权，最大限度降低国民经济对温室气体排放和能耗的依赖程度，形成绿色低碳发展的长期竞争力；四是从全球气候体制的参与者逐渐转变成为建设者和主导者，以此为平台之一促进提高中国在全球治理规则制定与实施中的地位、主导权和话语权；五是树立负责任、公平正义的良好形象。

Preface

Recently, the CPC Central Committee put forward a clear call to participate in global governance. In October 2015, the Fifth Plenary Session of the 18th Central Committee of the Communist Party of China proposed that "China will take an active part in global economic governance, promote the development of international economic order in the direction of equality, fairness, cooperation and win-win, and expedite the free trade zone strategy; actively undertake international responsibilities and liabilities, take active part in negotiations in response to global climate change, and take active part in the 2030 Agenda for Sustainable Development".

On October 12th, 2015, President Xi Jinping presided the 27th collective learning of the Political Bureau of the CPC Central Committee, stressing on global governance layout and system problem that, "We participate in global governance for the radical reason that we must realize the striving objective of two century, and realize the Chinese Dream of national revitalization. We must take the opportunity and chance, face the challenges in a good manner, stabilize domestic and international situations, drive the global governance system in a more reasonable and fairer direction, and create more favorable conditions for Chinese development and world peace."

On June 12th, 2015, Premier Li Keqiang pointed out at National Leading Group for Climate Change, Energy Conservation and Emissions Reduction Meeting that, "Active response to climate change is not only important for China to secure economy, energy resources, ecology, food safety and people's lives and properties from any loss or damage, and to improve sustainable development, but also an undertaking to take in-depth part in global governance, create human common destiny and drive joint development. As a responsible great nation, China will insist on the principle of joint and separate responsibilities, the principle of fairness and the principle of respective capabilities, and take the international obligations according to its national status, development stage and actual competence, China will continue to improve the ability to mitigate and adapt to climate change and make its own best efforts to promote global green and low-carbon transition and innovation of development path.

The discourses of party and state leaders have a clear judgment on global trend of green low-carbon development, present China's participation in top-level design of global climate governance, and constitute China's great contribution to promoting global climate governance, driving China's green low-carbon transformation, and innovating development concept.

We are in an important historical era. With the globalization progress over the past twenty years, the trend of revolution and development of science and technology and industries in new energy, information technology, new materials and other fields and worldwide green and low-carbon development has become increasingly evident. During this period, we have also experienced seven years' difficult and slow recovery after the global financial crisis. World economic development is in a profound adjustment period to identify new reorientation and find a new growth engine. The link between national development and world economy and global environment is increasingly close, making the global governance issues related to the management of global order and global public goods more prominent. Global climate governance, which is widely related to global politics, economy, resources and energy, security and other governance content, has also been a hot topic for heads of states in international politics. As for climate change, at the platform of 2012-2015 Ad Hoc Working Group on the Durban Platform, countries in the world conducted intense negotiations around the results of the 2015 Climate Conference in Paris and actively carried out climate diplomacy at the level of heads of state or government so as to improve their voice and influence in the whole global governance through their impact on global climate governance. Active participation in global climate governance has comprehensive spillover effects on participation in global governance, which is of great significance.

Meanwhile, China's social and economic development has entered a new normal. The whole country has formed a broad consensus, that is, accelerating the transformation of economic development mode, adjusting economic structure, deepening reform and making efforts to overcome the "middle income trap" has become the only way for China to realize the China dream, namely achieving the "two century goals" and achieving the great rejuvenation of the Chinese nation, which is embodied as a more democratic and prosperous country, richer and better life for the people and more healthy and beautiful ecological environment. This is highly consistent with the goals of active response to global climate change. China actively responds to climate change not because someone else wants us to do so, but because we ourselves want to do so. The interests of the Chinese nation are highly consistent with the interests of mankind in dealing with climate change, which is an important starting point for building a community of human interests.

Since the Chinese government announced its goals of response to climate change before 2020 in the Copenhagen UN Climate Change Conference 2009, China has adopted a series of major policies to mitigate and adapt to climate change and it has made noteworthy achievements. In June 2015, the Chinese government submitted the post-2020 INDC documents to the United Nations, showing the strong determination and pragmatic attitude of a responsible big developing country to deal with climate change. In recent years, in the process of multilateral and bilateral cooperation in international climate, especially in a series of international negotiations on climate change, China is playing an increasingly positive role as a global climate governance builder upholding justice, daring to assume responsibilities and actively promoting cooperation. China is devoted to the construction of global climate governance with a positive and confident strategic posture.

Currently, a major event for the construction of global climate governance is to conduct negotiations of the 2015 Paris Agreement at the Durban Platform according to the requirements of Durban resolution of the conference of parties of the 2011 United Nation Framework Convention on Climate Change so as to form an international legal agreement aimed at strengthening the fulfillment of the Convention with the principles of the Convention as guide under the Convention and make the Climate Conference in Paris a landmark climate governance conference. The Convention is the basis for global climate governance in moral, political and legal aspects and the main channel to carry out international climate cooperation under the current world order and global governance framework. The principle of "common but differentiated responsibilities", equity and respective capabilities established in the Convention is the fundamental principle that should be followed by developed countries and developing countries to participate in global climate governance. Although the Convention has been concluded for more than two decades and the strategic pattern of the world has undergone profound changes, there is no fundamental change in the historical responsibilities, development stage and respective capabilities of developed countries and developing countries in the world and the disparities in development level even show an expanding trend; there are still two basic camps, namely developed countries and developing countries, when dealing with international political and economic affairs within the United Nations system. From the perspective of the pattern of accumulative emissions, which is the basis of historical responsibility for climate change, developed countries are still major contributors to emissions. Even in 2012, in terms of accumulative emissions of carbon dioxide and other main greenhouse gases related to energy and cement production since the industrial revolution in 1750, the developed countries included in Appendix I of the Convention contributed nearly 70% of global accumulative emissions with less than 20% of the world population. The historical accumulative emissions of the United States and Europe have reached 5 times of China. According to the estimates based on research into relevant scenarios, even by 2030, the accumulative emissions of the United States and Europe will be 2.6 times of China. All these indicate that the basis of the fundamental principle established in the Convention still exists and the principle of the Convention still applies to guiding the negotiations of the Paris Agreement.

The construction of a global climate governance structure must be at the height of responsible for the future of mankind, fully reflect the demand for sustainable development of peoples of countries in different development stages, uphold justice and adhere to fairness. When defining the responsibilities of different countries to mitigate and adapt to climate change, responsibilities should be distinguished according to their historical responsibilities, development stage, respective capabilities and national conditions. Developed countries should take lead to achieve full and absolute quantified emission reduction within scope of the whole economy and provide developing countries with adequate and timely financial, technical and capability building support so as to help developing countries achieve low-carbon transition. With the financial and technical support of developed countries, developing countries should take positive action to address climate change and try to explore a low-carbon transition path suitable for national conditions, ultimately forming a cooperative and win-win global climate governance system through north-south cooperation and south-south cooperation.

As a developing country, in the process of participation in global climate governance, on one hand, China should correctly judge its development stage, capabilities and national conditions, consider sustainable development as the top priority, safeguard their own development interests together with developing countries, take the initiative to undertake responsibilities suitable for its development stage and capabilities and actively carry out south-south cooperation with developing countries; on the other hand, China should take the attitude of responsible for history and mankind, change its sustainable development responding to the challenges of climate change into opportunities, actively explore innovative development paths, closely link protecting global climate with the transforming development mode, quality upgrading, structural adjustment, deepening reform and other major strategic tasks in solving its own development and maximize win-win of response to climate change and economic and social development.

The joint team of National Center for Climate Change Strategy and International Cooperation led by Professor Zou Ji and "Programme of Energy & Climate Economics (PECE)" of Renmin University of China has actively engaged in research into global climate governance and policy advice in the past few years, many members have been personally involved in the technical support practices for multilateral negotiation process under the United Nations Framework Convention on Climate Change and multilateral and bilateral climate diplomacy, and they have done a lot of meaningful research by integrating theory with practice and made many useful suggestions for government policy decisions. In particular, since 2013, during their exchange and debate with international negotiation circles and think tanks, they have put forward the uniform interpretation and discussion of international climate policy with the conceptual framework of "innovation of human development path" and they have made great progress in reflecting the principle of equity, "common but differentiated responsibilities" and respective capabilities in the formation of climate negotiation positions, grasping the long-term development trends and goals of response to global climate change, accurately locating all major stakeholders of global climate governance, proposing solutions to climate negotiations, implementing and coordinating the strategic thoughts of the domestic and international situations, etc.

Although the discussion in the book is a little hasty and inadequate due to time constraints, its exploration in systematic discussion of global climate governance is of significance. We hope that more Chinese and foreign experts, scholars and think tanks can conduct long-term, continuous and in-depth research into the issue and provide more adequate intellectual support for the development of global climate governance.

It is particularly worth noting that many people in the author team are recent graduates and new employees, even doctoral students in school. They have received good academic training and have opportunities to personally participate in important practices, they are full of passion and vitality, and have brought hope to the development of China's climate change think tanks, with which I am very pleased.

That's all.

Special Representative of China's Climate Change Affairs
XIE Zhenhua

Foreword

Since the adoption of United Nations Framework Convention on Climate Change (UNFCCC) in 1992, China has been taking part in this crucial international process. Over the past twenty years, China has experienced dramatic changes in socioeconomic development, and the world strategic pattern has changed a lot accordingly. As China is becoming the world's second largest economy, it's gaining more and more influence and power in international issues. However, at the same time, the domestic Greenhouse Gases (GHGs) emissions have also risen rapidly with economic growth, which has made China become the country with the annual largest instantaneous emissions in the world. Therefore, whether it is in the aspects of international influence, GHGs annual emissions amount, or efforts and actions to accelerate domestic green low-carbon development transformation, China is playing an unprecedentedly important role in global climate governance based on UNFCCC. Compared to the performance in the 1990s, China is now making more efforts in leading the negotiation process, proposing detailed programmes, acting as a system designer and leader in the process of building global climate governance system. In the 27th collective learning session of the Political Bureau of the Central Committee of the Communist Party of China (CPC), General Secretary Xi Jinping proposed more fair and reasonable global governance. It is also concluded at the Fifth Plenary Session of the 18th Committee of the CPC that China will actively participate in global climate change negotiation and positively bear international responsibility. All these strategic decisions from CPC have laid a solid political foundation for China to be deeply involved in global climate governance and also created the conditions for using global climate governance as a starting point for China to fully participate in global governance.

National Center for Climate Change Strategy and International Cooperation (NCSC) is a government think-tank directly under the management of National Development and Reform Commission (NDRC). One of its important duties is to provide technical support for China in participating in international climate negotiation, which is implemented by the NCSC Department of International Cooperation. Under the guidance of its deputy director Professor Zou Ji, the NCSC Department of International Cooperation is actively working on issues such as multilateral negotiations of climate change, climate change international cooperation and other related bilateral and multilateral processes; conducting negotiation-related key strategic studies for NDRC, Ministry of Foreign Affairs and Ministry of Finance etc.; and providing direct technical support for Chinese delegation in international climate negotiation.

Programme of Energy & Climate Economics (PECE) from Renmin University of China has been focusing on studies on climate change policies for a long time. It has gained a lot of research experiences in global climate regime, low-carbon policies analysis, development of emissions reduction roadmap and low-carbon technology strategies, etc., with rich outcomes.

Most of the authors of this book are young researchers from the NCSC Department of International Cooperation and Renmin University of China. They have the opportunity to become witnesses and performers of China's participation in global climate governance. They have been directly involved in the negotiation of Ad Hoc Working Group on Long-term Cooperative Action (AWG-LCA), Ad Hoc Working Group on Kyoto Protocol (AWG-KP), Ad Hoc Working Group on the Durban Platform for Enhanced Action (AWG-DP) under the Convention and the two Subsidiary Bodies of UNFCCC. They have been playing important roles in the negotiation of specific issues such as Shared Vision, 2013-2015 Review, National Communication, Methodology of the Kyoto Protocol, Multilateral Evaluation, Work-stream 2 of the Durban Platform, as well as long-term goals, mitigation, global stocktake and technology issues of the Paris Agreement. Furthermore, they have participated in formulating domestic strategies, including draft of China's official Intended Nationally Determined Contribution (INDC), analyze and assess the INDCs of China and other Parties, conduct model inter-comparison studies of emissions scenarios of both China and the globe with international organizations and research institutes, provide analytical support for the Chinese government to develop short-term tactics and long-term strategies in the climate negotiation. In addition, they provided technical support for governmental review of the Fifth Assessment Report (AR5) of the Intergovernmental Panel on Climate Change (IPCC) Working Group III, carried out interpretation of IPCC AR5, assisted Chinese government in organizing domestic research institutes and experts to participate in the IPCC progress, and provided suggestions on the future work of IPCC to Chinese government.

They have also taken an active part in the bilateral communication and cooperation related to climate change, by providing technical support to bilateral intergovernmental communication between China and US, Europe, France, UK, Germany, Canada, Russia, Japan and South Korea, organizing and hosting dialogue with experts from BASIC countries and Like Minded Developing Countries, and from think tanks such as World Resources Institute, Center for American Progress, Harvard University, Institute for Sustainable Development and International Relations, Stockholm Environment Institute, South Center and so on. They have directly supported the formulation of the historical US-China Joint Statement on Climate Change as experts from think tank.

Furthermore, the NCSC Department of International Cooperation has been carrying out strategy and policy research on climate finance and technology. The related work includes research on climate change strategy of the New Development Bank BRICS (NDB), climate finance issues of G20, participating in the negotiation of Intergovernmental Committee of Experts on Sustainable Development Financing, assisting Chinese government in the negotiation with the board of the Green Climate Fund, conducting climate technology demands assessments in China, serving as the National Designated Entity of the Climate Technology Centre and Network (CTCN), participating in the negotiation of Technology Executive Committee (TEC), coordinating policy assessment of short-lived climate forces (including HFCs) technologies, etc. They have also been executing comparative studies on national climate change strategies and policies of major developed countries such as EU, Japan, Russia, Australia, as well as major developing countries such as Brazil, South Africa, India, Malaysia, Indonesia and so on. By analyzing the South-South cooperation mechanism of climate change, they have provided policy recommendations for Chinese government to establish the South-South Cooperation Fund on climate change.

Although these tasks are tedious and trivial, main authors of this book have devoted themselves to every detail with greatest enthusiasm and the sense of national pride and historical missions. Meanwhile, they have kept on analyzing, refining and summarizing the key issues like theoretical foundation, historical trend, information, cooperation framework, main governing bodies and the relation with global governance, etc., during their research and practice, with full advantage of information and intellectual resources of NCSC as a national platform. This book consists of these information, knowledge and experience of the authors in participating global climate governance.

This book, starting from the scientific issues of climate change, analyzes the particularity in response to climate change and its challenges to global governance and human development. Based on the review of global industrialization history and the process under the UNFCCC, this book elaborates the conceptual framework of global governance on climate based on innovation of human development paths in the context of sustainable development and equity. This book consists of 9 chapters. Chapter 1, which was accomplished by Wang Ke, Zou Ji, Qi Yue and Chen Ji, introduced the scientific findings regarding climate change and its particularity, as well as the overall requirements of global climate governance and the development constraints faced by various countries. Chapter 2, whose authors including Fu Sha, Cui Xueqin, Liu Junling, Zhang Xiaoli, and Zou Ji, analyzes the differences, changes and drivers based on the environmental Kuznets Curve and the review of emission trajectories of developed countries. Chapter 3, written by Cui Xueqin, Fu Sha and Zou Ji, focuses on the issues of global development and environmental equity, including the ethics basis and connotation of equity and equitable effort-sharing schemes. Chapter 4, finished by Fu Sha and Zou Ji, elaborates the connotation of innovation of human development paths and its application in the international climate regime. Chapter 5, accomplished by Qi Yue, Chen Ji, and Fu Sha, discusses the unique role of the UNFCCC in the course of international cooperation in response to climate change based on the review of its history and existing mechanisms. Chapter 6, completed by Chen Ji, Wang Ke, Liu Junling, Fu Sha and Zou Ji, examines the courses of global sustainable development, international trade, and cooperation on finance as well as their impacts, and further elaborates the spillover effect of climate change issue as a non-traditional security issue. Chapter 7, mainly delivered by Chen Ji and Qi Yue, analyzed the responsibilities, roles and appeals of major economies in the global climate regime, including the U.S., the European Union, and major developing countries. Chapter 8, accomplished mainly by Fu Sha and Zou Ji, focuses on China's role in the global governance on climate change. Finally, our understanding of global governance on climate change based on the concept of innovation of human development paths and its policy implications is summarized by Zou Ji in Chapter 9.

In addition to the contribution of the main authors, this book cannot be finished without the help of many institutions and colleagues. We would like to express our thankfulness to Mr. Xie Zhenhua in particular, the former vice chairman of National Development and Reform Commission and the present special representative on climate change affairs of China, who has a profound influence on the authors' points of view with his wisdom, mind, courage, numerous specific guidance and far sightedness. We would also give our sincere appreciation to the Climate Change Division of National Development and Reform Commission for providing the research platform and guidance, to Mr. Su Wei, the chief representative of Climate Change Negotiations and his team, for the precious support they generously offered to us. Our special appreciation also goes to the leaders of National Center for Climate Change Strategy and International Cooperation for their support. We would also like to thank the important contributions from Zhang Xiaohua, Hu Xiao, A'Rouna, and Liu Linwei in the previous researches, and the arduous work done by Xu Lili, Chen Yiying, Zhong Yang, Ma Hongyan, Sun Qian, and colleagues from Programme of Energy & Climate Economics (PECE) of Renmin University of China. China Clean Development Mechanism Fund Grants has provided strong support in writing and publishing of this book. We would also like to express our appreciation to the following projects: Bali Roadmap negotiations progress and prospects study, on the Durban platform major strategic issues and the overall strategy study, BASIC experts mechanism and developing countries think tank exchanges and cooperation platform for technical support for the project, after 2020 "fair, effective, win-win" international cooperation in the enforcement action arrangement design and China's countermeasure research, the bilateral practical cooperation in climate change technology support project, under the Convention 2013-2015 global long-term goals of the technical review support projects.

Due to author's time and capability limitation, there still exist some shortcomings and inevitable errors, and your critical opinions and suggestions are highly appreciated.

Zou Ji

ZOU Ji on behalf of the authors

Yu Yuan Tan, Beijing

late autumn of 2015

Summary

1. There has been an urgent need of developing a global climate governance system to manage the remaining global atmospheric carbon capacity, which has become increasingly scarce.

Based on the latest scientific findings, the fifth assessment report of the United Nations Intergovernmental Panel on Climate Change (IPCC), explores the correlation among temperature, greenhouse gas concentrations and historical cumulative emissions. The report states that if we still want to hold the increase in the global average temperature below 2°C above pre-industrial levels by the end of the 21st century, in order to minimize the risks from climate change, ambitious actions should be taken from now on to reduce the instantaneous greenhouse gas (GHG) emissions and then limit the cumulative emissions in the future. Under current proposals, global GHG emission should be reduced by 40-70% in 2050 compared with that in 2010, and zero-emission should be almost achieved by 2100. Obviously, the global atmospheric carbon capacity is becoming limited. Considering the global GHG emissions in 2010, the available quota of GHG emissions can only allow for another 17 to 31 years before 2050.

Although uncertainties exist in scientific assessment, it is clear that there is a top limit for the global GHG emissions according to science. It means that there will be a tight constraint of GHG emissions space for each country, which would have a tremendous and fundamental impact on the socioeconomic development of that country. In this regard, the increasingly scarce global atmospheric carbon capacity can be seen as the public wealth for the global citizens, and there is an urgent need to develop a global climate governance system of assigning and managing this space.

2. The externality of climate change, characterized with large temporal and spatial scales, brings enormous challenges to management for the public wealth of global climate.

Carbon dioxide and other greenhouse gases can remain in the atmosphere for a long time, thus have wide-ranging effects across the globe, whose externalities are transnational and intergenerational, bringing great uncertainty. These make the climate change beyond the spatial and temporal vision of extant normal decision in national and subnational level. In the absence of a super-sovereign governance body above national sovereignty, the formation of a constantly evolving and strengthening global climate governance structure can only be achieved via inter-governmental negotiations and transnational consultative process with sharing national values and development concepts. The process is slow, in conflict with the urgent requirement for addressing climate change, which brings huge challenges to global climate governance system. In addition, in current global political system, the management of externality with large spatial and temporal scales still requires national engagement as a key role, so the thinking and vision of national decision makers should extend across the globe and time.

3. Global climate governance system is being formed, with intensifying constraints on the greenhouse gas emissions.

In order to control the rights of greenhouse gas emissions and to build up a responsibility system of mitigation and adaptation, the international community in 1992 reached the United Nations Framework Convention on Climate Change (the Convention). As a political and legal foundation of the global climate governance system, the Convention has provided conceptual stipulation on all the key elements of global climate governance system, such as its objectives, key areas for cooperative actions, role and responsibilities of parties, mode mechanism for global cooperation and collaboration, which are endorsed and supported by, over 190 Parties both in terms of politics and international laws.

On the basis of the Convention, under the guidance of the principle of equity, the principle of common but differentiated responsibilities and the principle of respective capabilities, the global climate governance system is in constant improvement with intensifying constraints on greenhouse gas emissions. The Kyoto Protocol signed afterwards as well as all the decisions of the Conference of the Parties, together with the agreement possibly signed in the Paris Conference of 2015, have been all aiming at continuously enhancing the implementation of the Convention at the operational level and in different historical periods.

4. The principle of equity and the principle of common but differentiated responsibilities remain the cornerstone of global climate governance.

Sovereign states are the major entities in global climate governance and their position and responsibility are usually determined by their historical responsibility, development stage, capability and national circumstance. Since WWII and especially the enactment of the Convention, although the composition of global population, energy use and annual emissions between developed and developing world has changed drastically, the huge gap remains in economic growth, development stages and capabilities. Given its smaller share of population, developed world remains a larger greenhouse emitter both in terms of cumulative and per capita emissions. Therefore, insisting on the dichotomy between developing and developed countries as well as the principle of equity, the principle of common but differentiated responsibilities and the principle of respective capabilities of the Convention, is still vital - rather than outdated - in the construction of global climate governance system. The process of the Convention over two decades has also proven that the principle of common but differentiated responsibilities will not hinder the efforts in low-carbon transition in emerging economies, such as China. On the contrary, it has promoted their exploration on innovative low-carbon development pathways in accordance with their own historical responsibilities, development stages, capabilities and national circumstances.

Equity, effectiveness and multiple wins are the three basic criteria to evaluate whether global climate governance system is successful or not. As the fundamental principle of global climate governance, equity should be reflected on the basis of assessing four key elements of each country, namely historical responsibilities, development stages, capabilities and national circumstances. Since the Industrial Revolution, developed countries have emitted over 70% of cumulative CO_2 in to the atmosphere. The cumulative emissions of the US and Europe are four times larger than that of China. Undoubtedly, developed countries should take the major historical responsibility for global climate change at present and a long time in the future. Developed countries' well-constructed infrastructure, abundant materials, strong competitiveness, and dominant position in global governance at present are formed on the basis of consuming a large part of global common carbon resources in the past. Moreover, through investment, trade, spread of values, knowledge and technology in the ongoing globalization process, developed countries have imposed their carbon intensive development pattern and outsourced carbon emissions to developing countries.

For developing countries, expanding employment, reducing poverty, constructing infrastructure, improving social security system and enlarging middle-income class remain the top priorities of socioeconomic development. Their industrialization and urbanization will still be the major concerns of economic development in the foreseeable future, while facing unprecedented constraint of carbon emissions. Their modernization through the traditional carbon intensive development pathway set by developed countries hence will fundamentally conflict with this carbon constraint.

Developed and developing countries have common but differentiated responsibilities in addressing climate change since they have divergent historical responsibilities, development stages, capabilities and national circumstances. On one hand, developed countries shall speed up domestic de-carbonization and achieve deep emissions reduction in order to leave more space of carbon emission for the developing countries in transformation;

on the other hand, they shall provide financial, technological and capacity building support to enable low-carbon transformation of developing countries. Developing countries need to change their view on development that should be in line with achieving low-carbon transformation. This transformation can be made by promoting optimal allocation of productive elements as well as advanced technologies and low-carbon energies, improving well-designed infrastructure and appropriate layout of production capacities while ensuring the continuous increase of human development index.

Hence, common but differentiated responsibilities is the operationalization of climate equity in global climate governance, and the fundamental criterion to determine the adaptation and mitigation targets, corresponding responsibilities/rights to provide/receive financial and technology transfer support and the requirement for transparency of different Parties under the Convention. Without a fair foundation, the recognition to the legitimacy and the motion to take part in the global climate governance of developing countries will be severely eroded. As a result, the political mutual trust of multilateral cooperation will be undermined, and effective climate change strategies and policies will be hard to be formulated. Hence, to what extent is the essential principle of common but differentiated responsibilities being reflected in the agreement, will be a significant symbol to judge whether the 2015 Paris Conference is successful or not, and a sign to check in which direction of global climate governance is going in the future.

5. Controlling global GHG emissions and Breaking the Deadlock in the global climate governance through Innovation of Development Path.

An empirical examination on historical carbon dioxide emission tracks of major developed and developing countries since the industrial revolution would reveal that CO_2 emissions from fossil fuel combustion and cement production in major developed countries have already peaked after their industrialization and urbanization, yet with different peaking characteristics. The United States, Canada and Australia reach the peak with CO_2 emission per capita as high as about 20 tons; while that of European Union and Japan is around 10 tons. This shows that in the historical practices, there are two kinds of emissions trajectories with quite different peaking level ("the U.S, Canada, Australia"and "European and Japan" mode), and it is possible to achieve similar per capita income level with much less per capita CO_2 emissions.

In examining the correlation between CO_2 emission per capita and GDP per capita of an economy that has reached emission peak since the industrial revolution, one can find that no economy by far can avoid the Kuznets Curve phenomenon, meaning that energy consumption related carbon dioxide emissions per capita "first rises and then goes down" as level of development, or more precisely GDP per capita, increases, although with different characteristics in different countries. As a whole, developed countries peaked when its GDP per capita was around USD 20,000-25,000 (in 2010), indicating that an economy relies more on technology and service and less on fossil fuels consumption as it undergoes industrialization and urbanization. Most developing countries are still situated in the left half of the Kuznets Curve with greater possibility of increasing their emissions and their share of global emissions. In terms of historical cumulative emissions and emissions per capita, developing countries are still far lower than developed countries. However, developing countries can't follow the traditional development paths set by developed countries because of the national and global resources and emission budget constraints. Their choice among different development paths, namely the U.S.-Canada-Australia mode, EU-Japan mode and low-carbon innovative development path that is fundamentally different from that of developed countries, will exert great influence not only on the sustainable development of their own but also on global efforts in addressing climate change.

The essence of changing the emissions trajectory is to steer the development path. The connotation of development path is very rich, which is generally associated with a number of questions, such as what is the engine of economic growth, whether economic growth is made by higher elements efficiency or higher elements input, whether the income is mainly derived from investment or from consumption and net exports; and also some key variables, inter alia, population, income level, urbanization rate, share of carbon intensive energy or low (zero)-carbon energy in total primary energy, and energy intensity per unit of GDP which is affected by industrial products structure and technological progress. All of these will be reflected in the development of industrial sectors and infrastructure construction. With the increase of income level, the share of service industry in the economy will increase, while the domestic market for energy intensive or carbon intensive manufactory industry will become saturated, and the share will decrease. Technological advancement within manufactory industry, together with the extension of industrial chain, the change in product structure and the upgrading in value added, will result in the decrease of energy consumption and emission intensity.

It is necessary for developing countries to skip the conventional development model led by developed countries and initiate an innovative development path that is characterized by low-carbon as well as earlier and lower emissions peak. The concept of innovation of development path is not an empty academic concept but is distilled from the reflections upon the industrialization history of developed countries. Developing countries are able to peak earlier and lower by influencing the emission trajectory and innovating development paths with proper policies and institutional arrangements. Lowering the energy intensity per unit of GDP and decarbonization of energy structure will be key measures.

In general, the innovation of development paths starts with the belief that countries in different development stages need to take positive actions and explore options of development path innovation that suits their own development stage. This conceptual framework focuses on drivers that can influence the development of emission trajectories and paths and aims at breaking the deadlock in negotiations regarding global climate governance from a more substantial, practical and concrete aspect. Overall, this conceptual framework has constructive policy implications as it focuses on substantial actions and measures and identifies respective directions and emphases of actions of developed and developing countries under the general framework of sustainable development in consideration of innovation and transition of development paths.

The concept of innovation of development path can further help confirm that it is reasonable and necessary to insist on differentiation between developed and developing countries. Apart from cumulative emissions, developed countries are also responsible for the creation, domination, dissemination and reinforcement of the traditional carbon intensive development path. Most developing countries are in a disadvantageous position in global supply chain with high-energy consumption, high emission and low value-added. Although the emissions from developed countries are decreasing after completion of industrialization, they are still responsible for the emission increase trends and path dependency of developing countries in the context of globalization. Therefore, to change the future development path of developing countries is the problem of global system adjustment, which developed countries bear unshakable responsibility.

6. The UNFCCC is the main channel for global climate governance and the moral, political and legal basis of global climate governance.

The existing grouping between developed countries and developing countries does not only reflect political and economic reality of the current world, but also is the ongoing practice of setting up negotiating groups under the United Nations process for decades, which has profound historical and realistic origins and will last long. For a long period of time, the goal, the principles and the provisions of the UNFCCC will still be applicable. The objective of the UNFCCC reflects the idea of supporting sustainable development by protecting atmospheric environment. The UNFCCC introduced the principle of equity, the principle of common but differentiated responsibilities of developed countries and developing countries on mitigating and adapting climate change and the principle of respective capabilities. According to these principles and provisions, the common responsibility of all Parties includes, but not limited to, formulating national strategies and plans in response to climate change and submitting national GHG inventory and national communication. As for the differentiated responsibility, developed country Parties shall take lead to mitigate climate change by committing and implementing economy-wide absolute emission reduction targets and providing financial, technological and capacity building supports to developing country Parties, while the actions taken by developing country Parties shall be dependent on the availability of support from developed country Parties. Mitigation and adaptation have equal importance in response to climate change.

7. Setting up an appropriate goal for global climate governance and balancing the "bottom-up" and "top-down" approach.

The goal of the global climate governance should be a set of targets. The long-term global target should be to politically recognize the urgency and importance of combating climate change, guide the direction of global, regional and national low-carbon development, define the responsibilities of countries and the targets for various mechanisms and policies, and send clear signals to policy makers and investors. However, the simulated results with uncertainties should not be directly used as the quantitative basis for achieving the global targets as this may trigger the risks of "number-game" and "blame-game", increasing difficulties in the multilateral negotiation. In order to motivate all parties to keep the low-carbon development right on track, while practically promoting the transformation into low-carbon, the goal of global governance on climate should be based on the consensus of global low-carbon development and innovation of human development paths; it should include the economy-wide absolute emission reduction targets by developed countries and various relative targets by developing countries, such as the reduction of carbon emission intensity, the increase of non-fossil fuel consumption and the increase of forest carbon sink; and it should closely follow-up and report back the developments in technology and markets to accelerate the updating of the targets by stakeholders.

The debate on "top-down" versus "bottom-up" has been in the UNFCCC process for more than 20 years. After the US refusal to ratify the Kyoto Protocol, US domestic politics had underscored the influences in the multilateral process. The "top-down" approach faced with tremendous trouble in the absence of supra-national sovereignty. In the new round of climate negotiation, as a symbol of "bottom-up" approach, the "Intended Nationally Determined Contribution" was proposed and accepted by Parties in Warsaw Climate Change Conference of 2013. However, the debate regarding the stock-take or review process and its implications is still a key issue in negotiation and can hardly be determined until the last minute of Paris. Additionally, the legal forces of different contents and arrangements of the new agreement are also key issues to be addressed in global climate governance. It is also crucial to balance the "national determined contributions" with the arrangements of "global stock-take" and transparency, which aims at further enhancing the ambition. The global climate governance has to guarantee the environmental integrity and encourage the widest participation including the US and developing countries by combining the "top-down" and "bottom-up" approaches properly.

8. Dealing with the dialectical relationship between big power influence and extensive engagement is key to global climate governance, and the rights and interests of developing countries need to be respected and protected.

Decision making under the UNFCCC is based on consensus, which is the fundamental rule used by the United Nation bodies to deal with multilateral international affairs. It ensures equal right of Parties regardless of their size and economic development level, concerns of various stakeholders being heard and democracy with extensive engagement and transparency to enhance political mutual trust. In the process of the international climate progress, big powers have played a leading role and thus are of higher influence. To ensure the equity and effectiveness of the international climate progress, it is core to properly deal with the dialectical relationship between big power influence and extensive engagement. For historical reasons, developed countries are significantly well positioned in the rule-making process, which thereby raises a question as how to respect and protect the rights and interests of developing countries - particularly the least developed countries in Asia, Africa, Latin-America and the vulnerable small island developing countries. To enhance the capabilities of developing countries participating in the global climate governance, it is necessary to ensure proper institutional arrangement in place to meet the demands in finance, technology and capacity building of developing countries.

9. Global climate governance is a crucial component of global governance, which matters to the construction of human common destiny and global public wealth as well as global governance on politics, economy and security.

Strengthening global climate governance can help maintain global climate public wealth and realize human common interests and facilitate the construction of human common destiny. The core of this governance should be multiple-win cooperation. It is concerning the overall strategic target of human sustainable development as well as global governance on politics, economy, security and many other important aspects. Climate change has been identified as one of the 17 sustainable development goals put forward by the United Nations Sustainable Development Summit in September 2015. Climate change, with its nature of comprehensiveness, spans across the major areas on global governance:

for politics, it is particularly relevant to leading power, extent of voice and global influence; for economics, it is relevant to the trend and landscape of global economy, international trade and investment as well as financial standard and competitiveness; for the issue of security, it is also becoming an increasingly important part of Non-Traditional Security with regards to the tremendous losses caused by the negative impact of climate change.

10. Both developed and developing countries should explore their own innovation of development path and find the balance between competition and cooperation among each other, considering their different historical responsibilities, developmental stages, national capacities and circumstances.

Global climate governance has become a new boundary of international relations. Its influence can be as great as redefining every country's long-term competitiveness and thus changing the landscape of international competition. Its historical evolution and the need of innovative reform have been reflecting the dynamism of global political power. All countries in global climate governance have taken actions to different extents. They are trying to maintain and promote their international competence and profile, in order to take the lead in global climate governance. Developed countries (groups), such as US and EU, are competing for the leadership in governance patterns, elements, development pace, agenda setting, rules making, moral and legal basis, and other important decisions in global climate governance.

Since developed countries lead the way to the post-industrial age, their energy structures were generally optimized, and their industry structures were adjusted by globalization. Carbon intensive industries have been mostly moved to developing countries, while developed countries remain at the top of the industrial chain and maintain their technology edges. As a result, developed countries achieved continuous decrease of carbon intensity. Also, stable population and steady income growth enable developed countries to achieve significant GHG emissions reduction. Developed countries play a key role in leading the global low-carbon development.

However, developed countries with 20% of world's population still account for nearly 70% of world's cumulative emissions. It is estimated that developed countries' cumulative emission will remain larger than developing countries by 2030. The emissions per capita of USA, Canada, Australia, remain 15-20 tons per person, still far more than the world average. A large gap still exists between developed countries' aggregate efforts of emissions reduction and what is required by the IPCC fourth assessment report, namely reducing 25%-40% emissions by 2020 compared with the 1990 level.

Financial and technical supports from developed countries to developing countries are still at a slow pace, far from the requirements for addressing climate change. Developed countries' efforts still fall short of the expectations of developing countries and also cannot meet the requirements for dealing with global climate change. During the negations on global climate governance arrangement after 2020, some major developed countries make efforts to deviate from the principle of the Convention and tend to rewrite the rules of governance blatantly.

With their rise over two decades, emerging developing countries continue enhancing actions to addressing climate change. They have achieved strategic constantly and brought the issue of dealing with climate change into their socioeconomic development strategy, under the guidance of the principles and provisions of the Convention. Besides, according to the Cancun Decisions and the Paris Agreement in process, emerging developing countries set relative emissions targets and national climate change adaptation strategies based on respective national abilities and conditions. Although these efforts help developing countries to explore the new low-carbon development path, and also get marked achievements, lots of tough challenges still remains.

All in all, although developed and developing countries have different historical responsibilities, developmental stages, country capacities and national conditions, they have formed the strategic consensus of the low-carbon development path, international cooperation needs and some major issues. In the process of further development of the global climate governance, the consensus on achieving an innovative development pathways needs to be further condensed, and based on the different abilities and conditions, more targeted strategy of path innovation, goals, actions and policies, measures and supporting system of developed and developing countries in different development stages need to be recognized.

11. China's current development stage and its new global strategic positioning determine that innovation of development path is China's inherent requirement.

After more than thirty years of rapid development since the opening-up, China as the biggest developing country, which is undergoing rapid industrialization and urbanization and towards a higher stage of development, has achieved great achievements. China's total economic output now ranks the second of the world and its GDP per capita has increased to around USD7,800. China has almost accomplished its task of poverty alleviation and is now entering into a new stage to fully establishment of a well-off society by 2020.

From the perspective of growth economics, the inputs of elements like labor, capital, energy, land, water, mineral products and environmental capacity have made a major contribution to the rapid economic growth over three decades. Upgrading economic development to "new normal" condition means that China's traditional development mode relied on the quantity of input of elements has come to an end. In the economic take-off stage, by over consuming natural resources and environmental capacity to complete the primitive capital accumulation, this process has result in enormous social and environmental consequences. As the coal based energy endowment becomes a new comparative disadvantage under increasingly tight global carbon constraints, if China failed in transiting to new economic development mode driven by innovation and promoting technology and industrial revaluation characterized by low-carbon timely, China is very likely to fall into "middle income trap" restricted by resources and environment, which will seriously affect the realization of the two century goals. On the other hand, with the increase of the stock of social wealth and income level, it is possible to shift to a new development stage under which the economic growth mainly relies on the efficiency of inputs, thus alleviating the social consequences caused by overdraft of natural recourses.

China's foremost strategic priorities in the next decades continue to focus on increasing incomes, moving towards more advantageous development stage, and generally raising the standard of living. Meanwhile, China's traditional input-heavy growth model is no longer sustainable and making China faces the risks falling into the "middle-income trap".

In the context of comprehensive deepening reform and promoting the construction of ecological civilization, exploration and development of an innovative development path has become a prioritized theoretical and practical mission for China. Low-carbon development as an efficient and intensive economic development model, to foster and promote new economic growth point and model so as to improve the quality and effectiveness of economic growth by the application of new technologies and regulations that match new resources and environmental constraints, will become a powerful driving force to promote China's economic restructuring and development.

China needs to explore an innovative low-carbon development path which is better than the traditional ones of developed countries, and to upgrade its growth to a "new normal" condition: low-carbon development as a key driving force to improve elements productivity and achieve transition, and to achieve the innovation of development path by increasing energy efficiency, transforming energy mix, upgrading industrial structure, optimizing income structure and improving human resources. Innovation of development path and implementation of low-carbon strategy shall be understood as a way to achieve China's two century goals and to establish China's comparative advantages in global low-carbon trends and low-carbon technology revolution.

12. China's innovative development pathway is of considerable importance to the global green and low-carbon transformation.

Having experienced 30 years of rapid growth and 20 years of intensifying globalization, China now plays a pivotal role in the world economy. Therefore, in analyzing China's new development phase and the economic transition, one has to consider China's position in the world economy as a whole. China's innovative development path is of great importance for global transition towards green and low-carbon economic development.

Since the industrial revolution, the global cumulative greenhouse gas emissions pattern is still the same as it is contributed predominantly by developed countries. While the annual instantaneous emissions pattern has remarkably changed and will eventually alter the cumulative emissions pattern over the long term. China's rapid growth of total carbon emissions and emissions per capita has been increasingly noticeable. Although there is still disparity between China and the US/ EU in cumulative greenhouse gas emission, with the high GDP, considerable energy consumption, fast economy development and further acceleration of the China's industrialization and urbanization, it is inevitable that China's CO_2 emission growth will continue and eventually raise the international share of historical cumulative emissions. For this reason, the future carbon emission in China is critical to the global emission pattern and pathways. China's achievement of managing and controlling emissions through innovative development pathway will be essential for the world to achieve carbon emissions cut goals.

Along with the economic transition to a cleaner, more efficient and low-carbon development, there will be an enormous domestic market created in China for clean energy technology, products and services. On top of that, there will be large-scale demands for developed countries on regulations, technologies, standards and advanced management experiences. China, with the world's leading manufacturing capacity and high-quality production capacity, with the enormous domestic market and relative abundance of capital, can strengthen the collaborations with the US, EU and other countries throughout the low-carbon technology industry chains by means of bilateral and multilateral cooperation activities. Parties in the cooperation can optimize the global industrial layout and the specialized division of labor, enhance the R&D ability in low-carbon technology development and the innovation capacity for developing business models. Meanwhile, the cooperation can accelerate the cost reduction of low-carbon technology application, and help expand the market shares of low-carbon technologies and products worldwide. All of these outcomes will further help lower the carbon emissions, while advancing social prosperity and employment to achieve a re-balanced global economy. Moreover, the low-cost and low-carbon solutions will promote the renovation of infrastructure facilities in developed countries, reducing the difficulties for these countries to achieve a low-carbon transition.

As the world's largest developing country, China can provide other developing countries with referential demonstrations of its innovative development pathway and experience-sharing on its low-carbon sustainable development. By offering supports via the channel of South-South Cooperation, China can help these countries avoid the dependence on and Lock-in Effects of conventional high-carbon pathways, while stepping toward a "low-pollution, low-emission" but highly efficient & innovative development path. By self-transitioning, China will leverage low-carbon transition activities on a global scale, and will make significant contributions to the reshaping of a new global development path.

13. China's participation in global climate governance should aim to realizing the transition from a participant to a builder of the system while insisting on its positioning as a developing country.

Within today's world economy, energy and climate change system, China's basic position is a developing country with a unique status. Even though China has become the largest carbon emitter and the second-largest economy in the world, its main development indicators and the ongoing bifurcation of rural and urban economic structures demonstrate that China is still a developing country. This implies significant differences between China and developed countries in terms of their respective stages of development, development needs, historical responsibilities, and overall capacity. In addition, China also faces specific national circumstances like coal based energy resources endowment, vast landmass, limited stock of resources per capita in terms of both quantity and quality, etc. China needs to fully consider its developing stage, capacity and national circumstances while participating in global climate governance.

Climate change has become an important issue concerning international morality, global responsibility, national image and national development rights. China, as a developing country responsible for the entire world, on one hand shall respect historical facts, adhere to its positioning as a developing country and the principle of "common but differentiated responsibilities", require developed countries to fulfill their responsibilities in line with the principles and provisions of the Convention, e.g. to take lead and set example in mitigation, and to provide adequate technical, financial supports to developing countries, etc.; on the other hand, China shall be fully aware of changes in its development stage and situations it faces in the new period, recognizing that addressing climate change and innovating development path are something we need to do rather than something asked to do. In China's efforts to move its economy towards a more advanced stage, active participation in global climate governance is the internal requirements in respect of adjusting structure, overcoming middle-income trap, complying with the global trends of low-carbon development and technological revolution, and improving international low-carbon competence.

Based on its national circumstances and objective positioning, China should participate in the international climate process in an active and pragmatic way within the framework of sustainable development and based on the principles of Convention, play a constructive role as a responsible great country, undertake international responsibility to tackle climate change, which shall reflect its development stage, overall capacity, positioning and national circumstances, and transform its role from a participant to a builder to achieve strategic turn from "homeopathy" to "proactivity", ensuring an equitable, effective and win-win global climate regime.

In general, China should consider the following objectives in a comprehensive way in its participation in global climate governance:

(1) To guarantee reasonable essential emission space before peaking for China's smooth completion of industrialization, urbanization, and development path transition;

(2) To strive for an "equitable, effective and multiple win" international climate regime and support actions to address climate change, so as to minimize the negative impacts of climate change;

(3) To minimize domestic economic dependence on GHG emissions and energy consumption and cultivate long-term competitiveness in green and low-carbon development by pursuing lower and earlier peaking of emissions;

(4) To transform from a participant in the global international regime into a builder and leader, and use this as one of the platforms to improve China's position, leadership and voice in setting and implementing rules of global governance;

(5) To establish a positive image of China as a responsible nation with fairness and justice.

缩略词 Abbreviations

U.S.-ACEF	Africa Clean Energy Finance	美国 - 非洲清洁能源基金
ADP	Ad Hoc Working Group on the Durban Platform for Enhanced Action	德班增强行动平台
AGBM	Ad Hoc Group on the Berlin Mandate	柏林授权特设工作组
AIP	Africa Infrastructure Program	非洲基础设施项目
APEC	Asia-Pacific Economic Cooperation	亚太经济合作组织
AR5	Fifth Assessment Report	第五次评估报告
ARRA	American Recovery and Reinvestment Act	美国复苏和再投资法案
BAU	Business as Usual	基准情景
BECCS	Bioenergy with Carbon Capture and Storage	生物能源与碳捕集和封存
BR	Biennial Report	两年报
BUR	Biennial Update Report	两年更新年报
CAT	Climate Action Tracker	气候行动跟踪
CBDR	Common but Differentiated Responsibilities	共同但有区别的责任
RC	Respective Capabilities	各自能力
CCS	Carbon Capture and Storage	碳捕集和封存
CDIAC	Carbon Dioxide Information Analysis Center	二氧化碳信息分析中心
CD	Concentric Differentiation	同心圆
CDM	Clean Development Mechanism	清洁发展机制
CGE	Computable General Equilibrium	可计算一般均衡
CIF	Climate Investment Funds	气候投资基金
CMP	Conference of the Parties Serving as the Meeting of the Parties to the Kyoto Protocol	《京都议定书》缔约方会议
CO_2eq	Carbon Dioxide Equivalent	二氧化碳当量

COP	Conference of the Parties	缔约方大会
CRF	Common Report form	统一报告表格
CTCN	Climate Technology Centre & Network	气候技术中心与网络
DOE	Department of Energy	美国能源部
EGA	Environmental Goods Agreement	环境产品协议
EIA	U.S. Energy Information Administration	美国能源信息署
EKC	Environmental Kuznets Curve	环境库兹涅茨曲线
EPA	Environmental Protection Agency	美国环境保护署
ERF	Equity Reference Frameworks	公平参考体系
ERT	Expert Review Team	专家评审小组
ERU	Emission Reduction Unit	减排单位
EU ETS	European Union Emissions Trading System	欧盟排放交易系统
FCPF	Forest Carbon Partnership Facility	森林碳伙伴基金
G20	Group of Twenty	二十国集团
G8	Group of Eight	八国集团
GATT	General Agreement on Tariffs and Trade	关税及贸易总协定
GCCI	Global Climate Change Initiative	全球气候变化倡议
GCF	Green Climate Fund	绿色气候基金
GDP	Gross Domestic Product	国内生产总值
GEF	Global Environment Facility	全球环境基金
GHG	Greenhouse Gases	温室气体
GMI	Global Methane Initiative	全球甲烷行动倡议
GTAP	Global Trade Analysis Project	全球贸易分析模型
HDI	Human Development Index	人类发展指数
HFCs	Hydrofluorocarbons	氢氟碳化物
IDPs	Innovation of Development Paths	发展路径创新
IAR	International Assessment and Review	国际评估与审评
ICA	International Consultation and Analysis	国际磋商与分析

IDDRI	Institute for Sustainable Development and International Relations	法国巴黎可持续发展与国际关系研究所
IEA	International Energy Agency	国际能源署
INC	Intergovernmental Negotiating Committee	政府间谈判委员会
INDC	Intended Nationally Determined Contributions	国家自主减排贡献
IPCC	Intergovernmental Panel on Climate Change	政府间气候变化专门委员会
KP	Kyoto Protocol	京都议定书
LEG	Least Developed Countries Expert Group	最不发达国家专家组
LULUCF	Land Use, Land-Use Change and Forestry	土地利用、土地利用变化和林业
MCC	Millennium Challenge Corporation	美国千年挑战公司
MRV	Measurable, Reportable and Verifiable	可测量、可报告、可核实
NAMAs	Nationally Appropriate Mitigation Actions	国家适当减缓行动
NAPAs	National Adaptation Programmes of Action	国家适应行动方案
NAPs	National Adaptation Plans	国家适应计划
NASA	National Aeronautics and Space Administration	美国国家航空航天局
NCSC	National Center for Climate Change Strategy and International Cooperation	国家应对气候变化战略研究和国际合作中心
NIR	National Inventory Report	国家温室气体清单报告
NTS	Non-Traditional Security	非传统安全
ODA	Official Development Assistance	政府开发援助
OECD	Organization for Economic Co-operation and Development	经济合作与发展组织
OPIC	Overseas Private Investment Corporation	海外私人投资公司
ORNL	Oak Ridge National Laboratory	橡树岭国家实验室
PBL	Netherlands Environmental Assessment Agency	荷兰环境评估署
PECE	Programme of Energy and Climate Economics	能源与气候经济学项目组
PIK	Potsdam Institute for Climate Impact Research	德国波茨坦气候影响研究所
PM2.5	Fine Particulate Matter 2.5	直径小于或等于 2.5 微米的颗粒物

PPP	Purchasing Power Parity	购买力平价
PTA	Preferential Trade Agreement	优惠贸易协定
PTC	Production Tax Credit	生产税减免
R&D	Research & Development	研发
RCEP	Regional Comprehensive Economic Partnership	区域全面经济伙伴关系
RCPs	Representative Concentration Pathways	典型浓度路径
REDD-plus	Reducing Emissions from Deforestation and Forest Degradation and the Role of Conservation, Sustainable Management of Forests and Enhancement of Forest Carbon Stocks in Developing Countries	减少发展中国家毁林和森林退化造成的碳排放加养护、森林可持续经营以及增加森林碳储量
RUC	Renmin University of China	中国人民大学
SBI	Subsidiary Body for Implementation	附属履行机构
SBSTA	Subsidiary Body for Scientific and Technological Advice	附属科技咨询机构
TEC	Technology Executive Committee	技术执行委员会
TPP	Trans-Pacific Partnership	跨太平洋伙伴关系协议
TTE	Team of Technical Expert	技术专家小组
TTIP	The Transatlantic Trade and Investment Partnership	跨大西洋贸易与投资伙伴协议
UN	United Nations	联合国
UNDP	United Nations Development Programme	联合国开发计划署
UNEP	United Nations Environment Programme	联合国环境规划署
UNFCCC	United Nations Framework Convention on Climate Change	联合国气候变化框架公约
USAID	United States Agency for International Development	美国国际开发署
WB	World Bank	世界银行
WG3	Working Group III	第三工作组
WRI	World Resources Institute	世界资源研究所
WTO	World Trade Organization	世界贸易组织
WTP	Willingness to Pay	支付意愿

表目录

图目录

Tables

Figures

第一章 全球大气碳容量约束与全球气候公共财富管理

Chapter One Constraints of Global Carbon Capacity and Management of Global Public Goods

摘要

政府间气候变化专门委员会（Intergovernmental Panel on Climate Change，IPCC）第五次评估报告基于最新科学进展，进一步建立了温度、浓度、累积排放量之间的对应关系，提出为了较大可能实现2°C温升目标，全球2050年温室气体排放量相比2010年需要减少40%—70%，到2100年实现近零排放。虽然存在科学不确定性，但是气候科学评估结论仍然对全球温室气体排放提出了上限要求，世界各国未来排放空间将受到极为现实的紧约束，这将给各国的社会经济长期发展带来巨大而深远的影响。

温室气体在大气层留存的寿命周期长且影响范围涉及全球，同时具有全球跨国外部性和跨代外部性的问题特征，并存在很大的不确定性。这不仅使得气候变化问题超越了国家范围内现有决策主体的常规决策时空视野，并且由于目前还不存在一个凌驾于各国主权之上的纯粹超主权的治理主体，只能通过跨国谈判和协商进程，通过各国共享的价值观和发展理念等，形成不断演进和强化的全球气候治理结构，这给全球气候治理带来了巨大的挑战。此外，大时空尺度的外部性管理，要求国家主体发挥关键作用，国家决策者的决策思维和视野需要扩展至全球并跨越时间长河。

科学评估还表明，自工业革命以来，发达国家以比较少的人口占比占有了近70%的二氧化碳累积排放量，美欧迄今的累积排放量是中国的五倍。

为了对全球温室气体排放权益进行管理并构造减排责任体系，国际社会达成了《联合国气候变化框架公约》（以下简称《公约》）。自1992年《公约》达成以来，在共同但有区别的责任和各自能力原则指导下，作为全球治理体系重要组成部分的全球气候治理机制逐渐形成并不断完善，对温室气体排放的机制性约束不断强化。在该机制约束下，全球应对气候变化经历了从一个笼统的道义概念到政治共识，再到针对主权国家的具有法律约束力指标的巨大转变。2012年联合国气候变化谈判开启了德班平台进程，旨在全面加强履行《公约》的行动。目前各方正紧锣密鼓地就2015年巴黎协议进行谈判，以明确2020年以后全球减排机制和模式。

“公平、有效、共赢”是评价全球气候治理体系是否成功的三个基本标准。全球气候治理体系需要解决超国家主权的气候财富的公平分配、避免免费搭车、规则调整以体现全球力量格局的变化、在考虑历史责任前提下的剩余碳排放空间的分配以及在发达国家自身也存在不同发展模式和路径竞争的情况下如何推动全球发展路径创新等问题。

为了共同应对气候变化对全球带来的挑战，世界各国需要形成命运共同体，团结协作。但是由于历史责任以及当前能力的差异，发达国家和发展中国家负有共同但有区别的责任。因此发达国家对全球低碳转型负有更大的责任，也有能力发挥领导作用，作出更大的贡献。这种贡献一方面体现在自身的迅速和深度减排以腾出更大的碳排放空间用于支持发展中国家转型，另一方面则体现为利用其自身在资金、技术上的优势和在治理上的能力，积极支持发展中国家实现低碳转型和跨越。发展中国家则需要转变发展观，在保证实现人类发展指数持续上升的前提下，以新的要素组合、新的技术、新的能源结构和新的基础设施设计和生产力布局等，实现发展路径的创新和跃迁。全球气候治理体系能否公平的分配各国权益和责任，是该体系能否实现有效共赢的关键所在。

第一节　全球气候变化科学评估主要结论及全球低碳转型的含义

1896年，科学家就已经意识到了温室气体浓度增加将导致全球变暖，并初步估算出大气二氧化碳浓度加倍时全球温升5℃—6℃的结果。最近20多年来，全球温升将导致的严峻后果也引发了全球科学界对这一问题的广泛关注，很多来自不同学科、采用不同方法得到的研究成果都验证了温室气体在大气中累积浓度上升将导致全球表面温度升高这一结论。虽然由于气候系统的复杂性，相关研究仍有显著的局限性和不确定性，但为了防止可能产生的、不可逆的后果，人们必须立即采取行动，应对气候变化也被提上了国际政治议程和各国发展议程。

为了促进和加强气候变化领域的“科学－政治”互动，1988年政府间气候变化专门委员会（IPCC）正式成立，旨在评估气候变化科学研究的最新进展。IPCC迄今为止已发布了五次评估报告，为全球气候治理提供了更加全面和准确的科学依据。

一、全球气候变化及其带来的风险

IPCC第五次评估报告第一工作组报告指出，基于多个观测数据库的统计结果表明：全球变暖是一个正在发生的客观事实。最近的三个十年比1850之后任何一个年代都更温暖，1983—2012年，北半球经历了1400年以来最热的时期。全球范围内几乎所有的地区都经历了温升过程，1880—2012年，全球表面平均温升达到了0.85℃。对海水温度、海平面高度以及冰冻圈的观测结果也印证了全球变暖的趋势。海洋储存了生态系统中绝大部分的能量，1971—2010 年期间，全球 90% 以上的能量在海洋中累积，这导致水深 75 米以内的浅层海温上升了 0.11℃。在过去的 20 年中，格陵兰和南极的冰盖持续消融，全球范围内冰川进一步萎缩，北极和北半球的冰雪覆盖面积持续缩小。19 世纪中叶以来海平面高度上升的速度比过去两千年都要快，1901—2010 年，全球海平面高度上升了 0.19 米。

气候系统变化给全球生态系统和社会经济发展造成严重威胁，包括海平面上升、洪涝和风暴潮、农业生产力降低、极端气候事件、海洋和陆地生态系统服务功能丧失以及流行疾病和其他健康风险等危害。IPCC 第五次评估第二工作组报告指出，很多地区的降水变化和冰雪消融正在改变水文系统，并影响到水资源量和水质；部分生物物种的地理分布、季节性活动、迁徙模式等都发生了改变；温升导致粮食产量降低、人类健康出现不良状况；极端气候事件频繁发生；气候灾害也影响人们的生活，特别是贫困人口的生计，并使一些地区的暴力冲突加剧。按照目前的温升趋势，到 2100 年全球表面平均温度相比工业化前将升高约 4℃，在这种情况下，全球将面临气候变化带来的巨大的系统性风险，包括大量物种的灭绝、全球性和地区性的粮食紧缺、恶劣天气和极端气候事件以及由此带来的局部冲突等社会问题，气候变化已经成为全球可持续发展最大的威胁之一。

二、引起全球气候变化的原因

气候系统的变化主要是由地球能源收支平衡的改变引起的。1750 年爆发工业革命以来温室气体浓度增加是引起辐射强迫变化的主要原因，其中人为排放二氧化碳等温室气体的贡献最大。IPCC 第五次评估第三工作组报告明确指出："观测到的 1951—2010 年全球平均表面温度升高的一半以上极有可能是由温室气体浓度的人为增加和其他人为强迫共同导致的。"因此，减缓气候变化的根本就是控制温室气体排放。

1750—2011 年，全球人为累积二氧化碳排放量为 2 万亿吨，其中 30% 被海洋吸收，30% 被自然陆地生态系统吸收，剩下的 40% 留存在大气中。目前，二氧化碳、甲烷和氧化亚氮的大气浓度已经上升到了过去 800 万年中从未达到的水平，相比于工业化之前，二氧化碳的大气浓度升高了 40%。二氧化碳是最主要的人为温室气体，2010 年化石能源消费和水泥生产过程中排放的二氧化碳占全球温室气体排放的 65%，土地利用相关的二氧化碳排放占 11%。其他温室气体还包括甲烷、氧化亚氮和含氟气体等。自工业革命以来，主要的温室气体排放来自于列入《公约》附件一的发达国家。1751—2010 年期间，列入附件一的发达国家的累积二氧化碳排放占全球的 70%，与其较低的人口占比形成强烈的反差。

三、碳排放容量空间约束下的全球排放路径

《联合国气候变化框架公约》第 2 条明确了应对气候变化的目标是稳定大气中的温室气体浓度，防止气候系统危险的人为干扰，“使生态系统能够自然地适应气候变化、确保粮食生产免受威胁并使经济发展能够可持续地进行”，各方在谈判中对此目标进行了更深入的阐述。温升目标以及对应的浓度和排放目标是联合国气候谈判的核心问题，也是 IPCC 报告最核心的结论。2009 年哥本哈根气候大会以来，各国已就将全球至本世纪末相对 18 世纪中叶工业革命前温升水平控制在 2℃ 作为全球长期目标达成了政治共识。2009 年哥本哈根气候大会首次将 2℃ 温升目标写入协议，2010 年《坎昆协议》正式确认该目标。但是由于温升目标本身不能作为减排责任进行分解，真正对气候谈判和实际政策进程产生影响的，是基于温升目标而确定的温室气体减排目标。

自 IPCC 第四次评估报告以来，气候模式得到了显著改进，长期气候趋势模拟结果与观测趋势的一致性有所提高。IPCC 第五次评估报告基于对大量模式实验结果的综述，发现了温升和温室气体累积排放

之间存在着近似线性的关系，并将二者之间存在的系数定义为累积碳排放的瞬时气候响应（TCRE），进而计算出“将仅由人为 CO_2 排放引起的变暖限制在 2℃以内（相对于 1861—1880 年），需要将 1870 年以来的所有人为 CO_2 累积排放限制在约 3.6 万亿吨”。

第三工作组报告则基于全球模型组得到的 1200 组情景结果，对全球长期排放路径、技术需求、经济影响和投资需求进行了综合评估，并建立了温度、浓度、累积排放量以及 2050 年和 2100 年全球减排目标之间的对应关系，如表 1-1 所示。

从表 1-1 可以看出，为了较大可能地实现 2℃目标（more likely than not 以上水平，即＞50% 概率），到 2100 年全球温室气体浓度应该控制在 450—500 ppm CO_2eq 之间，其中 450 ppm CO_2eq 浓度（对应典型浓度路径情景 RCP2.6）下实现 2℃目标的可能性最大（likely，即概率水平为 66%—100%）。450 ppm CO_2eq 浓度下，全球 2050 年温室气体排放量相比 2010 年需要减少 41%—72%，到 2100 年的排放量相比 2010 年需要减少 78%—118%，也就是说排放量要接近于零。相应的，不考虑其他温室气体，2011—2050 年全球 CO_2 累积排放量为 5500 亿—13000 亿吨，2011—2100 年的 CO_2 累积排放量为 6300 亿—11800 亿吨。根据该报告前面给出的 2010 年全球 CO_2 排放量 380 亿吨，1971—2010 年累积排放 1.07 万亿吨的数据可以看出，2℃目标下，到 2050 年以及 2100 年的全球排放空间是很有限的。按照 2010 的排放水平，2050 年前全球剩余的碳排放空间仅够排放 17—31 年。

四、2030 年是全球实现长期减排目标的关键时间节点

2100 年 450ppm CO_2eq 浓度目标可以对应未来不同的减排路径。第一种是全球排放总量尽快达到排放峰值，然后开始总量下降，之后的减排速度可以相对缓和，并在 2100 年实现近零排放。第二种是全球暂缓采取大规模减排措施，全球实现排放峰值的时间推迟，在限定未

表 1-1 第五次评估第三工作组报告建立的温升控制目标和累积排放空间之间的应对关系（引自 AR5 WG3 决策者摘要表 1）

<table>
<tr><th>2100 年温室气体浓度（CO_2eq）</th><th rowspan="3">情景类别</th><th rowspan="3">所属 RCP（典型浓度路径）情景</th><th colspan="2">累计 CO_2 排放（$GtCO_2$）</th><th colspan="2">相对 2010 年年温室气体排放量的变化 (%)</th><th colspan="5">温度变化（相比 1850—1900 年水平）</th></tr>
<tr><th rowspan="2">浓度范围</th><th rowspan="2">2011—2050</th><th rowspan="2">2011—2100</th><th rowspan="2">2050</th><th rowspan="2">2100</th><th rowspan="2">2100 年温度变化（℃）</th><th colspan="4">21 世纪末低于此温度的可能性</th></tr>
<tr><th>1.5 ℃</th><th>2.0 ℃</th><th>3.0 ℃</th><th>4.0 ℃</th></tr>
<tr><td><430</td><td colspan="11">只有极少数模型研究了温室气体浓度低于 430ppCO_2eq 的情况</td></tr>
<tr><td>450(430—480)</td><td>全范围</td><td>RCP2.6</td><td>550—1300</td><td>630—1180</td><td>- 72 至 - 41</td><td>- 118 至 - 78</td><td>1.5—1.7(1.0—2.8)</td><td>不太可能</td><td>很可能</td><td rowspan="6">很可能</td><td rowspan="8">很可能</td></tr>
<tr><td rowspan="2">500(480—530)</td><td>从未超出浓度上限 530ppm CO_2eq</td><td></td><td>860—1180</td><td>960—1430</td><td>- 57 至 - 42</td><td>- 107 至 - 73</td><td>1.7—1.9(1.2—2.9)</td><td rowspan="6">不可能</td><td>较有可能</td></tr>
<tr><td>某些年超出浓度上限 530ppm CO_2eq</td><td></td><td>1130—1530</td><td>990—1550</td><td>- 55 至 - 25</td><td>- 114 至 - 90</td><td>1.8—2.0(1.2—3.3)</td><td>有可能</td></tr>
<tr><td rowspan="2">550(530—580)</td><td>从未超出浓度上限 530ppm CO_2eq</td><td></td><td>1070—1460</td><td>1240—2240</td><td>- 47 至 - 19</td><td>- 81 至 - 59</td><td>2.0—2.2(1.4—3.6)</td><td rowspan="3">不太可能</td></tr>
<tr><td>某些年超出浓度上限 530ppm CO_2eq</td><td></td><td>1420—1750</td><td>1170—2100</td><td>- 16 至 7</td><td>- 183 至 - 86</td><td>2.1—2.3(1.4—3.6)</td></tr>
<tr><td>（580—650）</td><td>全范围</td><td rowspan="2">RCP4.5</td><td>1260—1640</td><td>1870—2440</td><td>- 38 至 24</td><td>- 134 至 -50</td><td>2.3—2.6(1.5—4.2)</td></tr>
<tr><td>（650—720）</td><td>全范围</td><td>1310—1750</td><td>2570—3340</td><td>- 11 至 17</td><td>- 54 至 - 21</td><td>2.6—2.9(1.8—4.5)</td><td rowspan="2">不可能</td><td>较有可能</td></tr>
<tr><td>（720—1000）</td><td>全范围</td><td>RCP6.0</td><td>1570—1940</td><td>3620—4990</td><td>18 至 54</td><td>- 7 至 20</td><td>3.1—3.7(2.1—5.8)</td><td rowspan="2">不可能</td><td>不太可能</td></tr>
<tr><td>（>1000）</td><td>全范围</td><td>RCP8.5</td><td>1840—2310</td><td>5350—7010</td><td>52 至 95</td><td>74 至 178</td><td>4.1—4.8(2.8—7.8)</td><td>不可能</td><td>不可能</td><td>不可能</td></tr>
</table>

注：报告中所应用的不确定性术语所对应的概率水平：很可能（66%—100%），较有可能（>50%—100%），有可能（33%—66%），不太可能（0—<50%），不可能（0—33%）。

资料来源：IPCC, Climate Change 2014: Mitigation of climate change — Summary for policymakers. Table 1. http://www.ipcc.ch/pdf/assessment-report/ar5/wg3/ipcc_wg3_ar5_summary-for-policymakers.pdf

来累积排放的前提条件下，全球到峰值以后的排放总量要在较短的时间内加速下降。第三种则是全球实现排放峰值的时间推迟，而且在达到峰值以后排放总量仍然不能实现快速下降，则到 2070 年左右全球需要依赖大规模生物质能源以及碳捕集与封存技术（BECCS）等实现负排放。

各国已经在 2010 年的《坎昆协议》中明确了到 2020 年的温室气体减控排目标，当前又基于德班平台正在开展关于 2020 年以后尤其是 2030 年温室气体排放控制目标的谈判。第三工作组报告首先对《坎昆协议》中各国承诺目标作了分析，强调即使各国兑现了在《坎昆协议》中的减排承诺，与全球通过成本有效的方式实现 2℃目标所要求的减排力度仍然存在较大差距（如图 1-1 所示）。这意味着对全球 2020 年以后的减排力度和速度将有更高要求，更大程度上锁定在 2020 年前大规模建设的碳密集基础设施，后期将更多依赖于未经证实并存在巨大潜在风险的减排技术，中长期的减排成本也会更高，实现 2℃目标的可能性也将降低。报告的这一结论与联合国环境规划署（UNEP）发布的《排放差距报告 2014》的结论保持一致，支持了德班平台谈判中要求各国进一步提高 2020 年前减排力度这一议题。

报告还呼应当前德班平台谈判中各国承诺 2030 年减排目标这一最关键的议题，分析了如果要以成本有效的方式实现 2℃目标，全球 2030 年温室气体排放量需要控制在 300 亿—500 亿吨 CO_2eq 之间，也就是说相比 2010 年排放水平最高需要下降 40%，最低限度也需要回复到 2010 年的排放水平。这对各国提出了艰巨的任务。反之，如果 2030 年前全球排放超过 550 亿吨 CO_2eq，为了实现 2℃目标，2030 年以后的减排压力会进一步加大，需要在较短时期内实现排放陡降，远期全球甚至必须大幅度负排放，对低碳能源和近零碳能源的依赖将显著上升，甚至需要依赖风险较大的地质工程技术，因此，实现排放控制目标的不确定性增加，减排成本也将进一步加大。因此，根据报告的结论，2030 年前全球排放尽快达到峰值，是实现 2℃目标的必要前提。

图 1-1 2030 年不同的温室气体排放水平对应 2030 年以后的不同减排率以及低碳能源的不同水平

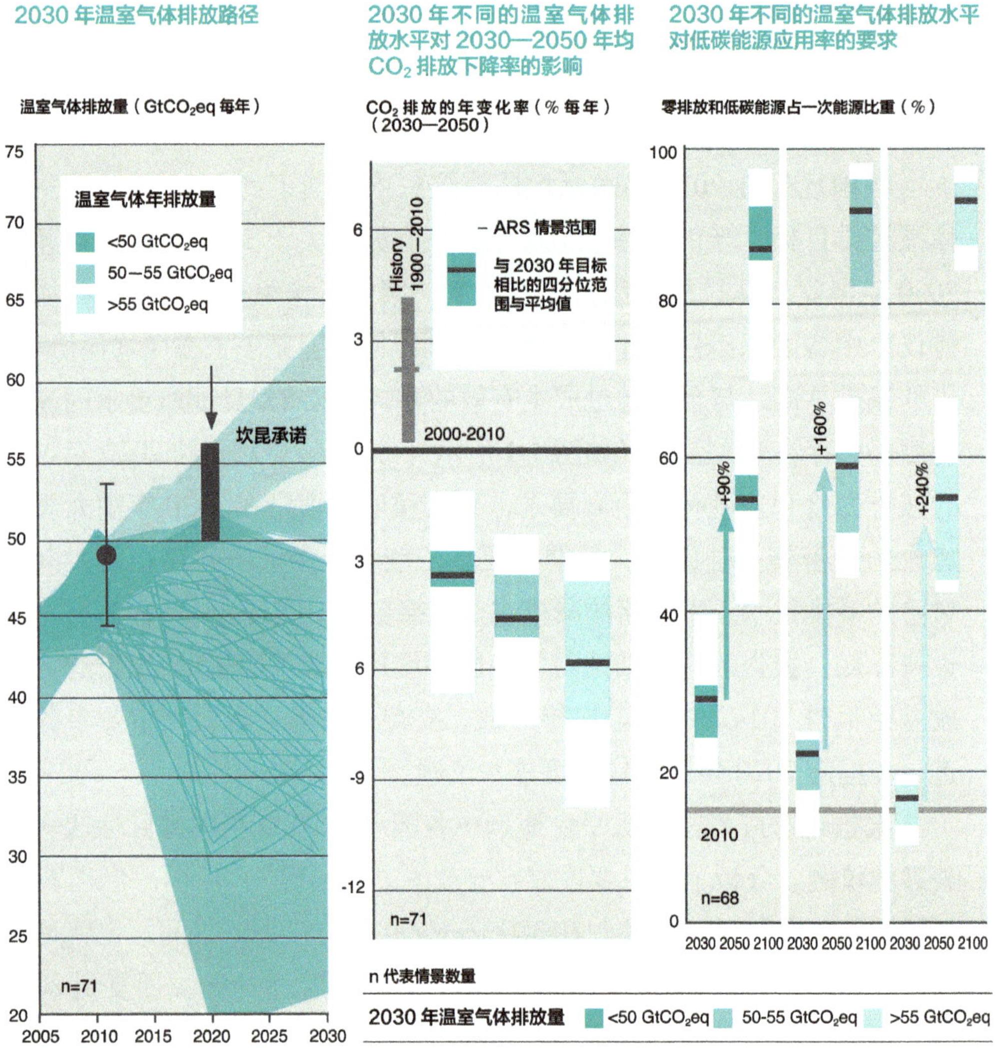

资料来源：IPCC, Climate Change 2014: Mitigation of climate change - Summary for policymakers. Figure 5. http://www.ipcc.ch/pdf/assessment-report/ar5/wg3/ipcc_wg3_ar5_summary-for-policymakers.pdf

五、应对气候变化所需的行动、措施和全球低碳转型

实现减排温室气体和减缓气候变化的目标需要跨部门的行动。根据 IPCC 第五次评估报告，从全球排放趋势的驱动因子来看，1971 年至 2010 年期间，人口和人均 GDP 是温室气体排放增长的主要驱动因子，同时单位 GDP 能源消耗持续下降，对排放增长起到了遏制的作用。但是，2000 年到 2010 年期间单位能耗碳强度从 1970 年到 2000 年期间的减排因子转变成为增排因子。从温室气体排放的来源看，2010 年供电和供热部门的排放占 25%，农业、林业和土地利用排放占 24%，建筑占 6.4%，交通占 14%，工业占 21%，其他能源相关部门占 9.6%。因此，长期来看，实现 2℃温控目标需要能源供应部门低碳化，全球能源供应部门的 CO_2 排放需要在 2040 年到 2070 年间比 2010 年下降至少 90%，要求低碳能源供应在电力供应中的比例由 2010 年的 30% 上升到 2050 年的 80%，到 2100 年彻底淘汰未配备碳捕集与封存（CCS）设备的化石燃料电厂。

然而，已经发生的和近期将要产生的气候变化影响无法通过减排措施消除，只能通过适应措施予以缓解。适应措施可减少生态系统和社会经济气候变化灾害脆弱性的影响，提高气候风险抵御力，降低气候变化带来的负面影响。但如果温升过高，气候风险将超出人类适应的能力范围。因此，应对气候变化需要平衡推进减缓和适应工作。

在过去 20 多年中，世界经济还经历了以互联网技术、现代通讯和交通技术、新能源技术、新材料技术等为标志的科技革命，由此给世界经济的面貌、商业模式和生产者与消费者的理念带来了巨大的变化。特别是经历了 2008 年的全球金融危机及之后的普遍衰退与复苏乏力，世界正处于一个调整和转型的时代。这个调整和转型要解决三个基本问题：一是全球新的增长点或增长动力在哪里？结构调整的方向和着力点是什么？二是对于处于不同发展阶段的发达与发展中国家经济体，其特定的调整重点、任务和背景是什么，与其他经济体的互动关系如

何？三是为了支撑和促进这样的结构调整和转型，需要在国际和国别层面上作哪些体制和政策上的改革？

应对气候变化对全球提出的碳排放约束，一方面是挑战，但其中也蕴含着机会。它和全球经济再平衡以及结构调整和转型的要求存在着某种吻合。二者都要求能源结构的转换，要求通过技术进步和结构调整而提高要素效率，要求增长的动力结构发生效率取向型的变迁以适应要素结构和稀缺程度的变化。在这个过程中，不仅对于其他局地污染物的控制具有协同效应，与低碳技术研发、示范和推广应用相关的巨额投资将成为驱动全球经济结构调整的重要因素，具有成为促进经济增长和就业新引擎的潜力，并带动全球发展模式的低碳转型。

第二节　气候变化问题的特殊性及对全球治理带来的挑战

气候系统是典型的复杂、高度非线性、开放的巨系统。根据 IPCC 科学评估报告，由于人类活动是造成近 50 年来气候变化的主要原因，因此在进行气候决策时，还需要将社会经济系统纳入决策视野，这进一步增加了气候决策的复杂性和不确定性。

由于温室气体在大气层留存的寿命周期长且影响范围涉及全球，气候变化问题具有最大的时空尺度，温室气体排放同时具有全球范围的跨国外部性和跨代外部性的特征。

一、气候问题的长时间尺度和跨代外部性

从跨代外部性看，占世界人口少数的发达国家经济体的人们，享受着其祖辈留下的丰厚物质遗产和在全球经济秩序中占据的有利地位，

这些物质遗产和有利地位都与其祖辈在现代化历史进程中先行累积排放的温室气体有关。这些发达国家自工业革命以来累积排放的温室气体，在人类总累积排放中依然占到多数比例，是造成今天气候变化的主要原因。同时，在当代全球化背景下，在国际经济体系中占据主导地位的发达国家，通过贸易、投资、技术转移和文化价值的传播与示范，将高碳的生产和生活方式转移传播到发展中国家，使后起的发展中国家形成高碳路径依赖。更进一步而言，尚未开启工业化进程因而累积排放寥寥的最不发达国家，在全球人口中占比可观，但在气候变化面前十分脆弱，适应能力有限，要为今天的全球气候变化承受诸多的源自发达国家历史累积排放的“损失与损害”。今天的发达国家是否应当承担更多应对全球气候变化的责任，而在自身减缓气候变化上率先垂范并对发展中国家减缓与适应气候变化提供充分的技术和资金援助呢？这些都给传统经济学研究以及相应的政治解决方案提出了新的要求。

气候变化的影响不是在当代完全显现出来的，大多数气候变化带来的负面影响，诸如海平面上升、气候异常等后果都是几十年甚至上百年才会发生。抑制气候变化的成本发生在当下，抑制气候变化的效益主要发生在遥远的未来。如何在代际的利益分配上进行权衡呢？在长时间尺度的气候变化问题中，一个典型的决策问题是，一个生命有限的投资者需要为五十年、百年甚至是几个世纪之后子孙后代的外部收益作出减缓或者适应的投资决策。按照现有理论，该决策者是基于一个标准的跨期费用效益分析框架而作出判断，在综合气候变化问题的伦理或价值判断基础上，从成本和收益两个角度评价应对气候变化目标和对策。成本角度，包括识别减缓和适应气候变化的技术选择，确定减缓和适应的程度，估计减缓和适应的成本以及分析成本在不同利益相关者之间的分担。而收益，则是减缓和适应气候变化措施所避免的未来气候变化损失。首先对气候变化在水资源、农业生产、生态系统、海岸带、公共健康等领域的物质影响利用环境经济学的评价方法开展经济损失评价并应用社会贴现率估出现值，然后评估减缓和适应气候变化的收益以及分析收益在不同利益相关者之间的分担。

问题在于，上述决策事实上超越了世界范围内现有决策主体的常规决策视野。费用和效益当如何具体衡量？必须将未来人类的成本和收益通过一定的贴现率转换成与当前的货币同质的价值量，然后进行成本效益大小的分析。贴现率的选择具有必要性，因为只有将未来的成本和由于采取行动带来的效益折算成当前的价值，才能够使人们在理性决策中选择最有利于人类福利的决定，实现人类福利最大化的目标。关键问题在于，长时间尺度内的费用和效益的量度，其贴现率当如何决定。

贴现率的选取首先是一个价值判断问题，受到社会伦理道德因素的影响，还受到人们对待风险和不确定性的态度影响。2006 年英国政府委托前世界银行首席经济学家尼古拉斯 · 斯特恩爵士推出的以“气候变化经济学”为主题的《斯特恩报告》，在全球引起广泛的关注和强烈的反响。斯特恩在该报告中给出了非常清晰的政策结论，即有力的及早行动所获得的效益大于成本，越早行动越主动，付出的代价越小，忽视气候变化问题将损害增长。应对气候变化从长期看是促进增长的正面战略（有利于增长），而且完全可以以一种不限制增长的方式实现。而迟缓的行动将导致在本世纪晚些时候或者在下个世纪出现经济活动的崩溃，其规模与 20 世纪上半叶的世界大战和经济大萧条相似并且这样的变化难以逆转。

以上结论除了对气候变化政治进程带来很大影响，也在学术上招致诺德豪斯等学者的批评。最具争议的问题就是斯特恩在报告中选取了非常低的贴现率。斯特恩关于立即采取调整和适应气候变化措施的结论与较低的贴现率的选取有紧密关系。贴现率反映了商品随时间贬值的速度，不同贴现率的选择在气候决策中存在巨大影响。诺德豪斯认为，贴现率大约为每年 6%，斯特恩则选取 1.4%。不同的贴现率导致两人的计算结果出现极大的差异。按照斯特恩的贴现率，今天价值

2470亿美元的商品，在一百年后大约价值1万亿美元；而根据诺德豪斯的贴现率，2108年价值1万亿美元的商品，今天仅值25亿美元。也就是说，对于任何商品，在一百年以后，斯特恩估算的价值几乎是诺德豪斯的100倍。或者说，今天投资所获得的百年后避免气候变化的价值（效益值）可以相差百倍。可见，由于气候变化的决策时间框架长达一百年左右，贴现率的选择对价值量的影响巨大。

最后，应对气候变化目标能否实现，很大程度上取决于相关技术的创新与应用。技术创新虽然也具有较长时间尺度，其生命周期大致分为研发、示范、市场开拓和商业化推广四个阶段。但是，相关技术战略和发展路线图从可操作性角度，一般展望到2050年，而且只有2030年左右的技术发展目标才相对细化。这是因为当前的决策者一般只能基于所能掌握的信息和资源进行决策，无法为不确定性过大的长期目标负责。因此，即使是时间尺度较长的技术创新决策问题，仍然无法与气候变化五十到一百年的决策时间框架尺度相比较，更何况五到十年的常规项目投资决策。由于气候变化问题与现有决策主体的决策时间框架尺度差距较大，而且气候变化长期目标所提供的政策信号，很大程度上作为人类社会发展的愿景，无法被当前的投资实践所体现。

二、气候变化的大空间尺度与全球外部性

与引致雾霾的细颗粒物质等局地污染物不同，各国排放的温室气体进入大气层参与全球大气循环，造成的全球变暖影响也具有全球意义。因此，气候变化问题的空间尺度特别大。按照气候科学原理，温升目标提出了对全球温室气体排放的上限约束，要求对全球大气层的温室气体容量进行管控。因此，气候变化已经演变为全球气候公共财富的管理问题。一个主权国家的决策主体需要处理全球气候利益和本国经济发展利益之间的关系。在一个主权国家内部，可以依据主权国家的强力，对环境产权作出强制性的定义，然后选择相关的政策手段，

如规制、财政与税收手段（近似庇古的解决方案）和排放贸易手段（即科斯的解决方案）、通过社会教育而改变观念意识进而改变需求的手段等。但对于全球气候公共财富的管理，目前还不存在一个凌驾于各国主权之上的纯粹超主权的治理模式，无法在国际层面像主权国家内部那样依靠国家强制权力解决国际外部性和全球公共财富管理的问题，只能通过跨国谈判和协商进程，通过各国共享的价值观和发展理念等，形成不断演进、强化的全球气候治理结构。这个全球气候治理结构的核心是国际责任体系的定义。通过谈判确定不同国家减控排责任的过程，在经济学意义上也是一个不断明晰和分配全球温室气体容量资源产权的过程。各方对这一产权分配的认识和立场，取决于对温室气体容量资源价值的认知和估价，以及此估价与其他方面（如重要双边外交关系、国家形象、与化石能源消耗相关的局地空气质量状况等）利弊得失的比较。对温室气体容量资源产权价值的估计，还会受到在特定发展阶段一个经济体的增长对温室气体排放的依赖程度、化石能源供求与价格长期变动趋势、相关技术变动趋势、政策环境变化预期、伦理偏好等诸多因素差异的影响。在保护全球气候公共财富，避免“免费搭车”的背景下，一个国家对自身责任的认定，还取决于对其他主要国家承担责任程度的认定和相互预期。因此，责任确定的过程必然是一个不断动态调整和演变的过程。这一过程首先是一个多边进程，其谈判进展会十分缓慢，有时也甚至十分脆弱、低效，但它却为其他小多边、双边进程提供伦理的、认知的、政治的和法律的基础，依然是全球治理的主渠道。在全球气候治理的发展过程中，还会遇到处理国际民主或广泛参与和大国影响力之间关系的实际问题。目前国际气候谈判的实践进程从某种意义上说快于理论研究进程，这也对创新的全球气候治理体制提出了新的要求。

三、气候变化问题的不确定性特征

气候变化问题还具有一个重要的特征，就是存在很大的不确定性以及巨大的甚至不可逆的风险。气候系统作为一种复杂巨系统，在其变化超过某个临界阈值以后，系统会进入正反馈状态，出现加速效应。一旦系统受到的外部干扰超过临界阈，破坏了系统的恢复能力，将造成系统的失衡，甚至可能崩溃，并且这种变化是不可逆转的。北极夏季海冰、西非季风、亚马逊热带雨林、北方森林、厄尔尼诺、格陵兰冰盖和西南极冰盖等正是地球气候系统中具有高度感敏性和指示作用的临界点。正是因为这些正反馈效应，使得国际社会特别强调预防性原则，《公约》就提出，"各缔约方应当采取预防措施，预测、防止或尽量减少引起气候变化的原因，并缓解其不利影响。当存在造成严重或不可逆转的损害的威胁时，不应当以科学上没有完全的确定性为理由推迟采取这类措施。"

考虑气候变化问题存在的不确定性以及存在大规模不可逆风险等特征，国际社会倡导建立气候变化风险评价框架，并基于此确定全球应对气候变化的减缓与适应目标。如本章第一节所述，以 IPCC 第五次评估报告为例，报告运用典型浓度路径情景（RCP）分析和累积碳排放与温升的近似线性关系这两种论述方法，按照温度、浓度、累积排放量之间的大致对应关系，建立了温升目标和累积排放空间之间的数量关系，并给出了不同温控目标下的累积排放空间范围和 2050 年排放目标范围。

但是由于受制于观测、方法、机理认识等局限，报告在具有重大政策含义的排放空间、长期排放目标问题上，只能给出十分宽泛的估值范围，而难以给出确切的数值。温升目标推断的长期排放目标和排放空间数值存在着巨大的不确定性，和应对气候变化社会经济技术措施的不确定性一道构成了双重不确定性。面对来自自然系统和社会经济系统叠加起来的这个双重不确定性，需要进一步发展处理不确定性的科学思维和正确方法，以一种辩证的眼光既重视不断探索这样一个目标趋势，形成不断逼近更精确目标的机制和过程，又不陷入对排放目标和排放空间数值的简单机械理解，不能将如此不确定的数字直接用于关于承诺目标的谈判。

第三节　全球气候管理体制的总体要求

自《公约》达成以来，全球气候治理机制逐渐形成并不断完善，对温室气体排放的机制性约束不断强化。在机制约束下，全球应对气候变化经历了从一个笼统的道义概念，到政治共识，再到主权国家具有法律约束力的量化指标的巨大转变。

《公约》第一次对全球应对气候变化行动提出了框架性目标，规定“将大气中温室气体的浓度稳定在防止气候系统受到危险的人为干扰的水平上。这一水平应当在足以使生态系统能够自然地适应气候变化、确保粮食生产免受威胁并使经济发展能够可持续地进行的时间范围内实现。”《公约》同时确立了发达国家和发展中国家应承担共同但有区别的责任，公平以及各自能力的原则，对发达国家缔约方提出了大规模减排温室气体的要求，并要求所有缔约方定期测量和报告温室气体排放情况，提交应对气候变化的国家方案，为后来的约束性全球气候治理体系建设奠定了良好的基础。《公约》还要求发达国家向发展中国家提供资金、技术和能力建设援助；要求平衡处理减缓与适应的关系，这进一步完善了应对气候变化全球治理的内容。

《京都议定书》在《公约》框架的指导下进一步细化了对发达国家减排温室气体的约束，大部分发达国家开始实施温室气体排放总量控制，建立全经济范围内绝对量化减排目标且具有法律约束力，并通过实施量化的审评和核算规则建立严格的遵约机制，保障量化减排目标的完成。从保障控制温室气体目标落实的角度来看，全球气候治理体制的约束性在《京都议定书》时期达到了顶峰。

哥本哈根会议前后，各缔约方逐步就到本世纪末将全球相比工业革命前，温升控制在2℃以内作为全球应对气候变化长期目标达成共识。尽管当时对2℃温升目标是否能够满足《公约》关于“将大气中温室气体的浓度稳定在防止气候系统受到危险的人为干扰的水平上”的要求缺乏广泛和深入的研究，但作为一项政治共识，为后续相关的科学研究设定了范围，为全球气候治理体制加强对主权国家温室气体排放的约束力，推动全球行动朝着更明确的方向努力奠定了基础。事实上，IPCC第五次评估报告的有关结论进一步明确了将全球温升控制在2℃以内反映了全球应对气候变化的现实需求。

在《坎昆协议》下，更大范围的缔约方接受了控制温室气体排放的制度约束。发展中国家提出了多样化的国家适宜减缓行动（NAMAs），发达国家则继续承担全经济范围量化减排指标。中国在这一时期提出了到2020年，单位国内生产总值二氧化碳排放比2005年下降40%—45%、非化石能源占一次能源消费比重达到15%左右、森林面积比2005年增加4000万公顷、森林蓄积量比2005年增加13亿立方米等目标。这些指标通过国民经济和社会发展五年规划以及专门的战略规划和政策措施形成国家层面的约束性指标体系，并逐步分解落实到地方和企业。

德班平台谈判以来，特别是华沙会议要求各缔约方提交国家自主贡献后（Intended Nationally Determinded Contribution INDC），未来自下而上全球气候治理模式基本形成。从目前的谈判进程看，巴黎会议将以各缔约方自主贡献为基础，定期盘点各方行动总体效果与2℃目标的差距，建立促进性和约束性兼备的全球气候治理体系，按照科学的要求和现实情况，保障各缔约方持续不断地朝实现《公约》目标的方向加速前进。

第四节　发达国家与发展中国家在全球气候变化中不同的地位和作用

如前所述，与全球气候变化直接联系的是与大气中温室气体浓度直接对应的累积排放量。从整个世界经济体系发展的历史看，自 1740 年以瓦特发明蒸汽机为标志的产业革命爆发以来，先行实现工业化的发达国家大量使用煤炭和石油等化石能源，以全球人口的少数占比形成了大约 70% 的能源使用相关二氧化碳累积排放。图 1-2 显示了中美欧累积排放的比较，它表明美、欧主要的累积排放发生在 1850 年以来，自 20 世纪 90 年代以来依然有显著的增长，2013 年至 2030 年间增速下降；而中国自 1991 年以来的累积排放显著增加，2013 年之后至 2030 年将呈现加速增加之势。截止到 2012 年，美欧自 1750 年以来的累积排放量之和是中国的 5 倍；到 2030 年，依然达到 2.6 倍。如果看人均累积排放量（见图 1-3），在 1751—2030 年期间，中国人均累积排放将达到 200 吨二氧化碳，美国超过 1200 吨，而欧盟超过 770 吨，美、欧分别是中国的 6 倍和 3.9 倍。

图 1-2　中美欧不同时期的历史累积排放对比

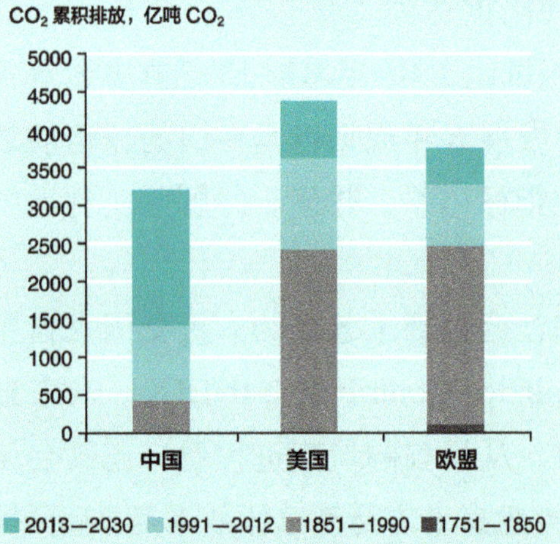

数据来源：仅包含能源相关 CO_2，1751—2012 年数据来自美国橡树岭国家实验室二氧化碳信息分析中心（CDIAC），2013—2030 年数据由国家气候战略中心基于各国国家自主贡献测算。

这些事实都表明：发达国家对今天及今后一个很长时期发生的全球气候变化负有主要的历史责任。尽管在上个世纪90年代缔结《联合国气候变化框架公约》之前，并不存在对国家碳排放行为予以约束的国际法律，“无法者不为过”这个适用于民商私法的准则似乎成立，但发达国家今天所享有的发达的基础设施和丰富的物质基础以及在全球所处的有利竞争地位和国际进程主导地位，都与其自工业革命以来（特别是1850年以来和二战以来）大量消耗化石能源而导致的温室气体累积排放有着千丝万缕的联系，以较低全球人口占比的发达国家今天的发达状态是建立在大量占用全球公共的碳容量资源的基础之上的。

图 1-3 中美欧人均 CO_2 历史累积排放比较

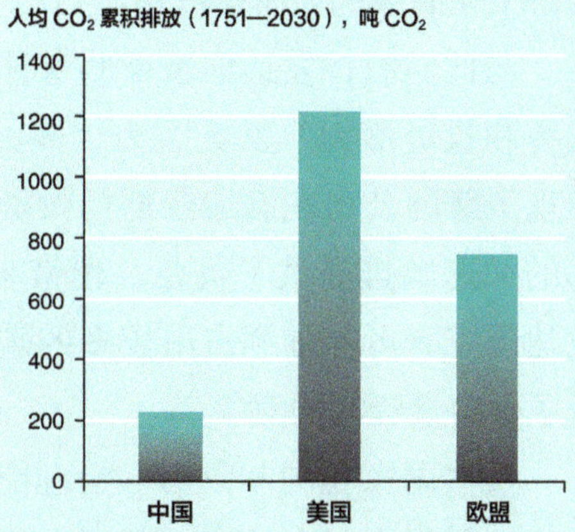

数据来源：仅包含能源相关 CO_2，1751—2012年数据来自美国橡树岭国家实验室二氧化碳信息分析中心（CDIAC），2013—2030年数据由国家气候战略中心基于各国国家自主贡献测算。

因此发达国家在全球命运共同体中，对全球低碳转型负有更大的责任，也有能力发挥领导作用，作出更大的贡献。这种贡献一方面体现在自身的深度和迅速减排以腾出更大的碳排放空间用于支持发展中国家的转型，另一方面则体现为利用其自身资金、技术的优势和在治理上的能力，积极支持发展中国家实现低碳转型和跨越。

进入后工业发展阶段后，发达国家的发展议程和增长动力发生变化，特别是在全球化背景下，实现了向新兴经济体发展中国家的高碳产业转移，相应的全球产业链的形式、产业结构以及盈利来源都发生了变化。在生产环节及总排放绝对量与占比下降的同时，消费的排放量（以建筑和交通部门为主）下降缓慢甚至继续上升，而以重化工为代表的制造业部门的排放主要发生在新兴经济体和其他启动工业化进程的发展中国家。在新的全球经济分工下，发展中国家还向发达国家出口工业制成品而将排放留给自己。

与此同时，发达国家面临着能源、建筑和交通等基础设施更新的任务和投资潜力。以美国为代表的发达国家还在启动再工业化进程，出现了跨国公司在制造业部门投资回流的迹象。这些都给发达国家内部的加速减排提供了机会，也带来了增排或迟缓减排的压力和挑战。发达国家有可能继续占用较多的碳容量资源，与后起的发展中国家形成了碳容量资源的竞争。

发达国家在国际政治、治理能力、技术研发、资金等方面具有较强的转型能力，在核能、可再生能源、燃气发电、能效、互联网和其他信息技术与材料技术等工业通用技术领域具有较强的研发能力，拥有大多数关键核心技术的知识产权。在金融的市场运作等方面发达国家也普遍具有支配地位和更强的影响力。如何将这些优势转化为保护全球碳容量资源的贡献，特别是在促进包括广大发展中国家在内的全球可持续发展方面，如何作出卓有成效的贡献，是发达国家面临的历史课题。

对发展中国家而言，人口众多的后起发展中国家在实现工业化、城镇化的过程中，受到了前所未有的全球范围碳排放制约。沿袭发达国家近现代经典高碳发展路径实现社会经济现代化与这个碳排放制约形成了趋势性的时代冲突。发展中国家面临的挑战是：

第一，所剩碳排放容量不足以支撑占全球人口大多数的发展中国家跟随发达国家历史上的高碳工业化路径而实现自身的现代化，需要发展路径的创新和跃迁。发展中国家需要转变发展观，在保证实现人类发展指数持续上升的前提下，以新的要素组合、新的技术、新的能源结构、新的基础设施设计和生产力布局，去支撑可持续发展的需要。

第二，实现全球低碳转型，一方面需要给发展中国家配置更多的现有碳容量资源，支持他们走过伴随工业化和城镇化的大规模基础设施建设的高碳发展阶段，保证他们在能力较弱的情况下实现平稳转型过渡。基于他们的路径依赖现状，需要给他们更长的一段时间实现过渡。发展中国家自身也应当抓紧寻找机会，实现跨越式的路径跃迁。

第三，需要建立发达国家与发展中国家的合作关系。发达国家要支持发展中国家打破高碳路径依赖，形成并发挥后发优势，在适应气候变化负面影响、低碳示范、技术开发与转让、贸易与投资政策协调、资金支持、人力资源开发等方面支持发展中国家的低碳转型。发展中国家需要建立新的低碳发展观，在可持续发展的框架下，主动应对气候变化，促进发展路径的创新。

第四，中国等主要发展中国家在工业化进程中面临的资源环境挑战，是其中等收入陷阱的重要内容，已经对其进一步的发展形成制约和威胁。控制碳排放，有可能带来改善空气质量、节约资源、加强能源安全和创造新的就业机会的共生效益。最大限度地实现共生效益，将提高减缓气候变化与可持续发展目标的一致性。加强适应气候变化的能力，有利于通过防灾减灾提高人民生命财产的安全保障，保护经济发展的果实或提高发展的实效。

第五节　全球气候公共财富管理面临的基本问题

自工业革命以来，特别是在过去一个半世纪里，温室气体容量作为全球公共财富，在发达国家大量消耗化石能源的工业化进程中，处于没有明晰的产权定义而自由取用（任意排放）的状态。过度使用温室气体容量资源，造成了今天和今后的全球气候变化。现在试图通过对公共财富明晰产权，通过主权国家政府之间的谈判以国家为单位确定排放限额，则面临着若干基本的问题：

一是全球范围内并不存在一个超越主权国家的合法强权强制力量，

去分配作为全球公共财富的温室气体容量资源。人类国际政治文明发展至今，解决国际和全球问题，主要还是通过谈判形成国际公约等国际法律，对国家行为规范进行约定。这样的规范和约定的严格程度尚很有限。这给解决公共财富保护和外部性内部化问题等标准做法的有效性带来了限制。

二是对免费搭车的忧虑。一些国家不减排或力度不够，会使另外一些国家觉得他们自身减排努力无效。由此，会形成互相的猜疑和以其他缔约方减排程度确定自身减排程度的博弈。

三是西方国家在法律体系、语言等方面的优势，使得他们在制定国际规则中具有先天的有利条件。现有国际气候规则中，需要更加公平地体现发展中国家的权益，发展中国家的话语权需要得到加强。这就对国际气候治理体系规则的调整提出了要求。

四是发达国家在工业化进程中先行占据了大量的温室气体容量资源，形成了今天的有利地位和享受的福利，他们将减排责任向发展中国家转嫁，造成了发达国家与发展中国家的矛盾。在全球气候责任体系中，如何体现历史责任，剩余的排放空间通过何种方式向发展中国家倾斜，以满足发展中国家的发展需求，并为发展中国家留足转型的空间，都是绕不过的难题。

五是发达国家之间也存在着不同的发展路径和能源与排放的不同比较优势，也存在着竞争。各主要发达国家的二氧化碳排放出现差异化的发展趋势，并代表了不同的发展模式和低碳转型要求。大体上可以将发达国家分为以美国、加拿大、澳洲为代表（人均 GDP 高，人均排放量也高，排放总量没有显著下降，出现暂时性峰值，但尚不稳定）和以欧盟及日本等国家为代表（人均 GDP 高，人均排放量较低，排放总量和人均排放量均已经达到峰值，并明显开始下降）的两种不同模式。而中国等发展中国家未来的发展路径，是沿着“高人均 GDP 高排放”的“美国模式”还是“高人均 GDP 较低排放”的“欧盟模式”抑或走出一条属于自己的新路径，对于其自身的社会经济可持续发展和全球应对气候变化的努力，都将具有十分重要的意义。而全球气候治理体系通过何种机制引导全球发展路径创新也是一项重要的时代命题。

综上所述，“公平、有效、共赢”是评价全球气候治理体系是否成功的三个基本标准。全球气候治理体系需要解决超国家主权的气候财富的公平分配、避免免费搭车、调整规则以体现全球力量格局的变化、在考虑历史责任前提下的剩余碳排放空间的分配以及在发达国家自身也存在不同发展模式和路径竞争的情况下如何推动全球发展路径的创新等问题。

作为现有全球治理体系一个有机组成部分的全球气候治理体系延续了二战以后所建立的全球安全、政治和经济治理体系，并不能适应全球化时代全球力量格局的变化，也不能适应由上述气候变化科学、体制与全球气候公共财富管理等问题带来的挑战。当然，政策实践的发展和创新往往会带来理论的重大突破和发展，气候变化问题的复杂性对现有学说提出挑战，相应也孕育了巨大的机会，从理论和实践上都提出了构建创新型全球气候治理体系的要求。

【参考文献】

[1] 邹骥，傅莎，陈济，等 . 基于“发展路径创新”的后 2020 年国际减排机制设计 [R]. 国家气候战略中心研究报告，2014.4.

[2] 邹骥，滕飞，傅莎 . 减缓气候变化社会经济评价研究的最新进展——对 IPCC 第五次评估第三工作组报告的评述 [J]. 气候变化研究进展，2014, 10(5): 313—322.

[3] 张晓华 . 应对气候变化科学研究进展——IPCC 第五次评估报告《综合报告》结论解读 .2014 .11 .19 .

[4] 张晓华，高云，祁悦 . IPCC 第五次评估报告第二工作组报告的主要结论对 2015 协议谈判的影响分析 [J]. 气候变化研究进展 2014, 10(3): 175—178.

[5] 张晓华，高云，祁悦，傅莎 . IPCC 第五次评估报告第一工作组主要结论对《联合国气候变化框架公约》进程的影响分析 [J]. 气候变化研究进展 2014, 10(1): 14—19.

[6] 张晓华，祁悦，傅莎 . IPCC 第五次评估报告对 2015 协议谈判的影响分析 [J]. 气候变化研究进展，2014, 36(12): 14—17, 25.

[7] IPCC, Climate Change 2013: Physic Science.
[8] IPCC, Climate Change 2014: Impacts, vulnerability and adaptation.
[9] IPCC, Climate Change 2014: Mitigation of climate change.
[10] IPCC, Climate Change 2014: Mitigation of Climate Change – Summary for Policymakers. http://www.ipcc.ch/pdf/assessment-report/ar5/wg3/ipcc_wg3_ar5_summary-for-policymakers.pdf.
[11] Stern, Nicholas. The Economics of Climate Change: The Stern Review [R]. Cambridge,
UK: Cambridge University Press, 2007.http://www.hmtreasury.gov.uk/independent_reviews/stern_review_economics_climate_change/sternreview_index.cfm.
[12] William Nordhaus.The Stern Review on the Economics of Climate Change[J].2007. http://www.econ.yale.edu/~nordhaus/homepage/stern-050307.pdf.

第二章 世界工业化历史进程的启示

Chapter Two

Revelation from Historic Course of the World Industrialization

摘要

本章在对主要先行工业化国家和发展中国家历史排放轨迹进行实证考察后发现，主要发达国家在完成工业化、城市化进程后，其能源相关 CO_2 加水泥排放总量普遍已达到峰值，但表现出不同的峰值特征，“美、加、澳”和“欧、日”的具体轨迹和峰值水平呈现不同的模式。美国、加拿大和澳大利亚的峰值水平较高，接近或超过 20 吨 CO_2/ 人。欧盟和日本等国家的峰值普遍在 10 吨 CO_2/ 人上下。

发展中国家普遍仍处于排放快速增长期，占世界排放比重不断上升，且这一态势在一定时期内还将持续，但从人均排放看，发展中国家的平均人均排放仍远低于发达国家。

从各国人均排放峰值与国内人均 GDP 关系来看，纵观工业革命以来的全球 CO_2 排放历史，目前还没有一个经济体能够摆脱随人均 GDP 和人类发展水平提高，人均 CO_2 排放水平“先增长后下降”的倒 U 型的库兹涅茨曲线现象，但在不同的国家会有不同的表现。这一现象表明的是一个经济体在工业化、城镇化进程中，在从不发达到发达的过渡过程中，经济质量不断提升，对化石燃料依赖下降，而对技术、服务依赖上升的过程。存在这一现象并不意味着发展中国家无所作为。发展中国家目前仍普遍处于曲线“左侧”，但已不具备延续发达国家老路的条件，未来是沿着“高发展水平高排放的美、澳、加模式”还是“相对高发展水平低排放的欧、日模式”的轨迹，抑或走出一条创新的不同于发达国家传统路径的更为低碳的发展路径，对其自身的社会经济可持续发展和全球应对气候变化的努力，都将具有十分重要的意义。发展中国家需要创新一条不同于美欧等发达国家的、更早达到更低峰值的发展道路。创新发展路径并不是空洞的学理概念，而是从对发达国家工业化历史进程的反思中提炼而成的。发展中国家通过适当的政策和体制管控措施并充分发挥后发优势，有望影响排放轨迹并实现发展路径创新，促成峰值更早更低到来。而要实现上述峰值，单位 GDP 能源强度下降和能源结构低碳化将是发展中国家未来排放下降的主要抓手。

第一节　主要先行工业化经济体历史排放轨迹的实证考察

累积排放是大气层中温室气体浓度上升、全球表面温升的直接原因，是界定气候变化历史责任的基本依据。但是累积排放又是自工业革命以来的瞬时（年度）排放的累加。在各经济体年度排放轨迹图中，轨迹线高度表示瞬时排放，轨迹线下方的面积大小则表示累积排放量。年度排放量的变化趋势将影响累积排放分布的未来走向。因此，需要同时考虑年度瞬时排放变化规律和累积排放量的分布情况。本章主要试图通过对发达国家、发展中国家历史排放轨迹及影响排放的驱动因子考察，为全球特别是发展中国家的转型发展提供借鉴。

从排放总量看，二氧化碳信息分析中心（CDIAC）、美国橡树岭国家实验室与国际能源署（IEA）提供的各国历史排放数据显示，美国、欧盟主要成员国、日本、澳大利亚、加拿大、俄罗斯等国家的能源相关 CO_2 加水泥生产排放总量已经达到峰值（本节下文中所指的关于 CO_2 排放的讨论均为能源相关 CO_2 加水泥工艺过程排放之和）。但是受经济周期和历史上不同驱动因素等的影响，各国表现出不同的峰值特征，是否达到了稳定峰值还需要结合经济周期变化进一步观察。

美国国内 CO_2 排放总量在 2007 年达到峰值，为 58.3 亿吨 CO_2，2008 年和 2009 年有所下降。在 1850—2000 年期间美国国内 CO_2 排放曾出现两次波动，分别在 1930 年和 1980 年，正值两次经济大危机，受经济下滑的影响，国内 CO_2 排放下降至波谷，之后随着经济复苏，排放再次攀升至新高并保持上涨趋势，而 2008 年和 2009 年美国国内排放总量再次下滑与金融危机的爆发时间一致（见图 2-1）。在 2010 年

图 2-1 美国历年CO_2排放

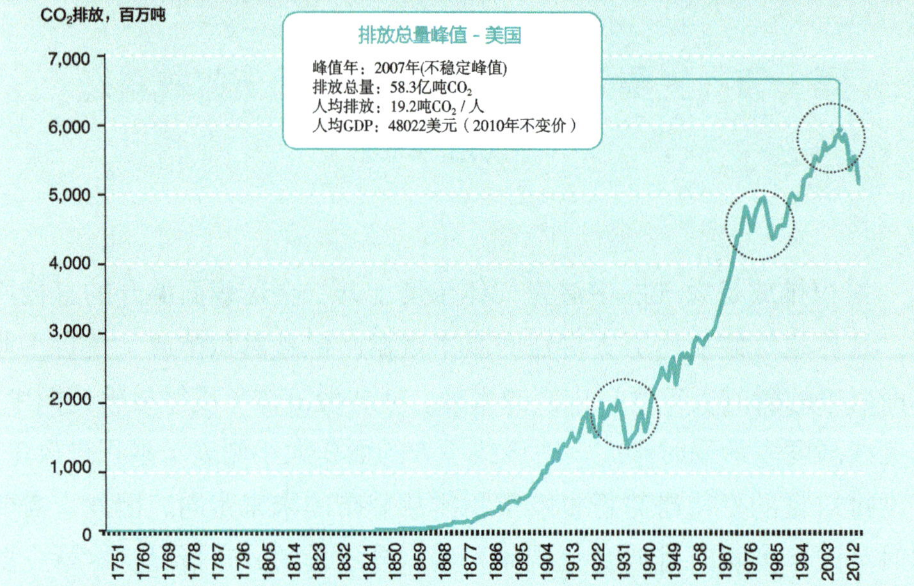

数据来源：

1. CO_2排放数据主要来自二氧化碳信息分析中心(Carbon Dioxide Information Analysis Center，CDIAC)，美国橡树岭国家实验室（Oak Ridge National Laboratory），2015，包含化石燃料燃烧的CO_2和水泥工艺过程的CO_2。
2. 人口数据来自联合国人口署(United Nations, Department of Economic and Social Affairs, Population Division 2014)。
3. GDP数据来自世界银行数据库2015（World Bank Database），为2010年美元不变价。
4. CDIAC的化石燃料燃烧的CO_2与IEA的化石燃料燃烧的CO_2数据基本一致，误差在1%左右。

排放短暂上升后，随着页岩气发展和交通、建筑部门能效标准的实施，美国排放又呈现较为稳定的持续下降趋势。

欧盟 28 国[①] CO_2 排放在 1979 年达到峰值，1990—2004 年为排放波动期，2004 年排放水平仅比 1990 年下降 1%，2005 年后欧盟排放总量下降趋势明显。英国排放峰值出现在 1971 年，为 6.6 亿吨，德国与法国的排放峰值年均为 1979 年，分别为 11.2 亿吨和 5.3 亿吨。波兰国

① 为 27 国数据，缺克罗地亚。

内的排放在 1987 年达到最高水平，此次排放峰值的出现也与经济危机相关。意大利的排放峰值出现在 2005 年，为 4.7 亿吨。英国、德国与法国国内 CO_2 排放在经历排放峰值后都出现了较为稳定的下降。上述五国的历史排放情况见图 2-2。

加拿大的排放总量峰值出现在 2005 年，为 5.6 亿吨。受金融危机的影响，澳大利亚的排放总量在 2009 年出现峰值，为 4.0 亿吨。日本

图 2-2 英国、德国、法国、波兰和意大利的历年 CO_2 排放

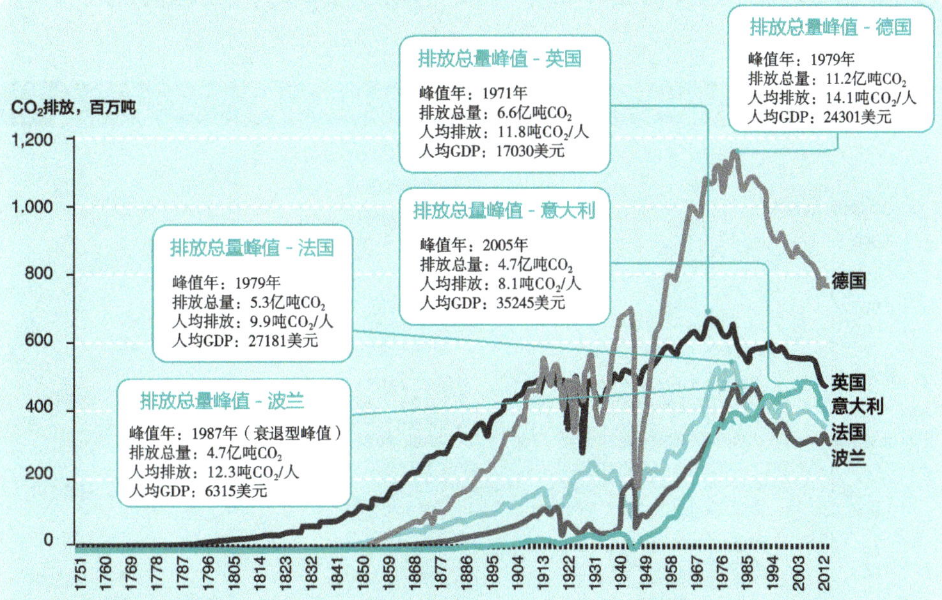

数据来源：

1. CO_2排放数据主要来自二氧化碳信息分析中心(Carbon Dioxide Information Analysis Center，CDIAC)，美国橡树岭国家实验室（Oak Ridge National Laboratory），2015，包含化石燃料燃烧的CO_2和水泥工艺过程的CO_2。
2. 德国1971—1990年数据来自国际能源署（IEA），CO_2 Emissions from Fuel Combustion 2011。
3. 人口数据来自联合国人口署(United Nations, Department of Economic and Social Affairs, Population Division 2014)。
4. GDP数据来自世界银行数据库2015（World Bank Database），为2010年美元不变价。
5. CDIAC的化石燃料燃烧的CO_2与IEA的化石燃料燃烧的CO_2数据基本一致，误差在1%左右。

国内排放总量在 2004 年前一直上升，2004 年达到最高值，为 12.6 亿吨，2005—2007 年间保持稳定，2008 年金融危机爆发后国内排放从 2007 年的 12.5 亿吨骤降为 2009 年的 11.0 亿吨，但随后受福岛核事故后关闭核电的影响，排放又在 2012 年逐渐回升至接近历史峰值水平。俄罗斯 CO_2 排放自 1988 年以来保持下降，但是这主要受当时国内“休克疗法”经济改革的影响，渡过经济转型期，随着国内经济的恢复，1999 年俄罗斯 CO_2 排放开始回升，从 1998 年的 15.3 亿吨增长至 2012 年的 18.3 亿吨。加拿大、澳大利亚、日本与俄罗斯国内 CO_2 排放趋势见图 2-3。

图 2-3 加拿大、澳大利亚、日本和俄罗斯历年 CO_2 排放

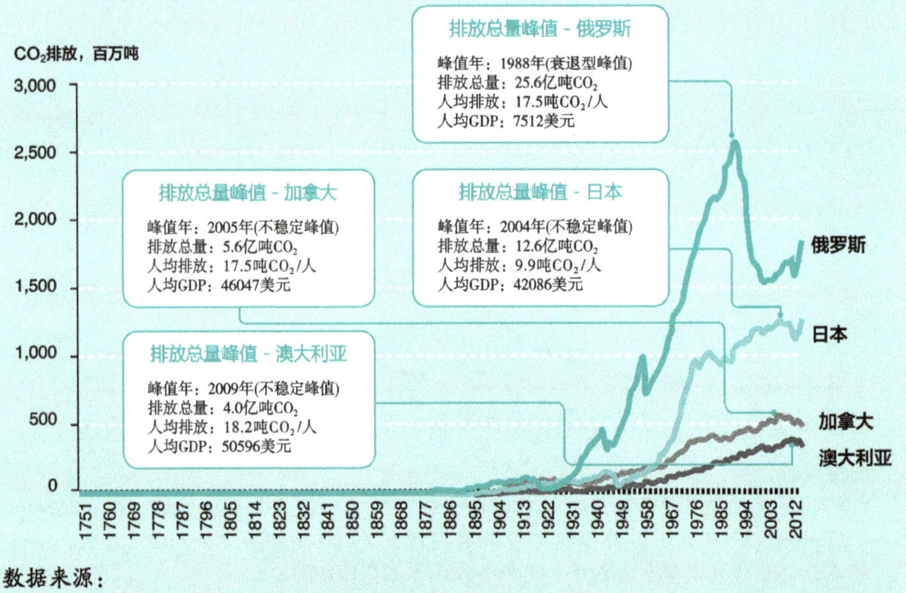

数据来源：

1. CO_2排放数据主要来自二氧化碳信息分析中心(Carbon Dioxide Information Analysis Center，CDIAC)，美国橡树岭国家实验室（Oak Ridge National Laboratory），2015，包含化石燃料燃烧的CO_2和水泥工艺过程的CO_2。
2. 人口数据来自联合国人口署(United Nations, Department of Economic and Social Affairs, Population Division 2014)。
3. GDP数据来自世界银行数据库2015（World Bank Database），为2010年美元不变价。
4. CDIAC的化石燃料燃烧的CO_2与IEA的化石燃料燃烧的CO_2数据基本一致，误差在1%左右。

从人均 CO_2 排放看，各国发展阶段与发展模式的不同导致各国人均 CO_2 排放峰值和排放趋势存在差异（见表 2-1）。

表 2-1 主要发达国家人均 CO_2 排放峰值

	人均 CO_2 排放峰值年	峰值年 CO_2 人均排放（吨 CO_2/ 人）
OECD	1979	11.80
附件一国家	2012	13.01
欧盟 27 国	1979	10.30
欧盟 15 国	1979	10.12
美国	1973	22.13
日本	2004	9.94
澳大利亚	1998	18.46
加拿大	1979	17.63
德国	1979	14.13
英国	1971	11.84
法国	1973	9.93

数据来源：
1. 二氧化碳信息分析中心(Carbon Dioxide Information Analysis Center，CDIAC)，美国橡树岭国家实验室（Oak Ridge National Laboratory），2015。
2. 人口数据来自联合国人口署(United Nations, Department of Economic and Social Affairs, Population Division 2014)。

美国人均 CO_2 排放在 1973 年达到峰值 22.1 吨 / 人，此后近 30 年时间一直处于高位平台期，没有显著的下降，并保持在 18 吨 / 人的排放水平。2010 年后呈现较为明显的下降趋势，2014 年已下降到 15 吨 / 人左右（见图 2-4）。

欧盟 28 国[①] 整体在 1979 年实现人均排放峰值，为 10 吨 / 人。英国和法国人均排放峰值分别出现在 1971 年和 1973 年，分别为 11.8 吨 / 人和 9.9 吨 / 人；德国在 1979 年达到人均排放峰值，为 14.1 吨 / 人；波兰的人均排放峰值在 1980 年，为 13.0 吨 / 人；意大利人均排放峰值为 8.1

① 为 27 国数据，缺克罗地亚数据。

图 2-4 美国历年人均CO_2排放

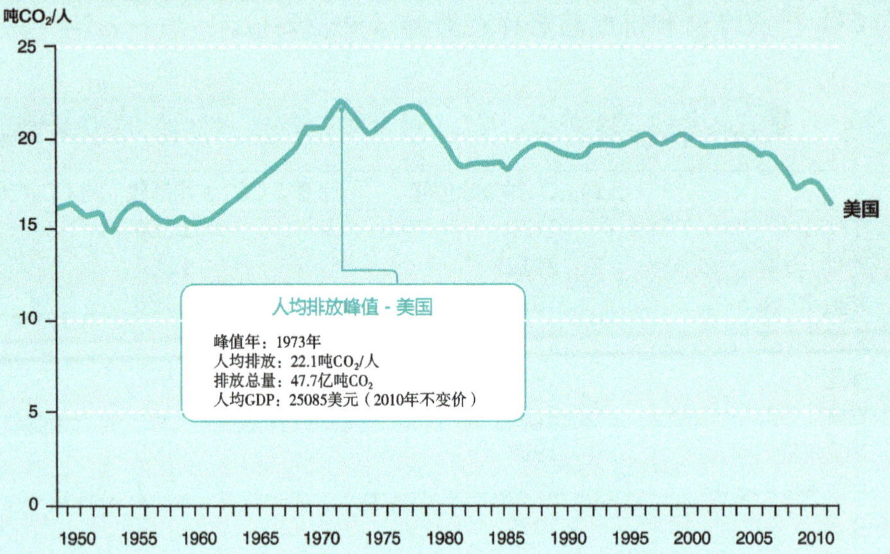

数据来源：
1. CO_2排放数据主要来自二氧化碳信息分析中心(Carbon Dioxide Information Analysis Center，CDIAC)，美国橡树岭国家实验室（Oak Ridge National Laboratory），2015，包含化石燃料燃烧的CO_2和水泥工艺过程的CO_2。
2. 人口数据来自联合国人口署(United Nations, Department of Economic and Social Affairs, Population Division 2014)。
3. GDP数据来自世界银行数据库2015（World Bank Database），为2010年美元不变价。
4. CDIAC的化石燃料燃烧的CO_2与IEA的化石燃料燃烧的CO_2数据基本一致，误差在1%左右。

吨 / 人，出现在 2004 年。在人均排放达到峰值后，英国、法国、德国、波兰和意大利人均排放呈逐年下降趋势（如图 2-5）。欧盟整体人均排放水平在 2013 年已下降到 7.4 吨 / 人左右。

加拿大人均排放趋势与美国类似，人均 CO_2 排放在 1979 年达到峰值 17.6 吨 / 人后一直保持在较高的水平，2012 年人均排放水平仍然为 14.3 吨 / 人水平；日本国内人均排放虽然在 2004 年达到最高值 9.9 吨 / 人，但此后一直维持在这个水平左右；澳大利亚国内人均排放尚未出现稳定峰值；俄罗斯国内人均排放从 1988 年最高值 17.5 吨 / 人迅速下降，并从 1999 年开始回升（见图 2-6）。

图 2-5 英国、德国、法国、波兰和意大利历年人均CO_2排放

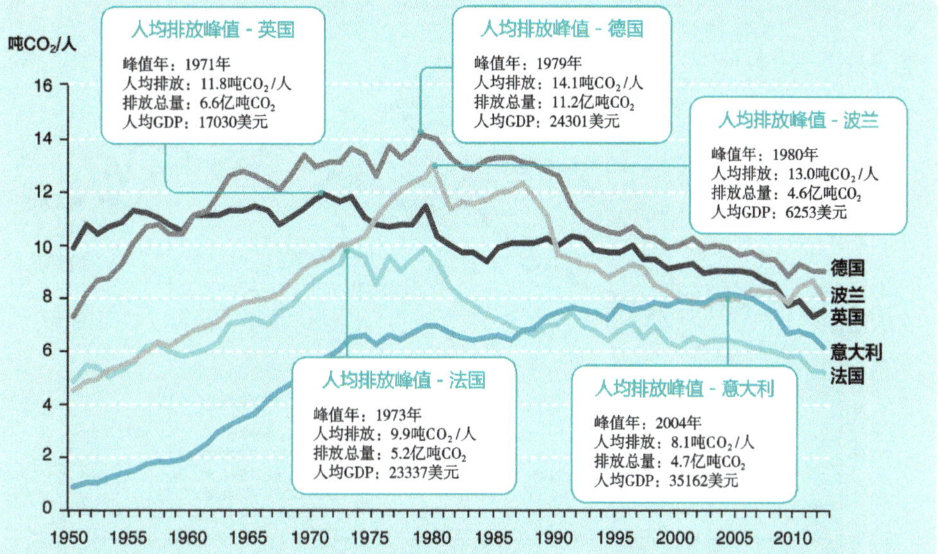

数据来源：

1. CO_2排放数据主要来自二氧化碳信息分析中心(Carbon Dioxide Information Analysis Center，CDIAC)，美国橡树岭国家实验室（Oak Ridge National Laboratory），2015，包含化石燃料燃烧的CO_2和水泥工艺过程的CO_2。
2. 德国1971—1990年数据来自国际能源署（IEA），CO_2 Emissions from Fuel Combustion 2011。
3. 人口数据来自联合国人口署(United Nations, Department of Economic and Social Affairs, Population Division 2014)。
4. GDP数据来自世界银行数据库2015（World Bank Database），为2010年美元不变价。
5. CDIAC的化石燃料燃烧的CO_2与IEA的化石燃料燃烧的CO_2数据基本一致，误差在1%左右。

发达国家在未来中短期内的排放绩效如何取决于经济何时复苏、与消费密切相关的交通建筑部门的减排效果、能源结构的进一步优化程度（如欧盟提高可再生能源、美国页岩气开发带来的实际影响程度）。

图 2-6 加拿大、澳大利亚、日本和俄罗斯历年人均CO_2排放

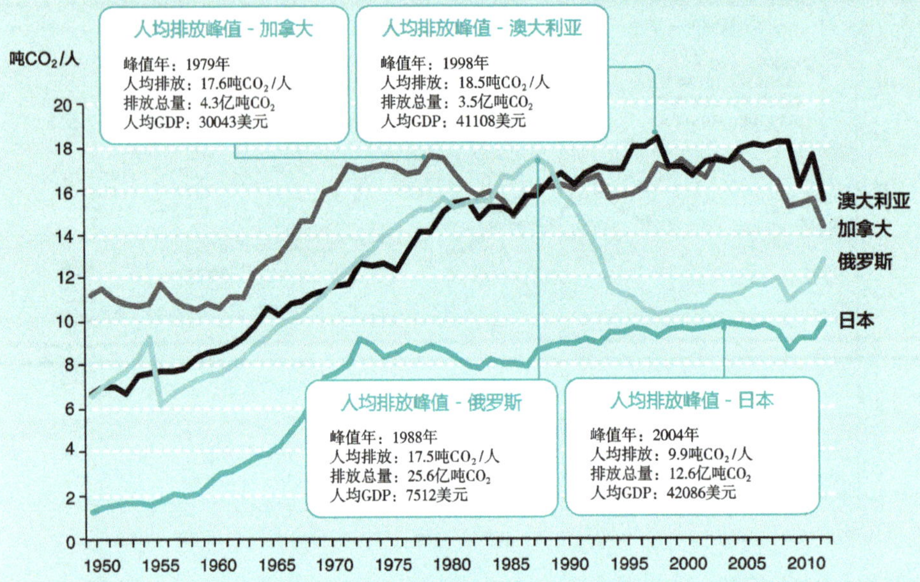

数据来源：

1. CO_2排放数据主要来自二氧化碳信息分析中心(Carbon Dioxide Information Analysis Center，CDIAC)，美国橡树岭国家实验室（Oak Ridge National Laboratory），2015，包含化石燃料燃烧的CO_2和水泥工艺过程的CO_2。
2. 人口数据来自联合国人口署(United Nations, Department of Economic and Social Affairs, Population Division 2014)。
3. GDP数据来自世界银行数据库2015（World Bank Database），为2010年美元不变价。
4. CDIAC的化石燃料燃烧的CO_2与IEA的化石燃料燃烧的CO_2数据基本一致，误差在1%左右。

第二节 发展中国家经济体历史排放轨迹的实证考察

从排放总量看，发展中国家的 CO_2 排放整体呈现快速上升态势，占世界 CO_2 总排放的比重不断上升，从 1990 年的 29% 上升到了 2012 年的 65%，其中中国、印度约占年度总排放的 1/3，而且这一态势还将持续。但从人均排放看，发展中国家的平均人均排放仍远低于发达国家，

历史累积排放总量占比和人均历史累积排放就更低。

中国、印度、南非、巴西作为发展中大国，国内排放总量快速上升，尤其是中国和印度（详情见图 2-7）。中国已成为世界第一大年度排放国，CO_2 排放从 1971 年的 8.8 亿吨迅速增长到 2012 年的 96.1 亿吨，增长 996%，在世界 CO_2 排放总量和增量中的比重越来越大，面临着人均 CO_2 排放赶上欧盟人均排放水平，能源活动和水泥生产过程的年度二氧化碳排放总量接近于欧盟、美国和日本碳排放量的总和，在 3—5 年内赶上 OECD 国家排放总量的态势。同时，受全球化背景下的产业分工以及以煤为主的能源资源禀赋等因素的影响，导致中国排放增速很快，在人均 GDP 4,400 美元时已经达到了法国人均 GDP 13,000 美元和意大利人均 GDP 18,000 美元时的人均 CO_2 排放水平。印度国

图 2-7 中国、印度、南非、巴西历年 CO_2 排放

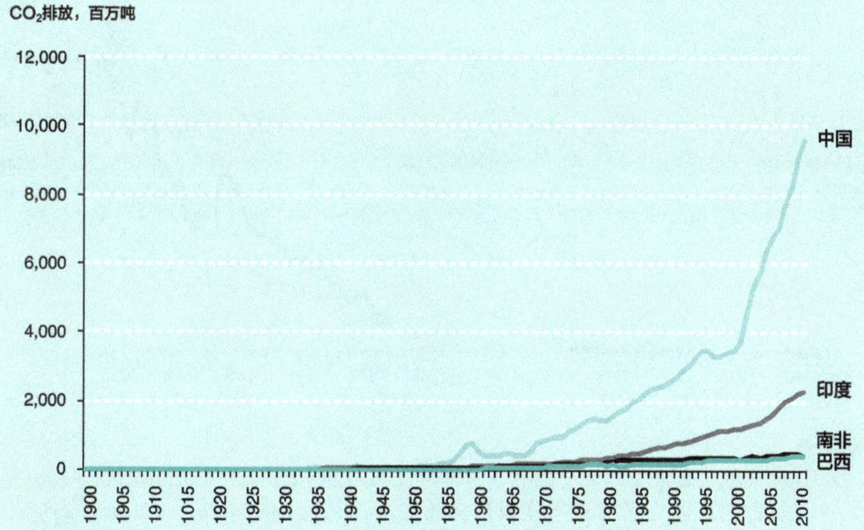

数据来源：

1. CO_2排放数据主要来自二氧化碳信息分析中心(Carbon Dioxide Information Analysis Center，CDIAC)，美国橡树岭国家实验室（Oak Ridge National Laboratory），2015，包含化石燃料燃烧的CO_2和水泥工艺过程的CO_2。
2. CDIAC的化石燃料燃烧的CO_2与IEA的化石燃料燃烧的CO_2数据基本一致，误差在1%左右。

内排放水平也从1971年的2.1亿吨增加到2012年的23.0亿吨，增长1015%。南非、巴西国内的排放水平分别从1971年的1.7亿吨和1.0亿吨上升至2012年的4.5亿吨和4.6亿吨，分别增长166%和342%。作为最大的几个发展中国家，中国、印度、南非、巴西国内的经济还有很大的发展空间和很强的发展需求，并在国际经济一体化、发达国家产业转移和国际贸易体系中担任重要角色，还有很高的排放需求，因此在未来短期内四国 CO_2 排放还很难出现峰值。

韩国[①]、东南亚国家、巴基斯坦、埃及、沙特、阿根廷、委内瑞拉等其他主要发展中国家要么是已经启动工业化进程的新兴经济体，要

图2-8 韩国、沙特、印尼、墨西哥历年 CO_2 排放

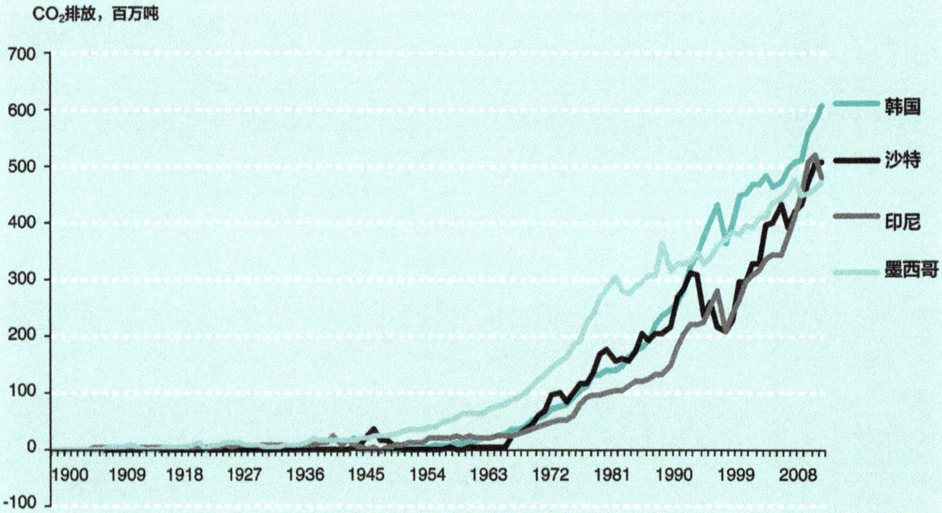

数据来源：
1. CO_2排放数据主要来自二氧化碳信息分析中心(Carbon Dioxide Information Analysis Center，CDIAC)，美国橡树岭国家实验室（Oak Ridge National Laboratory），2015，包含化石燃料燃烧的CO_2和水泥工艺过程的CO_2。
2. CDIAC的化石燃料燃烧的CO_2与IEA的化石燃料燃烧的CO_2数据基本一致，误差在1%左右

① 韩国、墨西哥已经是OECD成员国，但本文根据UNFCCC的划分，仍将其纳入发展中国家。

么是石油输出国，其排放总量自20世纪80年代以来进入急速上升阶段，还远未出现峰值和降低排放增速的迹象。2012年，韩国、沙特、印尼、墨西哥四国国内分别排放CO_2 6.1亿吨、5.1亿吨、4.8亿吨和4.7亿吨，与中国经历着类似的高碳发展阶段（如图2-8），此为这些国家有可能在争取排放空间问题上成为中国“立场相近国家”的共通利益和国情基础。基础四国中，中、印在能源结构、人均与总量排放上升趋势与增速等方面更为相似。

从人均排放看，中国、巴西、印度还未出现峰值，仍然在上升，到2012年中国、巴西、印度国内人均CO_2排放分别达到7.1吨/人、2.3吨/人和1.9吨/人；南非国内人均CO_2排放在2011年达到峰值10.3吨/人，随后下降（如图2-9）。

图2-9 中国、印度、南非、巴西历年人均CO_2排放

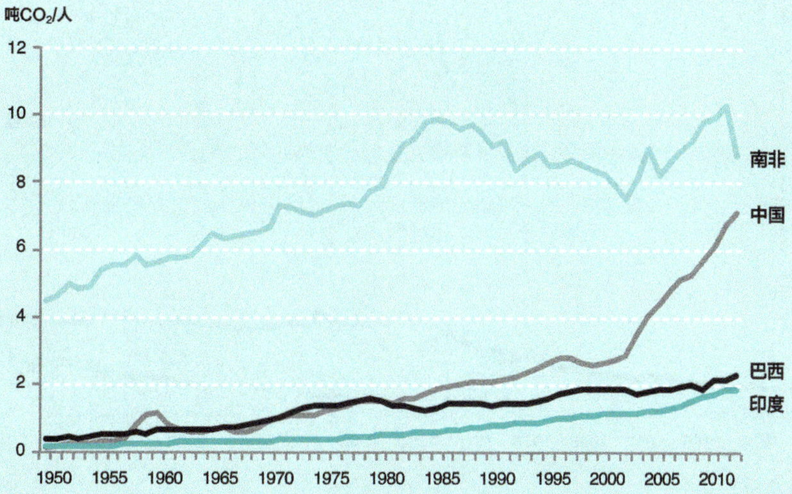

数据来源：
1. CO_2排放数据主要来自二氧化碳信息分析中心(Carbon Dioxide Information Analysis Center，CDIAC)，美国橡树岭国家实验室（Oak Ridge National Laboratory），2015，包含化石燃料燃烧的CO_2和水泥工艺过程的CO_2。
2. CDIAC的化石燃料燃烧的CO_2与IEA的化石燃料燃烧的CO_2数据基本一致，误差在1%左右。
3. 人口数据来自联合国人口署(United Nations, Department of Economic and Social Affairs, Population Division 2014)。

韩国、墨西哥、印尼人均 CO_2 排放仍然处于上升阶段（见图 2-10），2012 年三国国内人均排放分别达到 12.2 吨 / 人、4.3 吨 / 人和 2.1 吨 / 人；沙特阿拉伯国内人均排放受其国内经济主要依赖石油、经济结构单一影响，在 20 世纪 70 年代和 80 年代初期分别出现两次波动，在达到波峰后迅速下降，1998 年人均 CO_2 排放开始回升，到 2012 年达到 17.9 吨 / 人。

总而言之，步入工业化发展阶段的新兴经济体及石油输出国发展中国家，其年度排放总体依然处于上升趋势。

图 2-10 韩国、沙特、印尼、墨西哥历年人均 CO_2 排放

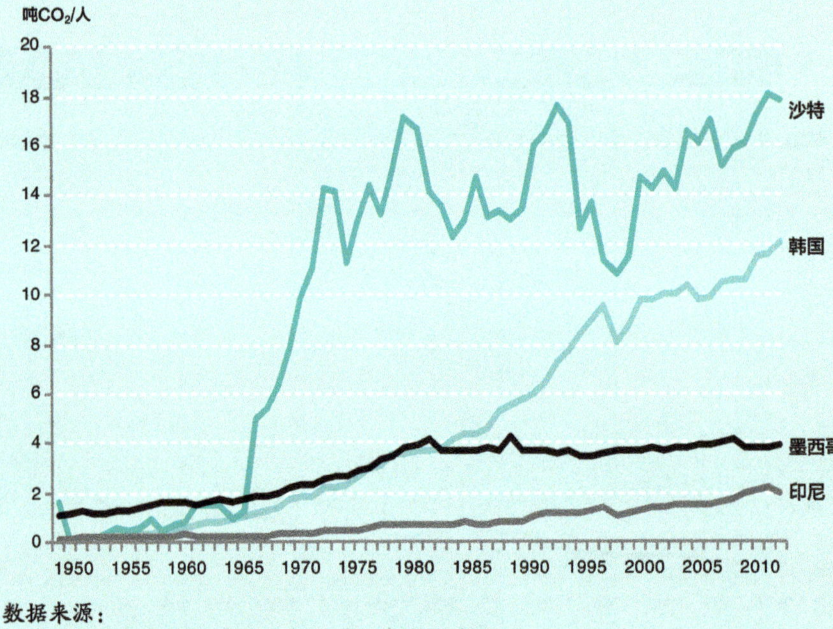

数据来源：
1. CO_2排放数据主要来自二氧化碳信息分析中心(Carbon Dioxide Information Analysis Center，CDIAC)，美国橡树岭国家实验室（Oak Ridge National Laboratory），2015，包含化石燃料燃烧的CO_2和水泥工艺过程的CO_2。
2. CDIAC的化石燃料燃烧的CO_2与IEA的化石燃料燃烧的CO_2数据基本一致，误差在1%左右。
3. 人口数据来自联合国人口署(United Nations, Department of Economic and Social Affairs, Population Division 2014)。

第三节　不同经济体历史排放和收入水平的关系

一、历史排放与收入之间的关系

如上文所述，主要发达国家基本在完成工业化、城市化进程后的后工业发展阶段才达到其排放峰值。OECD 国家整体、附件一国家整体、欧盟[①]整体及主要成员国、美国、加拿大等大体在 20 世纪 70 年代达到人均 CO_2 排放峰值，日本、澳大利亚大体在 20 世纪末和 21 世纪初达到人均 CO_2 排放峰值。其中“美、加、澳”和“欧、日”的具体轨迹和峰值水平呈现不同的模式。美国、加拿大和澳大利亚的人均峰值水平较高，接近或超过 20 吨 CO_2/ 人。欧盟和日本等国家的人均峰值普遍在 10 吨 CO_2/ 人上下。

从各国人均排放峰值与国内人均 GDP 关系来看，纵观工业革命以来的全球 CO_2 排放历史，目前还没有一个经济体能够摆脱随人均 GDP 和人类发展水平提高，人均 CO_2 排放水平“先增长后下降”的倒 U 型的库兹涅茨曲线现象。

在环境科学和环境经济学领域，有学者针对这种污染排放或环境退化程度与经济发展水平（通常以人均收入衡量）之间的倒 U 型关系，提出了环境库兹涅茨曲线（EKC）假说。EKC 假说认为，在经济发展的初期，污染排放和环境退化随着收入的增长而增长，而在超过某个收入水平的阈值之后，这一趋势将得到逆转，即收入水平的进一步提高将带来污染排放和环境损害的下降。

---

① 为 27 国数据，缺克罗地亚。

图 2-11 主要发达国家与发展中国家历史CO_2排放与收入之间的关系

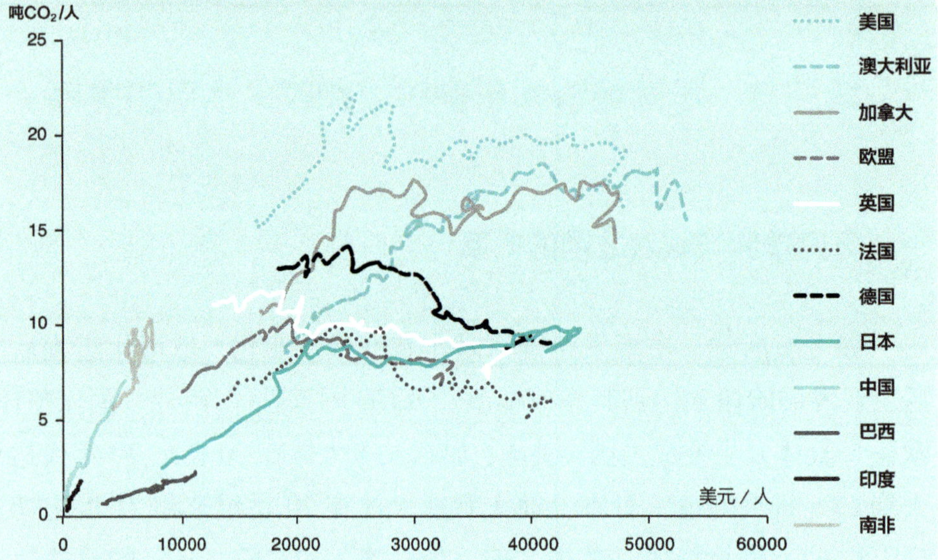

数据来源：

1. CO_2排放数据主要来自二氧化碳信息分析中心(Carbon Dioxide Information Analysis Center，CDIAC)，美国橡树岭国家实验室（Oak Ridge National Laboratory），2015，包含化石燃料燃烧的CO_2和水泥工艺过程的CO_2。
2. 德国1971—1990年数据来自国际能源署（IEA），CO_2 Emissions from Fuel Combustion 2011。
3. 人口数据来自联合国人口署(United Nations, Department of Economic and Social Affairs, Population Division 2014)
4. GDP数据来自世界银行数据库2015（World Bank Database），为2010年美元不变价。
5. CDIAC的化石燃料燃烧的CO_2与IEA的化石燃料燃烧的CO_2数据基本一致，误差在1%左右。

从历史趋势看（如图 2-11），美国国内人均 CO_2 排放达到峰值时对应的人均 GDP 水平为 2.5 万美元（2010 年价格，下同），欧盟整体人均峰值对应的人均 GDP 为 2.3 万美元，主要经济体如德国、法国、英国和意大利峰值年的人均 GDP 水平分别为 2.4 万美元、2.3 万美元、1.7 万美元和 3.5 万美元。澳大利亚和日本在国内人均 GDP 达到 4.2 万美元左右时实现人均排放峰值，加拿大虽然在 1979 年已经达到人均排放峰值，但是在人均收入超过 4 万美元时，人均排放还徘徊在峰值水平附近。中国、印度、巴西国内人均 CO_2 排放与人均 GDP 仍然在同步上升。

总体来看，世界主要发达国家实现国内人均排放峰值时对应的人均 GDP 水平有两种类别，一种为美国和欧盟，在人均 GDP 为 2 万—2.5 万美元左右时达到人均排放峰值，一种为日本、加拿大和澳大利亚，人均排放峰值对应的人均 GDP 水平为 4 万美元以上。

这一现象表明的是一个经济体在工业化、城镇化进程中，在从不发达到发达的过渡过程中，经济质量不断提升，对化石燃料依赖下降，而对技术、服务依赖上升的过程。以 EKC 曲线的峰值点为界，可以将 EKC 曲线分为左右两部分。曲线的左侧是“爬坡期”，此时随着收入上升，排放也将上升。位于曲线左侧的国家，通常还处于工业化和城市化进程中，收入水平还较低，并面临着消除贫困、改善人民生活水平等需求，需要大规模基础设施建设，因此经济发展对能源资源的依赖程度较高，尚未实现碳排放和经济发展的完全脱钩。越过 EKC 曲线的峰值拐点，将进入曲线右侧的“下降期”。以欧盟国家为代表的越过人均排放峰值的国家，通常已经进入后工业化时期，基础设施建设已基本完成，经济结构中制造业的比重较低，而服务业比重较高。此时经济发展对能源和排放的依赖程度较低，已经实现碳排放和经济发展的脱钩。因此，处于 EKC 曲线两侧的国家，对应着不同的发展阶段和社会经济进程，反映的经济活动性质、类型、优先议程依据是截然不同的，在经济学和伦理学上也是不同质的，不能简单、机械地对比它们的排放数据本身。对任何数据的解释，都离不开其对应的发展历史背景与国情背景。

二、影响排放与收入之间倒 U 型关系的驱动因素

历史排放和收入水平之间存在复杂的关系，而且在这些关系背后存在很多超越收入指标的驱动因素。

（一）发展阶段

借鉴马斯洛的需求层次理论，有研究认为，在发展的不同阶段，人们对于生存环境质量和舒适度的价值判断是不同的。在发展程度较

低的阶段，人们对温饱的关注较高，而对二氧化碳等温室气体排放带来的气候变化问题以及其他的环境质量问题没有太多关注，但随着发展水平的提高，温饱已不成问题，对于生存环境质量和舒适度（environmental amenities）的要求会提高。通常认为，环境质量的需求收入弹性大于 1，也就是说，在超过一定的收入水平阈值后，对高质量环境的支付意愿（willingness to pay, WTP）将以比收入水平更快的速度增长。因此在这一发展程度较高的阶段，人们将通过增加应对气候变化的防护性支出、加大对环保组织的捐赠、选择更低碳环保的产品、对减排的环境规制政策的出台施加压力等途径来减少二氧化碳的排放。人们将更容易认同保护环境的意识、政策，有更强的支付意愿和支付条件去支付更高的减排成本和适应成本。这种偏好或价值判断随发展阶段的变化更直接反映在局地环境问题上。在全球气候变化问题上不同发展阶段下的民意基础、普遍偏好倾向与支付意愿水平，也存在显著差异。

（二）能源结构

从能源结构的层面看，在经济发展初期，由于成本和技术条件的限制，能源结构较为单一，往往使用廉价的煤炭等高污染、高排放的能源进行生产，因此在该阶段，单位产出的污染强度较大，排放随着经济增长呈上升趋势。随着经济发展和技术进步，越来越多的清洁能源和技术得以推广和应用，能源结构得以升级，更加清洁化，从而在相同的产出和能耗下污染强度得以下降，出现污染排放与经济增长之间的脱钩。

（三）经济结构

从产出结构层面看，由于不同的经济部门之间的排放强度存在很大差别，因此当经济发展导致产业结构变化时，整体经济的排放强度将随之发生变化。通常认为，在发展的初期，经济的重心将从农业转向工业，尤其是高碳强度的重工业，从而带来排放增长。而在较高经济发展水平上高污染高排放的重工业占比甚至绝对产量将逐步下降，低碳强度的服务业占比逐步上升，从而使得单位经济产出的排放强度下降。即使在工业部门内部，产品结构也会升级，附加值上升，产业链延长，碳强度下降。

（四）技术进步

由技术进步带来的环境影响包括两个层面：一是技术进步提高全要素生产率，特别是提高能效，用更少的投入生产同样的产出，顺带产生更少的排放。尽管这类技术进步并不专门针对减排，但是也能够产生减排效应。二是随着排放控制和清洁技术的不断发展以及对传统技术的替代，将有效降低单位产出的排放，例如碳捕集、利用与封存技术（CCVS）。技术进步带来的这两方面影响都将有助于改善环境质量和控制排放。

（五）国际贸易和投资

国际贸易是解释历史排放和收入之间 EKC 曲线机理最重要的因素之一。随着一国收入的增长和环境损害的提高，更为严格的环境规制政策将会被实施，该国国内能耗污染密集型行业可能会通过国际贸易，或是国际投资的形式向国外——通常是低收入、低环境标准的发展中国家——转移。通过这种排放转移，发达国家自身的排放出现下降，同时可能带来发展中国家的排放上升。

（六）环境政策与治理的不断加强

政府的环境规制是缓解环境退化最重要的因素之一。当经济增长后，一国民众对于环境保护的公共意识将会提高，政府将会制定更有力度的环境规制政策，消除市场失效，防止环境退化进一步恶化。有研究估计了治理水平、政治体制、社会经济状况和教育投资等对 EKC 曲线的作用。同时有研究发现，如果一国的政治体制存在腐败、效率低下等问题时，EKC 曲线可能不存在或者拐点需要在很高的收入水平才能达到。

当然，对环境库兹涅茨曲线假说存在一定争议。许多研究从 EKC 模型的内生缺陷、模型解释力和计量方法等角度对 EKC 提出了批评。首先，EKC 的研究多采用简化式模型（reduced form model）形式，假设收入为外生变量，只考虑了经济对环境的单向影响，没有考虑环境损害对经济产出的反馈。其次，EKC 模型的解释力和可推广性也受到了部分研究的质疑。有研究认为没有证据表明发展中国家能够遵循发达国家的历史经验，也能越过拐点走上“收入上升，排放下降”的通道。再次，EKC 的实证研究使用的计量经济学方法，存在遗漏解释变量、环境和收入指标选取存在偏差等问题，但从前文所示的对主要经济体长时间尺度的实证排放数据分析看，这些争议并不影响收入和 CO_2 排放之间实证关系的存在。

三、对发展中国家的启示

尽管对于环境库兹涅茨曲线及其机理尚存在争议，但不能否定其经验证据的客观存在和政策含义。通过主要先行工业化经济体历史排放轨迹实证考察可看出，CO_2 历史排放与收入之间倒 U 型曲线关系在这些经济体确实存在。环境库兹涅茨曲线具有现实意义和政策含义，尤其对于发展中国家而言。

二氧化碳排放带来的问题不同于细颗粒物质（PM2.5）或二氧化硫等污染物，很难通过末端治理的方式进行削减，而是应该在发展过程中通过发展路径创新的方式解决。对于中国等发展中国家而言，在发展过程中不能一味依赖发达国家先污染后治理的传统道路，在新的历史时期，要开阔视野和思路，利用后发优势主动探索加速发展路径创新的可能和措施，在保证经济持续增长、人民生活水平持续提升的同时，使得碳排放与经济发展尽快脱钩，使得环境库兹涅茨曲线的拐点尽早、尽低到达，走出一条不同于美欧等发达国家的创新的发展路径。

另外，通过对主要先行工业化经济体历史排放轨迹的实证考察来看，CO_2 排放和收入之间倒 U 型曲线的具体形状在不同的国家会有不同的表现。如前文所述，从实现人均排放峰值时的排放水平来看，美国、加拿大和澳大利亚的峰值水平较高，接近或超过 20 吨 CO_2/ 人。而欧盟和日本等国家的峰值普遍在 10 吨 CO_2/ 人上下。从各国人均排放峰值与国内人均 GDP 关系来看，美国和欧盟在人均 GDP 为 2 万—2.5 万美元左右时达到人均排放峰值，日本、加拿大和澳大利亚人均排放峰值对应的人均 GDP 水平为 4 万美元以上。发展中国家未来是沿着“高发展水平高排放的美、澳、加模式”还是“相对高发展水平低排放的欧、日模式”的轨迹，抑或走出一条创新的不同于发达国家传统路径的更为低碳的发展路径，对其自身的社会经济可持续发展和全球应对气候变化的努力，都将具有十分重要的意义。

从全球剩余碳容量空间的客观约束来看，中国等发展中国家已经不具备走“美、加、澳”的“高发展高排放”的发展路径的条件，甚至以欧盟、日本为代表的“高发展、中等排放”的发展路径也难以为继。因此，以中国为代表的广大发展中国家，在争取发展权利、促进经济发展的同时，需要充分发挥后发优势，走出一条“高发展低排放”的创新发展路径，在保证经济持续发展的同时，促进 EKC 曲线的峰值尽早、尽低到来。

第四节　不同经济体历史排放变动的影响因素分析

若借用卡亚分解的概念框架，可从人口、人均 GDP、单位 GDP 能耗和单位能耗的 CO_2 排放几个方面，考察对 CO_2 排放的影响因素，数据详见表 2-2 和表 2-3。

欧盟主要成员国是仅有的几个达到长期稳定峰值的国家。分析其达到峰值后排放下降的主要驱动因子可以发现，欧盟国家 CO_2 排放降低主要依靠很低的人口增速（人口增速每年仅为 0.3% 左右，且城市化率普遍已高于 70%），较低的人均 GDP 增长速度（年均 2% 左右）以及单位 GDP 能源强度和单位能源碳强度的大幅下降，但各国情况略

表 2-2 各主要经济体 CO_2 排放影响因素对排放变化的贡献率（相对百分比）

国家（年份）	相对基年 CO_2 排放增长（%）	人口贡献（%）	人均 GDP 贡献（%）	单位 GDP 能耗贡献（%）	单位能源碳排放贡献（%）
法国（1979—2010）	－31.5	44.9	102.3	－62.7	－184.5
德国（1979—2010）	－30.9	12.4	130.7	－169.8	－73.2
英国（1973—2010）	－25.3	35.0	236.8	－294.3	－77.5
美国（1970—2007）	34.6	129.4	253.4	－245.1	－37.7
美国（2007—2010）	－5.7	45.0	－61.5	－59.0	－24.5
印度（1971—2009）	818.8	34.1	55.5	－23.6	34.0
日本（1970—2010）	48.1	50.9	223.0	－106.6	－67.4
中国（1970—2010）	968.1	20.8	126.1	－44.1	－2.8

注：负数表示排放下降或者是减排因素，正数表示排放上升或者是增排因素。根据 EIA 统计，由于可再生能源发展和页岩气开发，美国 2011—2012 年在经济增长的同时实现了 CO_2 排放下降。

数据来源：人口、GDP 和能源数据来自世界银行数据库（2012），其中中国能源数据来自能源统计年鉴。CO_2 排放数据来自 CDIAC。

表 2-3 各主要经济体 CO_2 排放影响因素对排放变化的贡献率（绝对值）

国家（年份）	相对基年 CO_2 排放增长（百万吨）	人口贡献（百万吨 CO_2）	人均 GDP 贡献（百万吨 CO_2）	单位 GDP 能耗贡献（百万吨 CO_2）	单位能源碳排放贡献（百万吨 CO_2）
法国（1979—2010）	－166.6	74.8	170.4	－104.4	－307.3
德国（1979—2010）	－341.1	42.1	445.7	－579.2	－249.6
英国（1973—2010）	－166.8	58.4	395.1	－491.0	－129.3
美国（1970—2007）	1498.1	1938.7	3796.1	－3671.7	－565.0
美国（2007—2010）	－334.4	150.5	－205.6	－197.4	－81.9
印度（1971—2009）	1685.6	575.1	935.4	－398.6	573.6
日本（1970—2010）	369.7	188.3	824.5	－394.1	－249.0
中国（1970—2010）	7469.4	1550.2	9422.4	－3291.2	－211.9

注：负数表示排放下降或者是减排因素，正数表示排放上升或者是增排因素。根据 EIA 统计，由于可再生能源发展和页岩气开发，美国 2011—2014 年在经济增长的同时实现了 CO_2 排放下降。

数据来源：人口、GDP 和能源数据来自世界银行数据库（2012），其中中国能源数据来自能源统计年鉴。CO_2 排放数据来自 CDIAC。

有不同。

能源结构低碳化是法国 CO_2 排放下降的主要因素。法国 2010 年与 1979 年相比，其单位 GDP 能源强度下降了 21%。单位能源碳强度因大量引入核能而降幅最大，达到 50%。

与法国不同，单位 GDP 能源强度下降（受产业结构调整和技术进步影响）是德国和英国 CO_2 排放下降的主要减排因素。德国单位 GDP 能源强度的降幅要大于单位能源碳强度降幅，2010 年相对 1979 年分别下降了 47% 和 24%。英国情况和德国类似，自 1973 年达到总量峰值后，2010 年的单位 GDP 能源强度和单位能源碳强度分别相对 1973 年下降了 58% 和 20%。

美国的情况比较复杂。虽然其人均排放在 1973 年就达到峰值，但受其人口增速较快的影响，其排放总量直至 2007 年才达到峰值。在 2007 年达到峰值后，2008 年、2009 年排放下降主要是受经济衰退的影响。而在 2010 年排放短暂上升后，根据美国能源信息署（EIA）统计，受可再生能源发展和页岩气开发等措施的影响，美国 2011—2014 年在经济增长的同时实现了 CO_2 排放下降。

人均 GDP 的高速增长是中国、印度等发展中国家 CO_2 排放高速增长的主要增排因素。和欧盟国家相比，中国由于经济的增长（2010 年的人均 GDP 相对 1970 年上升了 1884%），尽管 2010 年的单位 GDP 能源强度相对 1970 年下降了 65%，单位能源碳强度相对 1970 年下降了 7.5%，CO_2 排放量仍有大幅增长（2010 年的 CO_2 排放相对 1970 年上升了 968%）。印度情况和中国类似但也有区别，2009 年 CO_2 排放相对 1971 年上升了类似幅度，达到 819%，但引起排放增加的驱动因子与中国有所区别。除了人均 GDP 是重要的增排因素外，由于能源结构从传统生物质转向煤、石油等商品能源，能源结构高碳化也是重要的增排因素（2009 年单位能源碳强度相对 1971 年上升了 113%）。受上述两因素影响，尽管 2010 年的单位 GDP 能源强度相对 1970 年下降了 41%，CO_2 排放量仍有大幅增长。

第五节 小结

从上述分析中可以看出，纵观工业革命以来的全球 CO_2 排放历史，目前还没有一个主要经济体能够摆脱随人均GDP和人类发展水平提高，人均 CO_2 排放水平“先增长后下降”的倒U型的库兹涅茨曲线现象，但在不同的国家会有不同的表现，包括“相对高发展水平高排放的美、澳、加模式”或“相对高发展水平低排放的欧、日模式”。这一现象表明的是一个经济体在工业化、城镇化进程中，在从不发达到发达的过渡过程中，经济质量不断提升，对化石燃料依赖下降，而对技术、服务依赖上升的过程。存在这一现象并不意味着发展中国家可以无所作为。但发展中国家已不具备沿袭发达国家老路的条件，需要创新一条不同于美欧等发达国家的，更早达到更低峰值的发展道路。创新发展路径并不是空洞的学理概念，而是从对发达国家工业化历史进程的反思中提炼而成的。发展中国家通过适当的政策和体制管控措施并充分发挥后发优势，有望影响排放轨迹并实现发展路径创新，促成峰值更早更低地到来。而要实现上述峰值，单位GDP能源强度提高和能源结构低碳化将是发展中国家未来排放下降的主要抓手。

发展中国家的发展路径创新，如果能够得到发达国家的技术、资金援助，能够广泛开展南南合作，则实现的可能性和速度均可望大大提高。这一创新不仅是单一国别自主的事情，也不仅仅是后起发展中国家自主的事情，而是全球经济体作为一个整体的集体努力，需要全面开展国际合作。

此外，尽管中国2010年人均 CO_2 排放已超过法国、逼近意大利，但是中国和欧盟处于不同的发展阶段，欧盟国家普遍已经越过人均排放峰值，位于库兹涅茨曲线右侧的“下降期”，而中国还没有达到人均排放峰值，处于曲线左侧的“爬坡期”，两者对应着不同的发展阶段和社会经济进程，反映的经济活动性质、类型、依据是截然不同的，在经济学和伦理学上也是不同质的，不能简单相提并论。

【参考文献】

[1] Arrow, K., Bolin, B., Costanza, R., Dasgupta, P., Folke, C., Holling, C.S., Jansson, B-O., Levin, S., Maler, K-G., Perrings, C., Pimentel, D., 1995. Economic growth, carrying capacity, and the environment. Ecological Economics 15, 91–95.

[2] Baldwin, R., 1995. Does sustainability require growth? In: Goldin, I., Winters, L.A. (Eds.), The Economics of Sustainable Development. Cambridge Univ. Press, Cambridge, UK, pp. 19–47.

[3] Bhattarai M, Hammig M. Institutions and the environmental Kuznets curve for deforestation: a crosscountry analysis for Latin America, Africa and Asia[J]. World development, 2001, 29(6): 995-1010.

[4] Cole, M.A., 2004. Trade, the pollution haven hypothesis and the environmental Kuznets curve: examining the linkages. Ecological Economics 48, 71–81.

[5] Dasgupta, S., Laplante, B., Wang, H., Wheeler, D., 2002. Confronting the Environmental Kuznets Curve. Journal of Economic Perspectives 16 (1), 147–168.

[6] Dutt, K., 2009. Governance, institutions and the environment-income relationship: a cross-country study. Environment, Development and Sustainability 11, 705–723.

[7] International Energy Agency (IEA), 2011. CO_2 Emissions from Fuel Combustion 2011[EB/OL].
http://www.oecd-ilibrary.org/energy/co2-emissions-from-fuel-combustion-2011_co2_fuel-2011-en, 2015/09/20.

[8] IPCC, 2014: Climate Change 2014: Mitigation of Climate Change. Contribution of Working Group III to the Fifth Assessment Report of the Intergovernmental Panel on Climate Change [Edenhofer, O., R. Pichs-Madruga, Y. Sokona, E. Farahani, S. Kadner, K. Seyboth, A. Adler, I. Baum, S. Brunner, P. Eickemeier, B. Kriemann, J. Savolainen, S. Schlömer, C. von Stechow, T. Zwickel and J.C. Minx (eds.)]. Cambridge University Press, Cambridge, United Kingdom and New York, NY, USA.

[9] Kearsley A, Riddel M. A further inquiry into the Pollution Haven Hypothesis and the Environmental Kuznets Curve[J]. Ecological Economics, 2010, 69(4): 905-919.

[10] Leitao, A., 2010. Corruption and the Environmental Kuznets Curve: empirical evidence for sulphur. Ecological Economics 69, 2191–2201.

[11] Lopez, R., Mitra, S., 2000. Corruption, pollution, and the Kuznets environment curve. Journal of Environmental Economics and Management 40, 137–150.
[12] Oak Ridge National Laboratory, 2014. Carbon Dioxide Information Analysis Center (CDIAC) [EB/OL].
http://cdiac.ornl.gov/GCP/carbonbudget/2014/,2015/09/20
[13] Panayotou, T., 2003. Economic Growth and the Environment 2003, in Economic Survey of Europe: UNECE, no. 2, Chapter 2.
[14] Roberts, J.T., Grimes, R., 1997. Carbon intensity and economic development 1962–91: a brief exploration of the environmental Kuznets curve. World Development 25 (2), 191–198.
[15] Roca, J., 2003. Do individual preferences explain Environmental Kuznets Curve? Ecological Economics 45 (1), 3– 10.
[16] Selden T M, Song D. Environmental quality and development: is there a Kuznets curve for air pollution emissions?[J]. Journal of Environmental Economics and management, 1994, 27(2): 147-162.
[17] Suri V, Chapman D. Economic growth, trade and energy: implications for the environmental Kuznets curve[J]. Ecological economics, 1998, 25(2): 195-208.
[18] United Nations, Department of Economic and Social Affairs, Population Division, 2014. World Population Prospects [EB/OL].
http://esa.un.org/unpd/wpp/, 2015/09/20.
[19] Unruh, G.C., Moomaw, W.R., 1998. An alternative analysis of apparent EKC-type transitions. Ecological Economics 25, 221–229.
[20] World Bank, 2015. World Bank Database [EB/OL].
http://data.worldbank.org/, 2015/09/20.

第三章

全球发展与环境公平问题

Chapter Three

Global Development and Environmental Equity

气候公平的概念可界定为：所有国家、地区和个人都有平等地使用、享受气候系统资源的权利，也公平地分担稳定气候系统的义务和成本，在应对全球气候变化的背景下享受均等的实现可持续发展的权利和机会。

气候变化背景下的公平所蕴含的权利和机会的均等，包含生存权、发展权和环境资源产权等几个层面，具体包括基本生活需求得到满足和生活质量得到持续改善的权利和机会，得到安全、尊重进而最终实现自我的权利和机会，实现可持续发展的权利和机会，参与国际决策与共享信息的权利和机会。

气候公平对未来的全球气候治理提出了要求，是未来全球气候治理的核心要素，也应成为全球长期应对气候变化合作行动的重要指南。

历史责任、发展阶段、能力和国情是考量公平的四个核心要素。发达国家和发展中国家在所处的发展阶段、面临的发展需求、历史责任和现有能力等方面都存在巨大差别，要落实气候公平原则，就需要从这四个核心要素出发，综合考虑发达国家和发展中国家在这些要素上的差异。

“共同但有区别的责任”原则是在全球气候体制中落实公平的具体体现。《公约》订立二十年以来，国际形势和各国相对地位都发生了变化，在发展中国家内部出现了新兴发展中国家和其他发展中国家的细分。但是承认发展中国家内部的差异，并不能因此抹杀发达国家和发展中国家在历史责任、发展阶段、能力和国情等方面存在巨大区别的事实。

追求和实现“公平”和“共区”，并不意味着不行动，也不意味着追求均等的人均排放和历史累积排放，更不意味着发展中国家必然要重蹈发达国家在历史上的排放轨迹。发展中国家应在公平获取技术、资金、知识和教育机会的前提下，通过实现发展路径的低碳创新和变迁，以气候友好的方式实现脱贫、改善人民生活质量与安全状态的发展目标，在创新发展路径下实现各国公平获取可持续发展的权利和机会。

在全球减排方案设计中，主要通过努力分担框架，基于不同评价指标和方法将减排责任（或努力）分摊到各主要区域和主要经济体，实现气候公平原则的操作化。在努力分担方案中，通常被考虑的维度有责任、平等、能力和需求，以及成本有效性等。基于不同公平原则的努力分担方案，对于碳预算的地区分配结果有着重要影响。在历史责任和能力等指标权重较高的方案中，中东和非洲国家的减排力度可有所降低。同时，努力分担方案所基于的评价指标，其不同的计算方法，如衡量历史责任时起始年的选取，对分配方案的结果也会产生很大影响。但鉴于各方很难就具体指标和方法学达成一致，自上而下的公平责任分配方案很难在全球气候体制的构建过程中得到实际应用。

第一节 气候公平的伦理学基础

一、气候公平的概念与内涵

公平是指在道义上符合公正合理的（分配）原则。[①] 就哲学而言，公平体现为个人行为、体制或政策遵循某种社会道德规范及法律义务。就法律而言，公平意味着权利与责任的对等原则。本文所讨论的公平概念包含了正义（justice）、平等（equality）和公正（fairness/equity）等不同内涵，对这些用词的细微区别不作区分。

决定一项公共政策是否公平，有道义论和后果论两个维度。[②] 从道义论的角度讲，一个分配格局是否公平或公正取决于它在多大程度上尊重相关个人或群体的权利、义务、自由和平等地位。从后果论角度讲，决定一个政策或行动是否公平合理的因素是基于对其后果的判断，即所产生效用（utility）之大小。效果论和道义论的主要差别是前者更看重结果，对事件内在的道德质量不够关注；而后者更看重分配原则是否满足某种道德尺度，对事件所造成的后果和外部效应往往不够关注。前者围绕权利来谈分配，核心要素包括：生存权、发展权、平等权、自由表达意愿的权利、获取各种资源的能力、利益受损害者的索偿权（或损害方的赔偿 / 补偿责任）、各种需求的满足等；后者更关心分配结果是否是有效率的、合乎理性或符合预期的，核心要素包括效率（如成本效率、帕累托最优效率等）、效用（偏好或期望的实现）等。

气候公平是环境公平在气候变化领域的具体发展和体现。气候公

①② 郑艳，梁帆．气候公平原则与国际气候制度构建 [J]. 世界经济与政治，2011 (06)：69-90+158-159.

平概念的内涵可界定为：所有国家、地区和个人都有平等地使用、享受气候系统资源的权利，也公平地分担稳定气候系统的义务和成本，在应对全球气候变化的背景下享受均等的实现可持续发展的权利和机会。

气候公平是国际气候制度的基石。罗尔斯认为，“正义是社会制度的首要价值，正像真理是思想体系的首要价值一样。一种理论，无论它是多么精致和简洁，只要它不真实，就必须加以拒绝或修正；同样，某些法律和制度，不管它们如何有效率和有条理，只要它们不正义，就必须加以改造或废除。每个人都拥有一种基于正义的不可侵犯性，这种不可侵犯性即使以社会整体利益之名也不能逾越”。[①]

《从伦理角度看气候变化的布宜诺斯艾利斯宣言》明确指出，由于四方面原因，气候变化政策制定过程中，亟须考虑道德伦理问题：①只有在考虑道德伦理因素的前提下，国际社会才能在气候变化应对上做到公平；②在有关各种气候变化政策建议的科学和经济争议中，隐藏着深刻的公平和道德伦理问题；③为了克服目前阻碍国际气候谈判的各种障碍，有必要采取公平的气候变化政策措施；④公平的气候变化政策，可避免贫富差距进一步扩大，降低由于气候变化引发的粮食和水资源短缺，以及认为全球共有的大气的碳容量空间遭到不公平利用导致的潜在国际关系紧张。

二、气候公平的生存权、发展权和环境资源产权属性

公平是一个具有深刻伦理学内涵的概念，各方对公平有不同的理解，人类社会对公平原则尚未达成共识。一般而言，所谓“公平”，是基于世界体系中各个国家及其人民权利和机会均等的价值判断。在气

① 约翰·罗尔斯．正义论 [M]. 何怀宏，何包钢，廖申白，译．北京：中国社会科学出版社，1988.

候变化背景下，平等的权利和机会，体现为每个人都应享有与碳排放空间挂钩的均等的生存权、发展权和环境资源产权。

生存权就是国际人权公约上的“充分生活水准权”（the right to an adequate standard of living），即“人人有权为他自己和家庭获得充分的生活水准，包括足够的食物、衣着和住房，并能不断改进生活条件”。气候变化带来的负面影响，已经严重威胁了相当一部分人——尤其是贫困人口——的生存和基本生计，损害了其为自己和家庭获得适当生活水准的权利。应对气候变化的措施也可能对特点部门的人群（如煤矿工人）带来类似的影响。

发展权是在 20 世纪 80 年代提出并逐渐形成、充实和丰富起来的一项重要人权概念。1986 年通过的《发展权利宣言》指出，“每个人和所有各国人民均有权参与、促进并享受经济、社会、文化和政治发展，在这种发展中，所有人权和基本自由都能获得充分实现”。正如《发展权利宣言》所声明的，人是发展的中心主体。人类发展需要增强人的能力，扩大人的实际选择和实质自由。发达国家和发展中国家在发展需求上的侧重点不同。发展中国家的发展是指为了维持基本生活方面的社会生产力的发展，包括消除极端贫穷与饥饿、降低儿童死亡率、普及小学教育、改善产妇保健、促进男女平等并赋予妇女权利提供基本的社会保障等，而发达国家的发展则是指维持并提高现有生活的高标准。因此，满足基本生活的发展需求，应当优先于满足奢侈性消费的发展需求。

温室气体排放权，同时也是对碳排放空间这项环境容量资源的使用权。其权利主体被赋予了在一定时间、一定区域内排放一定数量的二氧化碳的权利，同时也被赋予了通过交易这种权利而从中获益的权利，属于环境资源产权的范畴。在较低发展水平上，这项权利与发展权有更多关联，部分反映了发展权利。保护全球气候的目标，使得碳排放空间有限，因此碳排放权这项环境资源产权也因其稀缺性而具有价值。对各国碳预算的初始分配，即是对这一特殊权利的初始界定和配置过程。由于发达国家在先行工业化过程中以不到 20% 人口先行占据了近半碳预算，对剩余碳预算的分配就与公平密切相关且具有极大伦理与政治敏感性。

因此在气候变化背景下，各个国家及其人民享有的均等权利和机会包含丰富的内涵，即基本生活需求得到满足和生活质量得到持续改善的权利和机会，得到安全、尊严进而最终实现自我的权利和机会，实现可持续发展的权利和机会，参与国际决策与共享信息的权利和机会。从天赋人权的意义上讲，这些权利是每个个人与生俱有的权利，这种平等的发展权利必须最终落实到每个个人身上。

第二节 气候公平的核心要素

诚如上文所言，气候公平体现为在全球应对气候变化背景下每个个体都享有平等的满足基本生活需求和持续改善生活质量的权利和机会，每个国家都享有平等的获取可持续发展的权利和机会。在造成气候变化的责任和受到气候变化的不利影响的问题上，各国之间存在巨大的不平等。气候公平对未来的全球气候治理提出了要求，是未来全球气候治理的核心要素，也应成为全球长期应对气候变化合作行动的重要指南。公平的气候体制有助于提高参与方的减排动力并确保协议的广泛参与和有效实施。

从国际气候谈判进程和研究进展看，历史责任、发展阶段、能力和国情是广为认可的考量公平的四个核心要素。发达国家和发展中国家在所处的发展阶段、面临的发展需求、历史责任、现有能力和国情等方面都存在巨大差别，综合考虑这些差异，对于落实公平原则具有重要的现实意义。“公平”的全球气候体制须全面反映各国的发展需求、历史责任、现有能力和国情。

一、历史责任

从引起气候变化的历史责任看，发达国家和发展中国家承担着不同的历史责任。工业革命以来的温室气体历史累积排放是衡量历史责任的基本指标。全球实现控制温升不超过2℃的目标将极大压缩未来全球的碳排放空间，世界各国都将面临排放空间不足的严峻挑战。尽管近年来发展中国家的碳排放快速增长，但从历史排放轨迹的全貌看，目前大气中由人类活动导致的温室气体存量绝大部分仍是由发达国家造成的，发达国家历史和当前的高碳排放，使全球排放空间分配不公成为既成事实，发达国家因此获得历史、现实与未来的既得利益，因而对此负有不可推卸的责任。从当前排放看，尽管发达国家排放量近年因经济衰退和其采取的减排措施有所下降，但人均排放水平依然处于高位，大大高于世界平均水平。2010年整个OECD国家以其全球18%的人口占到全球CO_2排放的38%，其排放总量在全球的占比远远高于其人口在全球的占比，人均CO_2排放仍远高于世界平均水平，更大大高于发展中国家的人均CO_2排放水平。从历史累积排放的角度看，自1751年到2010年间，由发达国家组成的附件一国家的能源相关CO_2累积排放占全球总累积排放的比重达到约70%。从人均历史累积排放看，1751—2010年全球平均人均能源相关CO_2累积排放为187吨CO_2/人，附件一国家为705.8吨CO_2/人，为世界平均水平的3.8倍。而非附件一国家仅为67.9吨CO_2/人，约为世界平均水平的1/3，附件一国家的1/10。发达国家自工业革命以来累积的CO_2排放已占全球的70%，对气候变化负有主要历史责任。虽然考虑森林和土地利用相关CO_2和其他温室气体后发达国家历史累积排放占全球比重有所降低，但不确定性非常大，特别是在区域层面，而且占比也达近50%。若考虑消费端的排放，即考虑贸易转移排放的影响，则发达国家的历史责任更高。

整体来看，《公约》实施20多年来，虽然发展中国家碳排放占全球的比例有所增长，但按累积排放量计算的各国历史责任的基本格局没有根本性改变。发达国家当前和今后相当长时期的历史累积排放和人均排放水平仍将高于发展中国家，已经并将进一步挤占和压缩发展中国家实现可持续发展所必需的碳排放空间，仍在继续扩大碳排放空间使用的不公平。

同时，发达国家除了负有累积排放的历史责任，还负有开创、主导、扩散和强化高碳发展路径的责任。在全球化背景下，发达国家通过对国际规则的影响优势及其海外投资、贸易联系，在全球供应链中的有利地位，对后起发展中国家的发展路径具有重要的示范效应和市场需求诱导效应。其技术规范、设计、商业模式、消费模式等都通过其海外投资和贸易的市场开发过程，对发展中国家的发展路径产生深刻影响，甚至形成路径复制和依赖，进而对发展中国家未来的高碳排放路径产生锁定效应。

二、发展阶段

从发展阶段看，发达国家和发展中国家目前仍处于不同的发展阶段，面临着不同的发展议程，应尊重不同发展阶段国家能源需求和 CO_2 排放强度不同的规律，保障发展中国家实现可持续发展必要的排放空间。经济发展、社会进步和环境保护，是世界可持续发展的“三大支柱”，三者互相依存。《公约》特别指出 ：“应当以统筹兼顾的方式把应对气候变化的行动与社会和经济发展协调起来，以免后者受到不利影响”，同时提出，“发展中国家为了迈向这一目标，其能源消耗将需要增长”。发达国家普遍已于 20 世纪 70 年代初完成工业化和城市化，进入后工业社会，其后 40 多年来能源消费和 CO_2 排放仍继续保持总体增长态势或稳定在相对高位水平。发达国家居于全球产业链的高端，在国际产业分工中已经将大部分高能耗高污染产业转移到新兴经济体国家，要素效率提高比要素投入增长对经济增长率作出的贡献更大，排放轨迹已越过排放峰值，处于环境库兹涅茨（EKC）曲线右侧的“下降期”，经济结构以第三产业为主，国内生产总值主要来自消费，温室气体排放主要来自建筑、交通等消费活动。

而发展中国家经济发展进程比发达国家滞后几十年到上百年，普遍仍处于工业化和城市化进程中，甚至还未启动工业化和城市化进程，仍处于 EKC 曲线左侧，未来仍需要满足较大的基础设施投资需求。重化工部门在未来很长阶段将依然占国民经济主导地位，在现有投入要素组合下有较强的增排需求。经济结构仍以二产甚至一产为主，国内生产总值主要来自投资，温室气体排放主要来自工业、农业和土地利用等部门的生存相关排放。发展中国家工业化和城市化发展阶段中能源消费的持续增长反映了世界各国发展进程中的共同规律，是实现可持续发展所必需的资源和生产要素需求，这是发展中国家不可剥夺的公平实现可持续发展的权利。

三、能力

从能力角度看，发达国家和发展中国家在资金、技术和软实力上的能力差距仍然显著存在。根据世界银行的统计，2012 年发达国家的人均 GDP(2010 年价格)仍然是发展中国家的 9 倍以上，与 1990 年相比，绝对量差距甚至在加大。大量核心技术仍掌握在发达国家跨国公司和国家实验室手中。发展中国家普遍仍处于全球产业链的中下游，在全球利益分配中处于劣势地位。根据联合国贸发署的统计，2008 年全球约有 8.2 万个跨国公司及其 82 万个分支机构，而大多数跨国公司来自于发达国家。前 100 名的跨国公司中来自发达国家的达到 73 家；前 5000 家跨国公司中，来自发达国家的占到 92%，并占 5000 家企业营业额的 90.9%。[①] 发达国家的企业在全球产业链中占据着高附加值环节，一方

① UNCTAD.World Investment Report[R].2009.unctad.org/en/Docs/wir2009-en.pdf.

面将更多的附加值留在了本国，另一方面将排放较高而附加值相对较低的生产环节转移到了发展中国家。而且除了传统能力，发达国家还在制定和实施国际规则的主导能力以及技术与知识的创新能力上具有优势。

四、国情

除去上述三个要素，公平还需充分考虑不同国家的特殊国情。《公约》3.2 条明确指出“应当充分考虑到发展中国家缔约方尤其是特别易受气候变化不利影响的那些发展中国家缔约方的具体需要和特殊情况，也应当充分考虑到那些按本公约必须承担不成比例或不正常负担的缔约方特别是发展中国家缔约方的具体需要和特殊性。”生态系统脆弱的国家，易遭自然灾害地区的国家，有低洼沿海地区的国家，有干旱和半干旱地区、森林地区和容易发生森林退化地区的国家，最贫穷国家和小岛屿国家等是气候损害和负面影响的主要承担者，需要在国际气候体系中基于“公平”原则对其予以特殊关注。同时，经济高度依赖于矿物燃料和相关的能源密集产品的生产、加工和出口所带来的收入，和 / 或高度依赖于这种燃料和产品的消费的国家，如石油输出国、能源资源禀赋以高碳化石燃料为主的国家，或者承担“世界工厂”职责，排放大多用于出口的国家，这些特殊国情都应在制定全球气候制度时得到充分考虑。

第三节 公平和“共区”原则在全球气候体制中的体现

一、公平获取可持续发展与“共同但有区别的责任”

全球应对气候变化的目标是保护地球大气环境，促进世界范围内的可持续发展。公平和可持续发展的理念和原则，是应对气候变化国际制度的核心和基础。《公约》3.1 条明确提出，“各缔约方应当在公平的基础上，并根据共同但有区别的责任和各自能力，为人类当代和后代的利益保护气候系统”。《公约》3.4 条也提出“各缔约方有权并且应当促进可持续发展”。由此可见，公平获得可持续发展是各缔约方在《公约》下的权利和义务，也应成为全球应对气候变化的共同愿景，成为应对气候变化国际制度改革和建设的基本原则。

“共同但有区别的责任”是在全球气候体制中反映公平的具体体现。在全球环境治理领域，“共同但有区别的责任”是一项重要的原则。“共同但有区别的责任”原则（以下简称“共区”原则）的理念可以追溯到 20 世纪 70 年代。1972 年《人类环境宣言》指出：发展中国家的环境问题大多是由于发展不足造成的。因此，发展中国家必须致力于发展工作，牢记他们的优先任务和保护及改善环境的必要性。“共区”原则的正式提出是在 1992 年里约联合国环境与发展大会上，并最终在《公约》以及后续的谈判中予以确立和体现。该原则首先来自于矫正正义（Corrective Justice）的主张，即“导致全球环境持续恶化的主要历史与现实原因是生产和消费的不可持续方式，因此造成这种损害的国家必须承担控制、减少和消除全球环境损害的责任，这种责任还必须与造成的损害相关，必须与他们各自的责任相一致”；其次来自于分配正义（Distribute Justice）的主张，即“历史上和目前全球温室气体排放的最大部分源自发达国家，发展中国家的人均排放仍相对较低，发展中国家在全球排放中所占的份额将会增加，以满足其社会和发展需要”。

“共同但有区别的责任”的含义有两层。第一层是“共同责任”，即全球所有国家，无论其大小、强弱、贫富，均对全球环境保护负有共同的责任，都应当积极参与其中。因此“共同责任”要求各国同心协力，携手应对全球气候变化问题。第二层是“区别责任”，是指不同国家在造成全球气候变化中的责任不同、应对气候变化的能力不同以及各自的发展阶段和国情不同，因此在承担全球气候保护上的义务也不同。

从上节分析中可以看出，《公约》订立 20 年以来，虽然发展中国家碳排放占全球的比重有所增长，但各国历史责任的基本格局没有改变，发达国家大大挤占了发展中国家的排放和发展空间，造成了历史的不公平，其减排力度和对发展中国家资金技术等支持不足，是在不断扩大和制造新的不公平。同发达国家相比，发展中国家在所处的发展阶段和能力上还有较大差距，且面临不同的国情。因此在国际气候体制中坚持“共区”原则仍有其合理性和必要性。

需要注意的是，强调公平和“共区”并不意味着不行动，事实上，发达国家和发展中国家都有采取强化行动应对气候变化的责任，只是需要坚持“共同但有区别的责任”原则，公平地承担应对气候变化的责任和义务。发达国家应率先深度减排，为发展中国家留出必要的排放空间，并向发展中国家提供资金、技术和能力建设支持，这是对其历史责任的后果和过多挤占发展中国家未来发展空间的一种补偿。发展中国家也要根据国情和自身能力，在可持续发展优先领域尽最大努力采取减缓行动。

同时，强调公平和“共区”并不意味着否认和回避发展中国家内部存在区别的事实，但发展中国家内部的差异并不能取代发达国家和发展中国家的区别。从发展路径可以看出，以中国为代表的新兴发展中国家与其他发展中国家（如最不发达国家、小岛屿发展中国家等）呈现出了不同的发展水平和发展需求。以中国为代表的新兴发展中国家，面临较强的工业化和城市化需求，目前正处于 EKC 曲线左侧的爬坡期，近期 CO_2 排放增速较快，且未来排放在一定时期内仍将继续上升。其他一些发展中国家如最不发达国家和一些小岛国，仍处于曲线的起点区，发展水平和排放水平都较低，还未进入起飞爬坡期，而且一些国家并没有工业化和城市化的需求。据此，在未来国际气候体制设计中，可在“两分法”前提下，适当考虑发展中国家内部的细分，接受新兴发展中国家和其他发展中国家的不同定位，在具体减排承诺方式和义务分担上，表现出适当的灵活性。当然，中国等新兴发展中国家与发达国家在历史责任、经济发展阶段、国情、资源禀赋、在全球供应链中的地位、经济结构和增长率来源、实际经济过程的内容等诸多方面，都存在着质的区别。这个区别是历史的，也是现实的，不容否认。

二、当前谈判中的主要公平方案

如何将《公约》的“共区”原则落地，区分不同的国家组别，并识别各类国家的“区别责任”，一直是《公约》进程下的国际气候谈判中的核心议题。在当前的2015协议谈判中，不同国家也提出了许多备选方案。

（一）自我区分方案

自我区分方案是对于不同国家，不预先作自上向下的区分，而是将这种区分留给每个国家自身，由各个国家自主决定其在全球应对气候变化中应负的责任和准备采取的行动。在华沙和利马气候大会上，提出并确认了各国以“国家自主减排贡献”形式将各国的目标和行动提交给《公约》秘书处，在一定程度上体现了“自我区分”的理念。自我区分方案的优点在于充分尊重了国家自主性，并具有高度灵活性，因此有助于扩大气候协议的参与范围。但是从气候公平角度来看，自我区分方案也存在很大的不足，它容易混淆发达国家与发展中国家之间的差别；无法保证各国减排目标的力度和公平性。对于各国，尤其是发达国家，其自主决定而非自上向下分担的目标，力度很可能与控制全球温升不超过2℃所要求的减排量相去甚远，也可能低于根据其造成全球气候变化的责任和应对气候变化的能力所应当承担的减排份额。

（二）动态调整国家分组方案

现有气候谈判中的国家区分，是在《公约》中以附件的形式将附件一和非附件一国家作了区分。部分国家认为，这种二分法过于僵化，没有考虑到《公约》制定20多年来各国地位间的巨大变化，已经不能完全反映世界局势的发展，并造成了《京都议定书》下具有定量减排义务的国家覆盖面过小，因此提出要以动态调整的方式进行国家区分。例如，埃塞俄比亚提出根据人均排放和收入对附件一和非附件一国家进行划分。美国提出增加附件 X 和附件 Y 的分类，分别由排放和经济增长趋势以及减排能力决定。这种对国家组别的动态划分，看似更精确更顺应变化，实质上从气候公平角度看也存在缺陷。从前文分析可知，《公约》订立 20 年以来，虽然发展中国家排放和能力有所增长，但各国历史责任的基本格局没有根本改变，而且承认发展中国家内部的差异，并不能因此抹杀发达国家和发展中国家存在巨大区别的事实。此外，在实际谈判中，很难就重新调整分组的标准和分组方案达成一致，缺少实际操作性。

（三）同心圆方案

针对 2015 气候协议，巴西提出了同心圆方案①。根据 INDC 下减排目标形式的差异，划分为三个同心圆，从内到外分别代表总体经济范围的绝对量减排目标，总体经济范围的强度、人均或偏离 BAU 的目标，以及部门的减排目标。所有附件一缔约方都应在同心圆的最内层。各缔约方应当在符合“共区”和各自能力原则的情况下朝着圆心方向移动，并且不允许从内部的圆向外部的圆倒退。同心圆方案着重考虑减缓，而对于国际气候协议的其他关键要素，如适应、资金、技术开发与转让、能力建设和透明度等问题缺乏考量。此外，对于哪些缔约方应当向圆心移动，以及如何移动等问题，也缺乏更细致的规定。

（四）公平参考框架

南非基于努力分担（burden-sharing）的方式，从《公约》的“共区”和各自能力原则出发，将这些原则操作化，提出了公平参考框架（Equity Reference Framework）②。该方案综合各国的历史排放责任、采取减缓行

① Brazil, Views of brazil on the elements of the new agreement under the convention applicable to all parties[M/OL]. 2014, http://www4.unfccc.int/submissions/Lists/OSPS ubmissionUpload/73_99_130602104651393682-BRAZIL%20ADP%20Elements.pdf.

② Winkler, H., Jayaraman, T., Pan, J., Oliviera, Adriano Santhiago de, Zhang, Y., Sant, G., Gonzalez Miguez, Jose Domingos, Letete, T., Marquard, A., Raubenheimber, S. (2011). Equitable access to sustainable development. Contribution to the body of scientific knowledge[R]. http://www.erc.uct.ac.za/Basic_Experts_Paper.pdf.

动的能力和发展需求三个指标，将全球剩余碳预算 / 全球减排责任在所有国家间进行分配。公平参考框架方案的问题在于不同的缔约方很难就衡量各国努力指标的选取达成共识，且指标的不同衡量方法对结果的影响很大。例如衡量历史责任时，时间起始点的选取对于责任划分影响很大。

三、在创新发展路径下实现气候公平

从上文的分析中可以看出，体现气候公平的核心是确保各方公平获取可持续发展的机会。在将碳排放与发展脱钩的历史进程中，平等的可持续发展权利与机会并不必然意味着均等的人均排放和历史累积排放，更不意味着发展中国家必然要重蹈发达国家在历史上的排放轨迹。实现这种公平的前提条件是发展中国家的经济得到大幅发展，当代科技革命和知识进步的成果能够被发展中国家及时分享。为此，一方面要确保发展中国家公平获取可持续发展的均等机会，促使广大发展中国家走上一条可持续的发展道路，在公平获取技术、资金、教育机会和知识的前提下，通过发展路径的低碳创新和变迁，以气候友好的方式去实现脱贫、改善人民生活质量与安全状态的发展目标。另一方面要确保发达国家基于其历史责任及在发展阶段、能力和在世界政治经济体系中占支配地位等方面的优势，遵循“共同但有区别的责任”原则，充分履行其减排责任，率先垂范，并向发展中国家提供技术、资金和能力建设支持，帮助发展中国家完成发展路径的创新和转型。

《公约》在创立之初就树立了这样的价值观，应当说这样的公平在世界范围内还远未实现，《公约》的精神在今天依然具有指导意义。各方只要充分按照《公约》所规定的各项原则去履行责任，就能实现公平。《公约》的原则是是否实现公平的判据。

第四节 基于气候公平不同原则的国际碳减排方案评价

一、基于气候公平不同原则的全球减排方案介绍

气候公平问题在全球减排方案中，主要体现为全球剩余的碳容量空间在地区间的不同分配，即通过努力分担框架 (Effort Sharing Framework)，基于不同评价指标和方法将减排责任 (或努力) 分摊到各主要区域和主要经济体，Höhne 等[①]通过总结现有全球减排方案研究，提出了四个在现有努力分担方案中通常被考虑的维度：责任、平等、能力和需求，以及成本有效性。

——责任：代表造成全球气候变化的历史责任，主要由一国的历史累积排放衡量。对于历史责任的计量起始年存在一定争议。一方认为应当以工业革命的起始年作为计量起点。另一方则认为应当从气候变化问题被人们知晓的时间，通常是 1990 年，作为计量起点。

——能力和发展需求：代表实施应对气候变化措施所需资源的可获取程度。各国的能力通常由 GDP(或人均 GDP) 或人类发展指数 (HDI) 来衡量。也有一些研究考虑了基本需求和发展权，强调一国在承担减排责任之前，有权利先满足基本生活和发展需求，对发展水平低于某一标准的人口或国家在一定程度上豁免减排责任。

——平等：强调每一个人不管其国籍、性别、年龄、能力和地位，都拥有权利排放同等数量的温室气体，亦即有权利获得同等数量的排放份额。在努力分担方案中，平等原则通常通过人均排放趋同来体现，

① Höhne N, den Elzen M, Escalante D. Regional GHG reduction targets based on effort sharing: a comparison of studies [J]. Climate Policy, 2014, 14 (1): 122-147.

趋同的路径和时间可以有所不同。

——成本有效性：根据各国的减排潜力来分配排放配额，减排潜力大的国家承担更大的减排份额。边际减排成本通常作为衡量成本有效性的指标。

基于历史责任、能力和需求、平等、成本有效等公平维度或不同维度的组合，现有研究提出了多种多样的全球减排方案。IPCC 第五次评估报告对这些努力分担研究进行了总结。报告基于文献考虑了如下努力分担方案（见表 3-1）：①考虑责任的分配方案，即考虑各国对全球历史排放和温升的共同但有区别的责任，如巴西案文；②考虑能力和发展需求的分配方案，即考虑各国不同的减缓气候变化能力或支付能力以及基本发展需求，如基于人均 GDP 分配；③考虑平等的分配方案，即考虑所有人具有平等的排污权利和不受污染的权利，如考虑人均排放趋同的紧缩趋同方案；④考虑成本有效原则的分配方案，即考虑各国减排潜力和成本最优的减排方案，如全球单一碳价方案；⑤综合考虑责任、能力和发展需求的方案，如温室气体排放发展权利方案(Greenhouse Development Rights, GDR)；⑥综合考虑责任和平等的方案，即碳预算和人均历史累积排放方案；⑦综合考虑责任、平等、能力和发展需求的分阶段方案 (Staged Approach)，包括多阶段方案 (M6)、共同但有区别趋同方案 (Common but Differentiated Convergence, CDC)、多指标趋同方案 (Multi-Criteria Convergence, MCC)、基于行业减排的 Triptych 方案、G8 方案和南北对话方案 (South-North Dialogue) 等。

二、不同方案的结果比较及评价

基于不同公平原则的努力分担方案，对于碳预算的地区分配结果有着重要影响。IPCC 第五次评估报告总结了成本最优情景和努力分担框架下各区域（OECD1990、经济转轨国家、亚洲、中东和非洲、拉丁美洲）2030 和 2050 年的减排要求（见图 3-1）。

表 3-1 全球减排方案

分类	考虑的要素：责任	能力和需求	平等	成本有效	描 述	方法应用	指标应用（举例）
责任	×				用造成气候变化的历史影响定义责任	巴西方案	（人均）累积排放
能力和需求		×			基于各国 GDP/HDI 水平或者 GDP 损失分配。有些方案也考虑排除了基本需求排放	各国能源强度 / 排放强度趋同； 发达国家 / 发展中国家承担同样的减排成本 / 损失（成本趋同方案）； 同样收入水平的人承担同样的减排义务	人均 GDP；HDI；单位 GDP 排放；GDP 贫困程度；国民收入分布
平等			×		人均排放趋同，以紧缩趋同方案为主，趋同年份不同。也有一些研究将基本可持续排放需求与紧缩趋同方案结合起来考虑	不同形式的人均排放趋同方案	人均排放
责任能力和/或需求	×	×			使用责任和能力作为分配依据	温室气体发展权； 责任，能力和可持续发展	人均 GDP； 国家收入分配； 贫困人口
人均累计排放均等	×		×		将全球累积碳排放空间基于不同指标分给特定国家的特定年份，如采用人均累积碳排放趋同的方案	碳预算； 人均累积排放均等	碳预算； 人均累计排放
分阶段的方法	×	×	×		包含各国在不同时间段采取有区分的承诺的方案，也包括基于部门的分配方案（如 Triptych 方案），以及等比例减排方案（祖父原则方案）	多阶段； 共同但有区别趋同； 技术水平趋同的部门减排方案	多指标混合； 不同部门的具体指标，如电力部门 2050 年可再生能源和非化石能源排放的比重
成本最优方案				×	基于各国的减排潜力进行分配	边际减排成本均等	全球单一碳价情景或者成本最优情景下的分配结果

资料来源： Höhne N, den Elzen M, Escalante D. Regional GHG reduction targets based on effort sharing: a comparison of studies [J]. Climate Policy, 2014, 14 (1): 122–147.

图 3-1 不同公平原则方案下 2030 年各地区减排要求

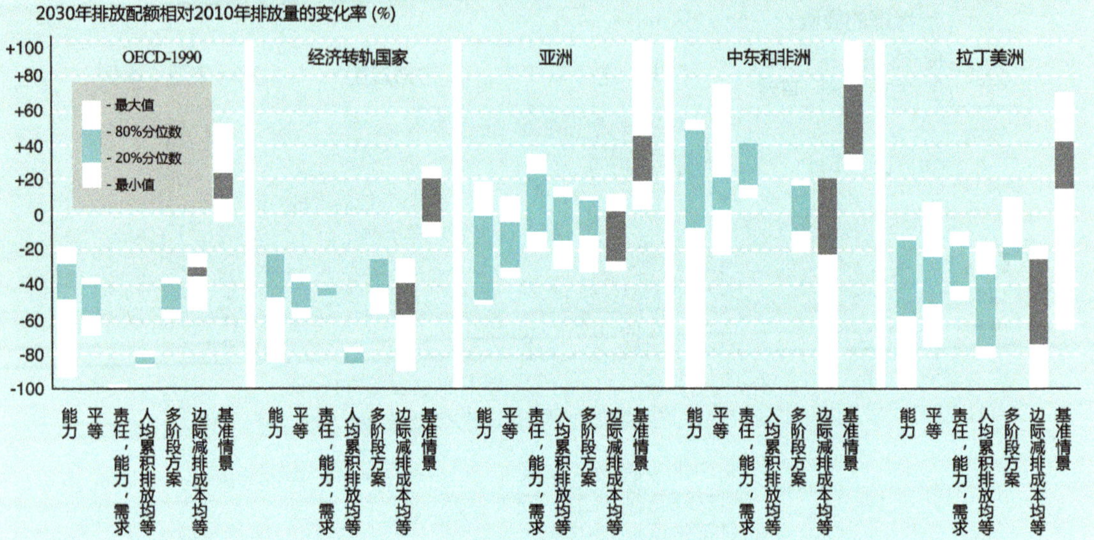

资料来源： IPCC. Climate change 2014: mitigation of climate change [M/OL]. Cambridge: Cambridge University Press, 2014 [2014-08-25]. http://www.IPCC.ch/report/ar5/wg3/.

在作为对比基准的成本最优情景下，为实现 430—530 ppm 浓度情景，南亚（主要是印度）2030 年的排放可大幅高于 2010 年水平，继续上升 50%—60%; 而东亚（主要是中国）则需要大幅低于 2010 年水平，下降 0—40%; OECD1990 国家 2030 年排放需要相对 2010 年下降约 32% (23%—40%); 经济转轨国家下降约 32%(18%—40%); 拉丁美洲 2030 年的排放需要相对 2010 年下降约 35% (16%—59%); 而中东和非洲 2030 年的排放则需略低于 2010 年水平。

而综合各种努力分担方案（包含考虑责任、能力、平等、发展需求等不同评价指标的不同分配方法，未考虑成本最优方案），碳预算的分配结果有所不同。为实现 450 ppm 目标，亚洲国家 2030 年的排放需要基本回到并略低于 2010 年的排放水平（下降 33%—上升 7%），OECD 国家 2030 年的排放需要在 2010 年的基础上下降约 50%(37%—75%)，经济转轨国家需要下降约 1/3 (28%—53%)。中东和非洲国家

2030 年的排放可略高于 2010 年水平（下降 7%—上升 24%），而拉丁美洲 2030 年的排放需低于 2010 年水平（下降 15%—49%）。

与成本最优情景相比，在努力分担框架下，受其历史责任和能力优势等影响，OECD1990 国家的减排力度需大幅增加，经济转轨国家和亚洲国家的减排力度有小幅增加，中东和非洲国家的减排力度可有所降低。

需要注意的是，如上述将全球排放空间和减排责任分摊到各区域需要基于一系列的评价指标和分配方法学。指标选取会对分配方案的结果产生很大影响。如在描述历史责任时，以 1750 年为起始年还是 1990 年为起始年，会对各国历史排放分布产生重要影响。① 而在 IPCC 报告总结努力分担方案结果所引用的文献中，在描述历史责任时，大多以 1990 年而非 1750 年作为计算历史累积排放的起始年。如 Baer，②Baer et al.，③Höhne & Moltmann，④ 在应用温室气体发展权方案时，取的都是 1990 年以来的历史责任。这将在很大程度上对分配结果产生影响。

总结现有的努力分担方案可以看出，多数方案仅考虑了历史责任、发展阶段和需求、能力、国情等气候公平的核心要素中的一部分，例如全球单一碳价方案仅考虑了成本有效原则，而忽视了各国在责任、能力、需求和发展阶段等方面的不同。温室气体发展权方案综合考虑了责任、能力和发展需求等公平要素，但忽视了气候公平蕴含的平等

---

① 傅莎，邹骥，张晓华，等 .IPCC 第五次评估报告历史排放趋势和未来减缓情景相关核心结论解读分析 [J]. 气候变化研究进展，2014, 10 (5): 323-330.

② Baer P. The greenhouse development rights framework for global burden sharing:reflection on principles and prospects[J]. Wiley Interdisciplinary Reviews: Climate Change, 2013, 4(1): 61-71.

③ Baer P, Fieldman G, Athanasiou T, et al. Greenhouse Development Rights: towards an equitable framework for global climate policy[J]. Cambridge Review of International Affairs, 2008, 21(4): 649-669.

④ Höhne N, Moltmann S. Distribution of emission allowances under the greenhouse development rights and other effort sharing approaches[J]. Germany: Heinrich-Böll-Stiftung, 2008.

的内涵，即人人都应当享有平等的排污权利和不受污染的权利。仅有少数方案，综合考虑了历史责任、发展阶段和需求、能力、国情等气候公平的所有核心要素，而且在具体选择实现公平原则的指标上也存在不公平的情况。例如在衡量历史责任时，选择 1990 年而非 1750 年作为起始年，实质上回避了发达国家和发展中国家在造成全球气候变化的责任上的巨大不平等。

因此，在实际全球气候体制构建过程中，鉴于各方很难就具体指标和方法学达成一致，自上而下的公平责任分配方案在政治上并不可行。区分问题可能是在国际气候体制中落实公平原则的关键。坚持发达国家和发展中国家区分仍有其合理性和必要性。在未来的国际气候体制中，可在“两分法”前提下，考虑小岛国和最不发达国家的特殊情况，在具体贡献方式和义务分担上，表现出适当的灵活性。

专栏：主要缔约方 INDC 目标公平性评估①

如何认识各国 INDC 的力度和公平性是目前谈判中的热点，也是各方评估的重点。近日，由生态公平组织和斯德哥尔摩环境研究所联合开展的气候公平参考项目发布报告，评估了主要《公约》缔约方国家自主贡献的力度和公平性。

研究结合《公约》要求，综合选取历史责任、能力和发展需求作为衡量公平的三个核心原则。其中，历史责任原则代表对造成全球气候变化问题责任越大的国家应当承担的减排义务越大；能力原则代表应对气候变化的能力越强的国家应当承担的减排义务越大；需求原则

① 资料来源：Civil Society Organizations. Fair shares: a civil society equity review of INDCs[R]. 2015. http://civilsocietyreview.org/wp-content/uploads/2015/11/CSO_FullReport.pdf.

代表各国满足基本生活和发展所必需的排放应当被充分保证。基于确定的公平原则，研究通过责任分担（Burden-sharing）的方法，提出了考虑不同历史责任起始年和不同发展需求衡量方式的三种公平分配方案（1850/高累进方案，1950/中累进方案和1990/低累进方案），据此对全球照常情景（BAU）排放与实现2℃目标所需排放之间的差值——即全球所需的减排努力——在主要缔约方之间进行公平分配，并对各国INDC目标和其应承担的“公平”减排责任进行比较。研究发现：

（1）现有国家自主贡献的力度总和仅为不到实现全球2℃温控目标要求的一半。根据2℃温控目标要求，到2030年，与照常情景相比，全球应减排350亿吨CO_2当量。然而，目前发达国家缔约方国家自主贡献总和到2030年的减排量约为60亿吨CO_2当量，发展中国家缔约方国家自主贡献总和减排量约为110亿吨CO_2当量（其中90亿吨为无条件的减排量，20亿吨为有条件的减排量）。目前所有缔约方国家自主

专栏图　公平分配方案与INDC目标的对比

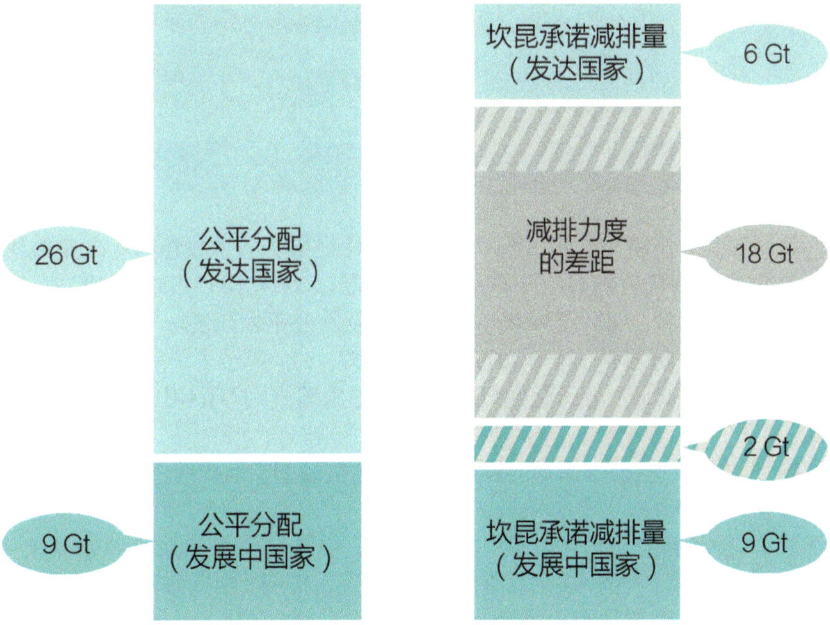

贡献总和减排量共计 170 亿吨 CO_2 当量，仅为 2℃温控目标要求的不到一半。

（2）发达国家缔约方国家自主贡献力度远未达到公平原则要求的水平。按照报告的量化分析，如果公平分配全球为实现 2℃目标所需的减排量，发达国家整体 2030 年应承担 110 吨 CO_2 当量的减排量，并为 150 吨 CO_2 当量减排量提供资金、技术和能力建设支持。而目前发达国家缔约方国家自主贡献总和仅能实现减排约 60 亿吨 CO_2 当量，与公平分配要求的减排量存在巨大差距。从个体情况看，美国和欧盟国家自主贡献力度约为公平分配减排要求的五分之一，日本约为十分之一。

中国国家自主贡献的力度和公平性值得称赞。按照报告的量化分析，如果公平分配全球为实现 2℃目标所需的减排量，从减排量看，中国 2030 年应在照常情景的基础上减排 33.7 亿—54.7 亿吨 CO_2 当量，而根据中国碳排放强度较 2005 年下降 60%—65% 的国家自主贡献目标计算，中国 2030 年将在照常情景的基础上减排 48.9 亿—65.1 亿吨 CO_2 当量，中国的减排力度最多可超过公平分配要求水平约 1 倍。

【参考文献】

[1] 傅莎，邹骥，张晓华，等 .IPCC 第五次评估报告历史排放趋势和未来减缓情景相关核心结论解读分析 [J]. 气候变化研究进展，2014, 10 (5): 323-330.

[2] 孙耀华，仲伟周 . 国际温室气体减排方案及其公平性研究——基于罗尔斯正义论的视角 [J]. 资源科学，2013 (07)：1430-1437.

[3] 杨通进 . 全球正义：分配温室气体排放权的伦理原则 [J]. 中国人民大学学报，2010 (02)：2-10.

[4] 郑艳，梁帆 . 气候公平原则与国际气候制度构建 [J]. 世界经济与政治，2011 (06)：69-90+158-159.

[5] 邹骥 .“经济转型与低碳发展”丛书总序：中国经济中长期发展的低碳化趋势：环境资源约束、效率与持续增长 [M]// 王克，邹骥 . 中国城市温室气体清单编制指南 . 北京：中国环境出版社，2014.

[6] Baer P, Fieldman G, Athanasiou T, et al. Greenhouse Development Rights: towards an equitable framework for global climate policy[J]. Cambridge Review of International Affairs, 2008, 21(4): 649-669.
[7] Baer P. The greenhouse development rights framework for global burden sharing: reflection on principles and prospects[J]. Wiley Interdisciplinary Reviews: Climate Change, 2013, 4(1): 61-71.
[8] Brazil, Views of brazil on the elements of the new agreement under the convention applicable to all parties[M/OL]. 2014, http://www4.unfccc.int/submissions/Lists/OSPS ubmission Upload/73_99_130602104651393682-BRAZIL%20ADP%20Elements.pdf.
[9] Den Elzen M G J, Oliver J G J, Höhne N, et al. Countries' contribution to climate change: effect of accounting for all greenhouse gases, recent trends, basic needs and the technological progress [J]. Climate Change, 2013,121(2):397-412.
[10] Höhne N, den Elzen M, Escalante D. Regional GHG reduction targets based on effort sharing: a comparison of studies [J]. Climate Policy, 2014, 14 (1): 122-147.
[11] Höhne N, Moltmann S. Distribution of emission allowances under the greenhouse development rights and other effort sharing approaches[R]. Germany: Heinrich-Böll-Stiftung, 2008.
[12] IPCC. Climate change 2014: mitigation of climate change [M/OL]. Cambridge: Cambridge University Press, 2014 [2014-08-25]. http:/www.IPCC.ch/report/ar5/wg3/.
[13] UNCTAD, World Investment Report [R], 2009. http://unctad.org/en/Docs/wir2009_en.pdf.
[14] Winkler, H., Jayaraman, T., Pan, J., Oliviera, Adriano Santhiago de, Zhang, Y., Sant, G., Gonzalez Miguez, Jose Domingos, Letete, T., Marquard, A., Raubenheimber, S. (2011). Equitable access to sustainable development. Contribution to the body of scientific knowledge[R]. http://www.erc.uct.ac.za/Basic_Experts_Paper.pdf.

第四章

人类发展路径创新：另辟蹊径的发展

Chapter Four

Innovation of Human Development Path: an Alternative

应对气候变化归根到底是发展问题，应该在发展进程中解决气候变化问题。要想改变排放的路径，必须转变发展的路径。发展路径创新的出发点是处于不同发展阶段的国家都要以积极的姿态行动起来，探寻与本国发展阶段相符的发展路径创新；其着眼点是影响排放轨迹和路径的发展驱动因素，以求在更实质、更实际、更具体的层面上打破全球气候治理谈判的僵局。从创新发展路径的角度入手设计 2020 年后国际气候体制，有助于全面、公平反映发达国家与发展中国家不同的责任、能力、国情和在不同发展阶段的不同重点议题，更好地将减排目标和措施与影响排放路径的驱动因子联系在一起，采取更有可行性、可控性的有效措施，并切实考虑到发达国家与发展中国家的不同利益关切，实现合作共赢。

创新发展路径的最终目的是以尽可能少的排放实现一定的发展水平。从增长函数看，创新发展路径就是要让经济增长更多地依赖要素效率的提高，而降低对要素投入扩张的依赖。而从卡亚分解看，创新发展路径则需要从人口（包括总量、增速和分布等）、GDP（包括总量、增速、结构等）、能源强度（包括产业 / 产品结构和能效水平等）及能源碳强度（包括能源结构及相应的资源禀赋和可获取性等）等驱动因子入手。

发达国家、新兴发展中国家和其他发展中国家在 2020 年后面临着不同的创新发展路径的任务。

对发达国家而言，其主要任务是发挥示范作用和市场需求导向作用，加速下降温室气体排放，率先垂范一条创新的低碳可持续发展示范路径，并通过向发展中国家提供资金、技术、能力建设支持及政策、经验交流，将上述示范路径复制、推广、扩散到发展中国家。发达国家的示范作用对于全球创新发展路径有着至关重要的作用。

对发展中国家特别是新兴发展中国家而言，有必要也有可能改变现有发展模式，通过发展路径的跃迁，开创一条比美欧等发达国家传统发展路径更为低碳、更早达到更低峰值的创新发展道路，但其实现转型的可能性在很大程度上取决于发达国家所给予的支持程度和示范程度。

从创新发展路径的角度出发，发达国家和发展中国家未来减排贡献可以包括如下四方面指标：量化转型程度指标；量化转型过程指标；定性指标和保障指标。发达国家、发展中国家应基于其责任、能力、发展阶段、发展需求和特定国情选择不同的贡献指标。

对发达国家而言，应延续其一贯的全经济领域绝对量化减排指标，同时为发展中国家发展路径的变迁提供资金、技术、能力建设支持。此外，发达国家还可承诺示范指标如特定部门的减排目标和特定技术的发展目标。

对新兴发展中国家而言，其主要任务是实现发展路径的变迁，争取尽早达到更低峰值。未来贡献可采用“意向性的量化转型程度目标（如峰值目标、强度指标、总量目标、人均目标）+更为严格的量化转型过程指标”的形式。具体的指标选择可以是灵活多样化的,根据本国国情选取。需要强调的是，这些行动需要得到来自发达国家在技术、资金、能力建设和优惠的贸易与投资条件等方面的充分支持。

对其他发展中国家而言，可以自主根据国内需求选择是否采取行动及具体的行动指标，同时需要承诺在接受国际社会支持时优先接受低碳技术、低碳投资、低碳基础设施建设等方面的援助。

第一节 发展路径创新的必要性和可能性

应对全球气候变化的挑战要求现有社会经济系统、经济发展、城市发展、能源生产和消费模式发生根本性改变 (Rotmans et al., 2001;IPCC, 2015)。[①②] 应对气候变化归根到底是发展问题，应该在发展进程中解决气候变化问题。除非气候变化政策更深入的嵌入潜在发展道路的改变，否则实现气候目标将是昂贵的和具有破坏性的 (Robinson et al., 2001; Swart, Robinson, & Cohen, 2003)。[③④] 这表明，有必要将气候政策与经济、社会、技术和环境相关的更广泛的可持续发展目标之间进行整合 (Bizikova, Burch, Cohen, & Robinson, 2010; Sathaye et al.,2007)。[⑤⑥] 只有在确保每个国家可持续发展机会和达到一定发展水平

① Rotmans, J., Kemp, R., & van Asselt, M. (2001). More evolution than revolution: Transition management in public policy. Foresight, 3(1), 15–31.

② IPCC, 2014. Climate Change 2014, Mitigation of Climate Change. Contribution ofWorking Group III to the Fifth Assessment Report of the Intergovernmental Panel on Climate Change. Cambridge University Press, Cambridge, United Kingdom and New York, NY, USA.

③ Robinson, J., Morita, T., Adegbulugbe, A., Alcamo, J., Herbert, D., Lebre La Rovere, E., Dadi, Z. (2001). Greenhouse gas emissions: Mitigation scenarios and implications. In B. Metz, O. Davidson, R. Swart, & J. Pan (Eds.), Climate change 2011: Mitigation (pp. 115–166). Cambridge: Cambridge University Press.

④ Swart, R., Robinson, J., & Cohen, S. (2003). Climate change and sustainable development: Expanding the options.Climate Policy, Special Issue on Climate Change and Sustainable Development, 3(Suppl. 1), S19–S40.

⑤ Bizikova, L., Burch, S., Cohen, S., & Robinson, J. (2010). A participatory integrated assessment approach to local climate change responses: Linking sustainable development with climate change adaptation & mitigation. In K. O'Brien, B. Kristoffersen, & A. St. Clair (Eds.), Climate change, ethics and human security (pp. 157–179). Cambridge: Cambridge University Press.

⑥ Sathaye, J., Najam, A., Cocklin, C., Heller, T., Lecocq, F., & Robinson, J. (2007). Sustainable development and mitigation. In B. Metz, O. Davidson, P. Bosch, R. Dave, & L. Meye (Eds.), Climate change 2007: Mitigation. Contribution of Working Group III to the Fourth Assessment Report of the Intergovernmental Panel on Climate Change (pp. 691–743). Cambridge: Cambridge University Press.

的前提下讨论应对气候变化问题，才能激励各个国家的广泛参与。通过减少发展中国家的发展机会实现减排的方式，很难获得各方的认可，也不可能真实实现。

要想改变排放的路径，必须转变发展的路径。这是对近现代发达国家实现工业化全过程所对应的温室气体排放路径进行实证研究后得出的基本结论。通过上文对工业化国家历史发展路径的实证探讨可以发现：主要先行工业化经济体普遍经历了随人均GDP提高，人均CO_2排放水平“先增长后下降”的倒U型的库兹涅茨曲线现象。发展中国家目前普遍仍处于曲线左侧，存在继续增排的需求。但与此同时，受全球实现2℃温控目标的碳排放空间约束以及发展中国家自身环境排放容量约束等因素的影响，发展中国家已不再具备延续发达国家传统高碳高排放发展路径的条件，即便是延续相对较为集约低碳的“欧、日”发展路径也将难以为继。而且发展中国家整体目前也不再具备发达国家通过将高碳产业转移到发展中国家实现结构转型和减排的条件。发展中国家必须创新一条区别于发达国家的更为低碳、集约、高效的发展路径。

另一方面，受技术进步、后发优势等因素的影响，发展中国家也具备了在发达国家先行示范和提供充分技术、资金和能力建设支持的基础上，开拓出一条创新发展路径的可能性。

李克强总理在2015年6月12日的国家应对气候变化及节能减排工作领导小组会议上明确指出，中国将为促进全球绿色低碳转型与发展路径创新尽自身最大努力。① 国家应对气候变化战略研究和国际合作中心的邹骥等自2013年起就研究提出“发展路径创新”的理论框架。②③④

① 李克强．着力推进提质升级发展 强化应对气候变化行动；http://www.ccchina.gov.cn/Detail.aspx?newsId=53255&TId=61.

② 邹骥．全球减排须着眼发展路径创新 [J]. 瞭望，2013.11.11.

③ 邹骥，傅莎．参与全球气候治理与实现发展路径创新 [N]. 经济日报，2015-7-1.

④ Zou Ji , Fu Sha. The challenges of the post-COP21 regime: interpreting CBDR in the INDC context[J], International Environmental Agreements: Politics, Law and Economics, Volume 15, No.3, September 2015, 10.1007/s10784-015-9303-8, available: http://link.springer.com/article/10.1007/s10784-015-9303-8.

发展路径创新的出发点是处于不同发展阶段的国家都要以积极的姿态行动起来，探寻与本国发展阶段相符的发展路径创新；其着眼点是影响排放轨迹和路径的发展驱动因素，以求在更实质、更实际、更具体的层面上打破全球气候治理谈判的僵局。这一概念框架着眼于实质的行动和措施，在可持续发展的总体框架下，围绕发展路径的创新和变迁，识别出发达国家与发展中国家各自行动的方向和重点，在政策含义上具有积极且重要的建设性意义。

将国际气候体制谈判的着眼点更多转向发展路径的创新，首先要解决一个公平获取可持续发展的问题，保证发展中国家有足够的机会实现发展路径的低碳创新与变迁并以此为国际气候体制谈判的前提，即占世界人口大多数的发展中国家在未来既要发展，又要低碳，二者缺一不可。从创新发展路径的角度入手设计 2020 年后国际气候体制，有助于全面、公平反映发达国家和发展中国家不同的责任、能力、国情和在不同发展阶段的不同重点发展议题，更好地将减排目标和措施与影响排放路径的驱动因子联系在一起、采取更有可行性、可控性的有效措施，并切实考虑到发达国家和发展中国家的不同利益关切，实现合作共赢。这一核心概念框架的内涵包括了发展驱动因子与排放的内在逻辑关系，深入到问题表象背后，触及问题本质，试图从社会经济系统内的动因去解释给地球气候系统带来的影响。

当然，将气候谈判引向发展路径创新和变迁层面，发展中国家也会面临转型阵痛、国际援助条件不兑现、克服“路径依赖”不成功等风险，要专门设计安全防范机制，防止路径转换中的硬着陆。

第二节　发展路径创新的内涵

发展路径可被定义为随着时间发展，决定人与自然系统互动的技术、经济、社会、制度、文化方面的一系列复杂安排 (Sathaye et al., 2007)。① 一个经济体的发展路径受到众多因素的影响，由包括结构和效率、技术、资金、能力、制度、意识等各个方面，由其经济增长来源特征、资源及要素投入组合、经济结构、基础设施、产业组织、技术体系选择、体制政策及社会治理等诸多方面的特征所决定。发达国家传统的排放路径，主要取决于其传统的发展路径，即经济增长的来源与动力、不同层次的产业与产品结构及在产品和服务供应链上的位置，技术体系，能源结构，建筑、交通和能源等基础设施的设计选择及城市形态的规划和与能源、排放相关的战略、体制与政策选择等。发展路径的变迁是一个复杂的过程（Frantzeskaki, Gorissen, & Loorbach, 2012)②，需从不同治理层次对转型进行管理。在本文中，创新发展路径的最终目的是以尽可能少的排放实现一定的发展水平，即通过经济增长模式、能源生产和消费模式、土地利用模式等的转型来减少发展对碳排放空间的依赖。

从增长函数看，经济增长率等于要素效率增长率和要素投入增长率之和，即可以分为要素规模的产出贡献及要素效率的产出贡献，如

---

① Sathaye, J., Najam, A., Cocklin, C., Heller, T., Lecocq, F., & Robinson, J. (2007). Sustainable development and mitigation. In B. Metz, O. Davidson, P. Bosch, R. Dave, & L. Meye (Eds.), Climate change 2007: Mitigation. Contribution of Working Group III to the Fourth Assessment Report of the Intergovernmental Panel on Climate Change (pp. 691–743). Cambridge: Cambridge University Press.

② Frantzeskaki, N., Loorbach, D., & Meadowcroft, J. (2012). Governing transitions to sustainability: Transition management as a governance approach towards pursuing sustainability. International Journal of Sustainable Development,15, 19–36.

公式（1）所示。如果把能源和碳排放（即碳排放容量资源）作为要素，则创新发展路径主要是指改变经济增长的来源，就是要让经济增长更多地依赖要素效率的提高，而降低对要素投入扩张的依赖，尽可能地提高单位能源投入的产出效率，以及单位碳排放的产出效率。在创新发展路径下，经济增长不再追求过高增速，而是以提升经济增长质量和效益为目标。同时，创新发展道路更有利于社会财富积累和发展成果共享，更符合经济社会发展的终极目标。发展路径转型的过程，也是优胜劣汰、不断激发经济增长活力的过程。在这一过程中，高碳排放、高污染行业发展会受到限制，继之而起的将是代表产业优化升级方向的战略性新兴产业和现代服务业，催生的将是新的经济增长点。这正是“转方式”之题中应有之意，也是经济体从欠发达向发达发展的必由之路和表征。

$$\Upsilon = \Upsilon_{\text{要素效率}} + \Upsilon_{\text{要素投入}} \qquad (1)$$

若借助卡亚分解分析影响发展路径的驱动因子，即从人口、人均GDP、单位GDP能耗和单位能源的CO_2排放几个方面，考察对CO_2排放的影响（见图4-1），可以发现，创新发展路径需从影响发展路径的诸多驱动因子入手，包括影响一个经济体温室气体排放路径的能源结构、能源强度（以及影响能源强度的能源效率或技术水平、三次产业结构、工业内部产业结构和行业内产品结构等）、人均收入水平与构

图 4-1 从卡亚分解的角度看“创新发展路径”

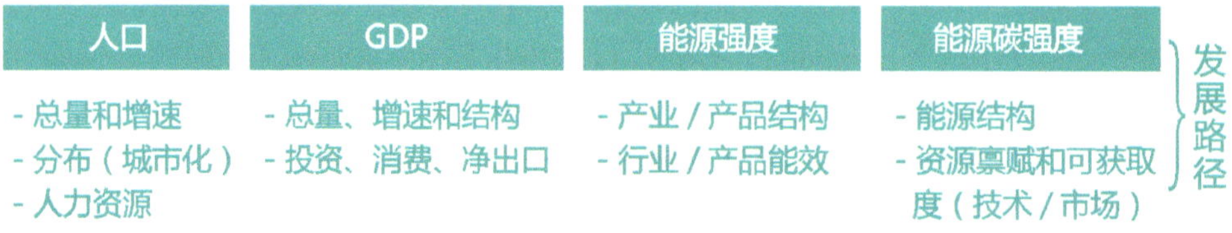

资料来源：邹骥，傅莎．基于“发展路径创新”的后 2020 年国际减排机制设计［R］．国家气候战略中心工作报告，2014.

成、人口数量和分布、基础设施乃至经济整体空间布局等诸多发展要素。对这些要素的综合调整，就是在一定排放约束下创新发展路径、优化配置资源、实现低碳可持续发展目标的过程。发展路径创新能够带来应对气候变化与发展的双赢。

需要注意的是，技术转型是促使发展路径转型发生的重要因素。然而，如果不考虑相应的公共政策、体制机制安排、公共/私营部门伙伴关系形成、消费者和市场偏好转变等方面的转型，技术转型将仅仅局限于某项技术，而不足以产生足够的发展路径变革，在有关发展路径创新的探讨中，有关社会、意识、文化层面的转变也非常重要 (Geels & Schot, 2007)[①]。总之，转型不是单纯的技术转型，而是主体行为、治理方法，以及价值观在发展路径中发展根本性的转型。

第三节　发展中国家和发达国家创新发展路径的任务

发达国家、新兴发展中国家和其他发展中国家都有创新发展路径的需求，但由于其所处的发展阶段的差异，在 2020 年后将面临不同的创新发展路径任务（如图 4-2 所示）。

发达国家已完成其工业化、城市化的发展进程，其主要任务是在既有的发展路径的基础上实现加速变迁，实现全经济范围内的绝对量化加速减排，确保发达国家基于其历史责任及在发展阶段、能力和在世界政治经济体系中占支配地位等方面的优势，遵循“共同但有区别责任”的原则，充分履行其减排责任，发挥率先垂范的作用，并向发展中国家提供技术、资金和能力建设支持，帮助促进发展中国家完成发展路径的创新和转型。

① Geels, F. W., & Schot, J. (2007). Typology of sociotechnical transition pathways. Research Policy, 36, 399–417.

图4-2 创新可持续发展路径

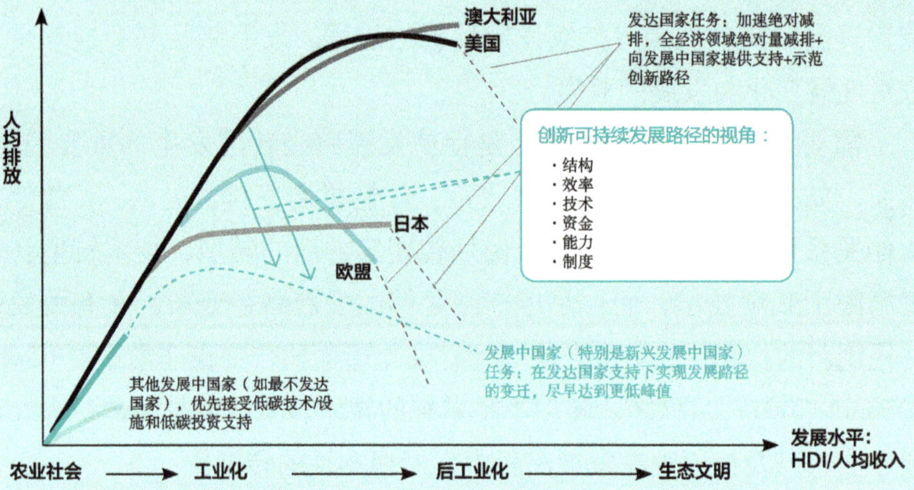

资料来源：Zou Ji , Fu Sha. The challenges of the post-COP21 regime: interpreting CBDR in the INDC context[J], International Environmental Agreements: Politics, Law and Economics, Volume 15, No.3, September 2015, 10.1007/s10784-015-9303-8, available: http://link.springer.com/article/10.1007/s10784-015-9303-8.

而发展中国家仍然处于发展进程中，主要任务是在确保公平获取可持续发展的均等机会的前提下，在公平获取技术、资金、教育机会和知识，共享当代科技革命和知识进步的成果的前提下，在发展进程中实现发展路径的跃迁，走上一条可持续的发展道路，以气候友好的方式实现脱贫、改善人民生活质量与安全状态的发展目标。

下文将对发达国家和发展中国家创新发展路径的具体任务进行阐释。

一、发达国家：示范与对世界未来发展趋势的引领

如前文所述，对发达国家而言，考虑到其发展阶段已进入后工业化阶段，有历史责任并具有较强的资金、技术、社会治理能力，且在

国际政治、经济、能源、环境、科技事务中居主导优势地位的现实，其主要任务是发挥示范作用和市场需求导向作用，加速下降温室气体排放，率先垂范一条创新的低碳可持续发展示范路径，并通过向发展中国家提供资金、技术、能力建设支持及政策、经验交流，将上述示范路径复制、推广、扩散到发展中国家。发达国家的示范作用对于全球创新发展路径有着至关重要的作用。考虑到发达国家其历史上和现实中在国际事务中所具有的主导、示范作用，对发展中国家在意识、发展观、价值观、发展道路与模式的巨大影响，发达国家不从根本上作出转型示范而让发展中国家作出改变，既不公平，也不可行。

具体而言，发达国家创新发展路径的重点任务包括：

(1) 基于其历史责任，通过消费模式和生活方式的转变和技术进步率先减排，尽快促成温室气体排放的大幅下降，保障发展中国家转型发展路径所需的必要增量排放空间。

(2) 充分利用其技术研发优势，强化对减缓、适应气候变化关键技术的创新，并通过消除阻碍技术转让障碍来帮助发展中国家获取气候友好技术，推动技术在发展中国家的扩散和应用，为发展中国家应对气候变化提供科研和信息支撑，提高发展中国家应对气候变化的整体技术水平。

(3) 提供公共资金，并基于此撬动充足的私人资金，解决发展中国家应对气候变化面临的增量成本和资金需求，并通过培训、知识共享等方式为发展中国家提高应对气候变化能力提供能力建设。例如为发展中国家应对气候变化提供适当的国际融资机制和援助；帮助培训应对气候变化人才，并适当提供培训后的工作机会；直接派出应对气候变化专家工作对发展中国家人员进行就地培训和指导等。

(4) 与发展中国家交流共享应对气候变化的经验教训，为发展中国家制定应对气候变化国家战略提供咨询意见，提升发展中国家将应对气候变化纳入主流政策框架的能力。

二、中国和典型发展中国家：路径依赖与路径创新

如前文所述，对发展中国家而言，未来的发展路径是沿着“高发展水平高排放”还是“高发展水平低排放”的轨迹，或者走出一条创新的不同于发达国家传统路径的低碳发展路径，对其自身的社会经济可持续发展和全球应对气候变化的努力，都将具有十分重要的意义。考虑到后发优势、全球排放空间约束和国内资源环境约束，这些新兴发展中国家有必要也有可能改变现有发展模式，通过发展路径的跃迁，克服路径依赖，开创一条比美欧等发达国家传统发展路径更为低碳的、更早达到更低排放峰值的创新的发展路径。但是受发展中国家自身能力不足的限制，其实现转型、克服路径依赖的可能性在很大程度上取决于发达国家所给予的支持程度和示范程度。在强调发达国家示范、支持作用重要性之时，不断提高发展中国家自身的能力，是最终越过峰值阈值的根本所在。

通过本书第二章的卡亚分解研究可以发现，单位 GDP 能源强度和单位能源碳强度的大幅下降是那些已达到稳定峰值的发达国家实现生产端排放下降的主要因素。考虑到发展中国家未来的人口和人均 GDP 仍将持续增加，应着力通过调整产业结构和产品结构，延长产业链，提高产品增加值，提高能源效率，调整能源结构以及采用先进减碳技术等措施降低单位产出能源强度和单位能源碳强度。此外，发展中国家在制定创新发展路径战略时，还应充分考虑技术进步、基础设施的锁定效应和消费行为改变等主要驱动因子对减排的影响。

具体而言，发展中国家创新路径的任务包括：

（1）增长方式的低碳转型。通过改变产业发展方式，靠技术进步和创新驱动产业增长，促进传统产业的低碳转型，大力发展新型的绿色低碳经济，逐步实现经济增长和碳排放的脱钩。

（2）能源系统的低碳转型。通过实施能源生产和消费革命，构建以清洁、低碳为特征的能源供应体系，合理控制能源需求总量，大幅

提升能源利用效率和管理水平，逐步实现能源增长和碳排放的脱钩。

(3) 消费模式的低碳转型。通过转变人类生活和消费方式，在满足合理消费需求和提升生活品质的同时，通过广泛的宣传教育和积极的政策引导，形成绿色、低碳、节约的消费理念和生活方式，逐步实现消费增长和碳排放的脱钩。

(4) 城市化模式的低碳转型。未来主要的城市化进程和与之相伴随的基础设施存量的增长将主要发生在发展中国家。城市将成为发展中国家未来排放主体。城市形态、城市设计和城市连通性是决定城市温室气体长期排放水平的重要影响因素。为避免巨额锁定效应，发展中国家有必要探索低碳城镇化的可能性，通过改变新、扩建城市的城市形态，加强低碳基础设施建设，改善对土地利用模式与公共交通体系的设计和规划，实现低碳转型。

发展中国家应致力于通过上述四种转型形成新的国际竞争优势和经济增长点，推动发展由量到质的转变，逐步形成与国情相符的创新发展路径。

此外，发展中国家还应充分考虑减排行动的风险以及转型路径与其他社会经济发展优先选项间的关系，包括与社会保障、扶贫、就业、能源安全、粮食安全、局地环境污染、长期竞争力等的关系，通过统筹兼顾、多目标寻优，力争最大化协同效益，最小化风险。协调低碳发展目标与其他社会经济发展目标，最大化协同效应。

石油输出国国家是发展中国家中一个比较特殊的集团，承受着“没有发展的增长”之患，经济过度依靠石油出口，不具有完整多样的经济结构和可持续发展的能力，在全球产业链中依然处于低端，其发展路径创新的主要议题是经济多样性和克服发展路径对发达国家高碳消费与排放模式的依赖。

第四节 “创新发展路径”概念框架在后 2020 年国际减排机制中的具体落实

基于前述创新发展路径的概念框架，发达国家和发展中国家未来减排贡献可以在如下四类指标中分别选择承诺不同的指标：①量化转型总体进展指标：即全经济领域范围的减控排指标，包括基于特定年份的绝对量化减排指标、总量控制/峰值指标、人均指标等；②量化转型过程指标：即与影响转型的具体驱动因子相关联的指标，包括能源强度、能源结构（包括低碳能源比重/装机等）、具体行业指标（如能效、先进技术扩散率或产能占比等）和其他排放源或碳汇控制指标（包括碳汇/工艺过程/非 CO_2 温室气体控制指标等）；③定性指标：即相应的政策措施种类、体制或法律安排进展、能力发展指标；④支持保障指标：即发达国家向发展中国家提供的资金、技术和能力建设支持指标，用于考核发达国家帮助发展中国家转型发展路径的力度和质量。

以上四类指标具有不同特点和适用性。减排要求更为严格、经济发达成熟、变迁稳定、核算统计与治理能力较强的发达经济体同时适用于全经济范围绝对量化指标、量化转型过程指标和支持保障指标；而发展迅速但尚不成熟稳定、变化大且不确定性高、治理能力弱的发展中经济体则更适用于量化转型过程指标和定性指标。

关于全球长期目标，无论是温升目标还是碳预算目标或减排目标，由于存在着较大的科学不确定性，宜于正面承认并强调其在确定全球与国别低碳发展方向、提高行动紧迫性及给决策者趋势性合理预期等方面的重要作用，总体肯定，并且要积极倡导不断探索，但在用其数值直接作为目标决策依据时，应以决策科学中不确定性条件下决策的原理，持谨慎的态度，轻易不形成约束性的全球定量目标指标。

发达国家、新兴发展中国家和最不发达国家、小岛屿国家等其他发展中国家可在创新各自发展路径的共同愿景下，在图 4-3 所示的指标体系中，基于其历史责任、能力、发展阶段、发展需求和特定国情选择不同的“贡献”指标。

对发达国家而言，其承诺应延续其一贯的全经济领域绝对量化减排指标（力度水平参考公平、历史责任和科学要求），同时为发展中国家发展路径的变迁创新提供资金、技术、能力建设支持。此外，为更好地形成发展路径示范效应，发达国家还可承诺量化转型过程指标，如特定部门的减排目标和特定技术的发展目标，包括 R&D 投入目标等。对发达国家提出此“更高”要求，主要是考虑到它们的历史责任、此前的实践经验、综合能力优势，及在国际进程中起

图 4-3 “创新发展路径”在 2020 年后国际减排机制中的具体体现

创新发展路径

指标体系
- 量化转型总体进展指标：
  排放总量/峰值/人均/一定发展水平（HDI 或人均 GDP 等）下的人均排放水平；
- 量化转型过程指标：
  能源强度；
  能源结构：低碳能源比重/装机等；
  行业指标：如能效/先进技术扩散率等；碳汇/工艺过程/非 CO_2 温室气体；
- 定性指标：
  体制、政策、能力等
- 支持保障指标：
  资金、技术支持

→

减排贡献

贡献是什么？
- 发达国家：全经济领域的绝对量化减排目标 + 支持目标 + 特定行业/技术的示范目标；
- 有工业化、城市化需求的发展中国家（新兴发展中国家）：意向性的量化转型程度承诺（如峰值）+ 较为严格的量化转型过程承诺；基于发达国家支持，具体指标自主根据本国国情和发展需求选择；
- 其他发展中国家（AOSIS/LDC）：自主根据国内需求选择

评估和动态调整机制？
评估路径转型的进展，发达国家根据公平和科学要求强化在减排和支持两个方面的承诺和示范水平；发展中国家基于届时的最优实践水平自愿动态调整和更新目标，需要给予发展中国家一定的宽限期

←

全球长期目标

- 2°C政治共识
- 意向性的排放空间和长期减排目标（范围值）——基于 IPCC
- 定期评估，不断探索

资料来源：邹骥，傅莎 . 基于“发展路径创新”的后 2020 年国际减排机制设计［R］. 国家气候战略中心工作报告，2014.

主导作用的示范效应。

对发展中国家中的新兴发展中国家而言，其主要任务是实现发展路径的变迁，争取尽早达到更低峰值。考虑到其正处于更为复杂多变的社会经济转型情景中，社会经济系统的变化面临更大不确定性，不适宜将严格的全经济领域绝对量化减排指标作为控排管理的核心指标，而应将重点放在相对减排目标和不确定性更小、可操作性更强的转型过程指标上。但为了方便各缔约方理解其转型效果，其未来贡献可以采用“意向性的量化转型总体进展指标（如峰值目标、总量目标、人均目标）+更为确切的量化转型过程指标”的形式。意向性的转型总体进展量化指标给出了总体的努力方向，并展示了转型过程指标可能实现的转型效果。而转型过程指标则可与更为具体、可操作的转型行动安排相结合，给决策者、企业、投资者、消费者和公众更清晰的行动指导信号。具体的指标选择可以是灵活多样化的，根据本国国情选取。而且需要强调的是，这些行动需要得到来自发达国家在技术、资金、能力建设和优惠的贸易与投资条件等方面的充分支持。将意向性总体进展指标和具体的转型过程影响指标相结合的方式也是基于科学决策的考虑。当前的决策者一般基于当前所能掌握的信息和资源进行决策，无法为不确定性过大的长期目标负责。过程指标是行动导向的，可以为当今的决策者提供直接具体的行动指导，一旦对应能效、能源结构等各项驱动因子的行动推进，肯定会推动排放总量的控制，同时将对发展中国家未来发展总量和排放总量权益空间的限制风险控制在较低的程度。

对于其他发展中国家，如最不发达国家、小岛屿国家而言，可以自主根据国内需求选择是否采取行动及具体的行动指标（如定性指标），同时需要承诺在接受国际社会支持时优先接受低碳技术、低碳投资、低碳基础设施建设等方面的援助。

为增进各缔约方之间对各自贡献内容、力度的相互理解，及时掌握发展路径的“创新”进展程度和“变迁”程度，并根据发展路径转型能力的变动情况，如技术进步等更新信息确保转型的速度和力度，确保全球 2℃温控目标的实现，在各方确定 2015 协议下的贡献水平后，还可考虑建立相应的定期评估和动态调整机制。该机制的主要目的是参考 2℃目标给出的意向性的排放空间和长期减排目标，评估转型路径的进展并确保各缔约方尽其最大可能变迁其发展路径从而实现最大程度的减排。具体而言，机制将通过定期的路径审评促使发达国家根据公平和科学要求强化其量化承诺水平、示范效果和对发展中国家技术和资金支持的水平。同时发展中国家可以考虑在相应的发达国家支持下，基于届时的最优实践水平，如最优可得技术，自愿调整更新贡献力度。

需强调的是，发达国家的贡献水平并不必然受到其本国减排能力的限制。除了本国减排，发达国家还可以通过直接提供额外的资金、技术、能力或者清洁能源供应帮助发展中国家减排，向发展中国家直接购买碳排放份额，通过加速开发推广转让新技术和提高现有最优实践水平以提高发展中国家转型潜力等方式实现其减排承诺（如图 4-4 所示）。

图 4-4 发达国家 / 新兴发展中国家减排贡献示意

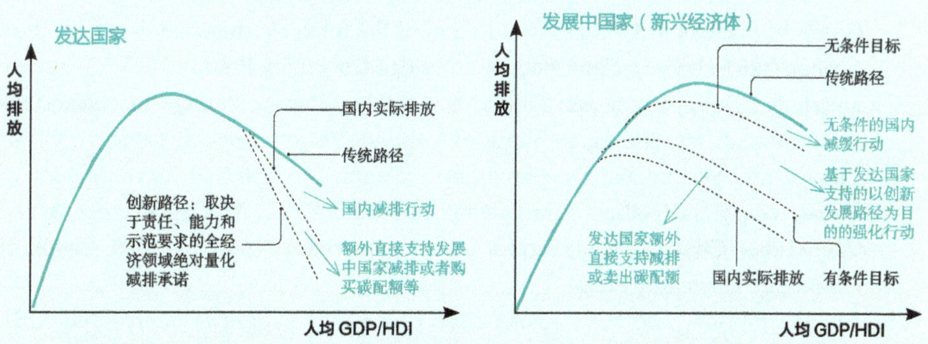

资料来源：邹骥，傅莎．基于“发展路径创新”的后 2020 年国际减排机制设计［R］．国家气候战略中心工作报告，2014.

【参考文献】

[1] 李克强．着力推进提质升级发展 强化应对气候变化行动．http://www.ccchina.gov.cn/Detail.aspx?newsId=53255&TId=61.

[2] 邹骥．全球减排须着眼发展路径创新 [J]. 瞭望，2013.11.11.

[3] 邹骥，傅莎．基于“发展路径创新”的后 2020 年国际减排机制设计 [R]. 国家气候战略中心工作报告，2014.

[4] 邹骥，傅莎．参与全球气候治理与实现发展路径创新 [N]. 经济日报，2015-7-1.

[5] Bizikova, L., Burch, S., Cohen, S., & Robinson, J. (2010). A participatory integrated assessment approach to local climate change responses: Linking sustainable.

[6] Frantzeskaki, N., Loorbach, D., & Meadowcroft, J. (2012). Governing transitions to sustainability: Transition management as a governance approach towards pursuing sustainability. International Journal of Sustainable Development,15, 19–36.

[7] Geels, F. W., & Schot, J. (2007). Typology of sociotechnical transition pathways. Research Policy, 36, 399–417.

[8] IPCC, 2014. Climate Change 2014, Mitigation of Climate Change. Contribution of Working Group Ⅲ to the Fifth Assessment Report of the Intergovernmental Panel on Climate Change. Cambridge University Press, Cambridge, United Kingdom and New York, NY, USA.

[9] Robinson, J., Morita, T., Adegbulugbe, A., Alcamo, J., Herbert, D., Lebre La Rovere, E., Dadi, Z. (2001). Greenhouse gas emissions: Mitigation scenarios and implications. In B. Metz, O. Davidson, R. Swart, & J. Pan (Eds.), Climate change 2011: Mitigation (pp. 115–166). Cambridge: Cambridge University Press.

[10] Rotmans, J., Kemp, R., & van Asselt, M. (2001). More evolution than revolution: Transition management in public policy. Foresight, 3(1), 15–31.

[11] Sathaye, J., Najam, A., Cocklin, C., Heller, T., Lecocq, F., & Robinson, J. (2007). Sustainable development and mitigation. In B. Metz, O. Davidson, P. Bosch, R. Dave, & L. Meye (Eds.), Climate change 2007: Mitigation. Contribution of Working Group Ⅲ to the Fourth Assessment Report of the Intergovernmental Panel on Climate Change (pp. 691–743). Cambridge: Cambridge University Press.

[12] Swart, R., Robinson, J., & Cohen, S. (2003). Climate change and sustainable development: Expanding the options.Climate Policy, Special Issue on Climate Change and Sustainable Development, 3(Suppl. 1), S19–S40. development with climate change adaptation & mitigation. In K. O’ Brien, B. Kristoffersen, & A. St. Clair (Eds.), Climate change, ethics and human security (pp. 157–179). Cambridge: Cambridge University Press.

[13] Zou Ji , Fu Sha. The challenges of the post-COP21 regime: interpreting CBDR in the INDC context[J], International Environmental Agreements: Politics, Law and Economics, Volume 15, No.3, September 2015, 10.1007/s10784-015-9303-8, available: http://link.springer.com/article/10.1007/s10784-015-9303-8 .

第五章

全球气候治理体系

——《联合国气候变化框架公约》

Chapter Five

Global Climate Governance
— the United Nation Framework Convention on Climate Change

摘要

气候变化是全球外部性和公共财富管理问题，需要所有国家积极行动、广泛合作。自20世纪80年代末，气候变化开始呈现在国际政治舞台上，全球气候治理取得了积极的成果。气候变化问题具有较大的时空尺度，且有显著的不确定性，一般的基于成本效益的决策无法指导应对气候变化工作，需要在国家层面开展对话以制定一个全球的行动目标，并作出承诺引导国内的行动及相关合作。因此，以主权国家为缔约方的《联合国气候变化框架公约》是全球应对气候变化的主渠道，它拥有197个缔约方，覆盖了超过全球90%以上的排放量，以科学为基础，制定了合作的原则，明确了行动的目标，识别了核心要素，经过20多年各方不懈的努力，建立了一系列运行机制和机构并开发了相关的工具，建立并增强了各方携手应对气候变化的互信基础，提高了各国各界应对气候变化的意识，是全球气候治理的核心，是推动全球应对气候变化行动与合作的基础。在《公约》进程中我们获得了很多有益的经验，达成一系列富有成效的成果，各国为实现目标制定的政策措施效果也开始显现，全球气候治理下各国的参与程度不断提高，也有更多的利益相关方参与到国际气候合作进程中来。为了在超主权力量缺位的客观现实中确保更广泛的参与，国际气候机制也开始从严格向松散转变，《公约》下的合作模式逐渐由“自上而下”向“自下而上”演变，在《公约》原则和规定的指导下，缔约方被赋予更多的自主权。然而，气候变化是一个长期而复杂的问题，关于应对气候变化国际合作的一些根本问题的讨论从未停止，需要在评估行动进展和相关科学、技术、政治、经济发展的基础上不断调整应对策略。正在谈判中的2015年巴黎气候协议，应注重落实《公约》的原则，巩固缔约方合作的信任基础，明确全球低碳转型和提高气候抵御力的行动方向，鼓励各方不断提高目标的力度，加强合作，携手应对气候变化带来的挑战。

第一节　全球气候治理的发展和演变

一、全球气候治理的基本框架

如前文所述，气候变化问题具有独特的特征，其影响具有较大尺度的时空属性，相关研究的不确定性极强，在个体和市场主体的决策过程中难以纳入这种超大时空尺度的问题，因此需要各国一道站在全人类福祉的高度，在考虑子孙后代可持续发展机会的背景下，制定应对气候变化的全球议程。因此，如图 5-1 所示，全球气候治理的根本

图 5-1　全球气候治理的概念框架

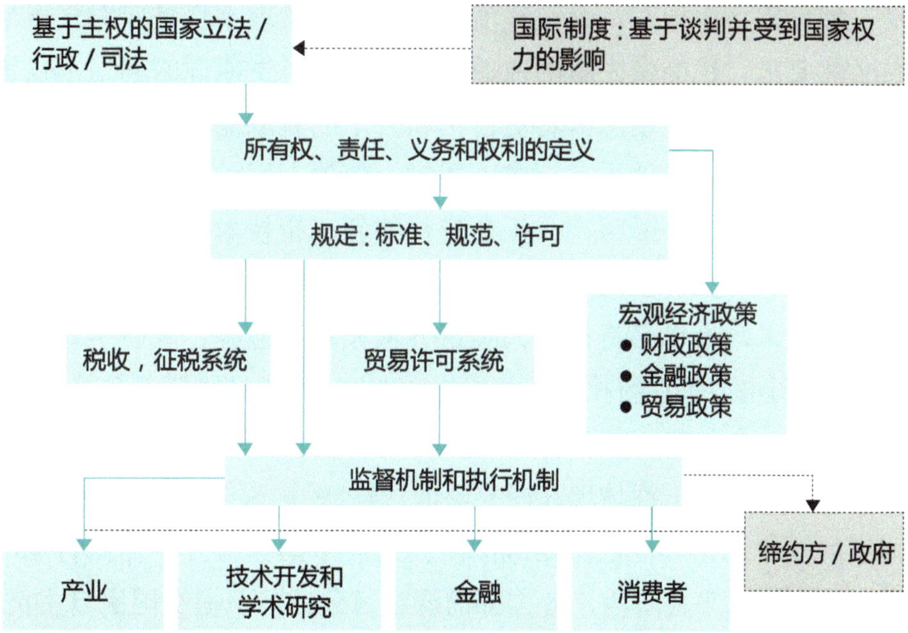

资料来源：邹骥. International Regine for Addressing Global Climate Externality. 坎昆气候大会哈佛大学边会发言.

是具有国家主权的缔约方推动的国际多边机制，这对于设定全球行动目标、明确行动与合作的原则、识别应对气候变化的关键问题具有决定性的作用，是全球气候治理不可缺少的核心和根本。

国家层面明确了行动目标及合作机制后，才有可能在国内制定相关的法律法规和政策措施，利用各种手段，不断推动应对气候变化的行动与合作。政策措施的时间尺度与市场主体和个人的决策和行为周期更加贴近，国家目标和政策也可激发地方政府、私营部门、社会团体以及学术界、媒体等的积极行动，促生一系列的倡议和联盟，为全球气候治理成果的落实提供了坚实基础。

二、全球气候治理的发展趋势

气候“阴谋论”在相当长的一段时间内都有广大的“市场”，人们认为减少化石能源消费必然产生额外成本，限制经济增长，“低碳发展”是一种负担。但现在，人们逐渐认识到低碳发展不仅是应对气候变化的必然要求，也能带来新的效益。比如，可再生能源的利用不仅带来良好的环境效益，也意味着更自由和公平的获取能源的机会。低碳转型已成为各国发展的内在诉求。同时，随着人们对环境与发展之间关系的认识与思考不断深入，正在进行的新一轮技术革命将使人类朝着气候友好和环境有益的方向发展，使人们以有限的资源实现更好的发展。新技术革命也必将打破人类过分依赖化石能源、威胁全球气候的困境，各国也力争在新技术革命中抢占制高点，维持国际竞争力。因此，近些年来，各方都更广泛和积极地参与到全球气候治理当中。

广泛参与一方面体现在国家层面，《公约》最初阶段只聚焦于发达国家的减排承诺，但巴厘路线图阶段发展中国家就开始作出行动的承诺，到了德班平台阶段，各方都需要在《公约》下提交国家自主贡献，全球各国共同参与已经成为定局。另一方面，广泛参与体现在非国家主体的行动与合作上，近年来涌现了众多的区域间、企业间和公民团体的合作倡议，在其他国际多边机制下气候变化也成为重要的话题。

全球治理中需要把握一个重要的平衡，即参与程度与国际机制的严格程度之间的平衡。IPCC 第五次评估报告第三工作组报告的结论更倾向于建立一个广泛参与但约束力并不很强的国际机制，认为这样能更好地推动全球应对气候变化进程。同时，也有一些国家仍然推崇具有严格约束力的国际机制，特别是减排机制。虽然基于现在的研究我们还无法准确找到这个平衡点，但在《公约》进程下确实呈现出各方广泛参与的同时机制约束力降低的趋势,《公约》最初建立的“自上而下”的体制模式逐渐演化为“自下而上”的体制模式。形成这一趋势的另外一个直接重要因素就是美国国会很难批准白宫政府向国际社会承诺的减排目标和援助发展中国家的资金预算。

这种变化显著体现在各方的“承诺”上。1992 年通过的《联合国气候变化框架公约》明确提出了“共同但有区别的责任”原则，为气候变化国际合作奠定了基础。此后的《公约》谈判进程就一直在为探索具体有效的合作模式而努力，人们对合作模式的理解也不断发生变化，这也反映在用于描述各方责任的具体用词的变化上。回顾《公约》20 年的进程，此次“贡献”一词的使用意味着气候变化的国际气候体制逐渐从严格走向松散。《公约》最早不仅明确了应对气候变化国际合作的目标和原则，还规定了各缔约方的相关承诺（commitment）来具体界定各方在应对气候变化问题上的责任，此处的“承诺”有着更多强制性义务的含义。《公约》指出，各国本着“共同但有区别的责任”以及具体的国情开展应对气候变化的行动，发达国家应率先应对气候变化及其不利影响，附件一缔约方应采取措施限制人为温室气体排放并保护碳汇，以减缓气候变化，作为附件二的发达国家缔约方向发展中国家缔约方提供新的额外的资金和无害环境的技术支持。此后，1997 年的《京都议定书》进一步量化了附件一缔约方的减排承诺，建立了附件一国家温室气体排放的报告、审评和遵约审查模式。然而，自从美国宣布拒绝核准《京都议定书》，整个气候变化进程就进入了一个低潮期，特别是发达国家开始逐渐回避“承诺”一词的使用，2007年通过的“巴厘行动计划”就开始改用“承诺 / 行动”（commitment/action）的表述。2009 年哥本哈根大会和 2010 年坎昆大会期间，相关用词又被改为“许诺”（pledges）。2011 年启动新一轮谈判的德班平台授权则是使用了“加强行动”（enhanced action）来描述未来的国际合作。此次在华沙，各方明确放弃了“承诺”一词的使用，而使用“贡献”（contribution）代替。用词上的变化显而易见，目标的约束力不断被弱化，各方愈加倾向于灵活和松散的国际合作机制。

从 20 世纪 70 年代人们逐渐建立起气候变化科学认识以来，人类活动引起气候变化的证据越来越确凿，应对气候变化行动的重要性和紧迫性不断增强。然而，是什么原因使得气候变化国际体制从最开始的严格逐渐走向松散？首先，国际政治经济形势和温室气体排放格局发生了变化。“金砖国家”等新兴发展中国家经济增长势头强劲，而一些发达国家国内却面临很多社会经济问题，特别是 2008 年经济危机之后，发达国家承受了复苏经济的巨大压力，新兴发展中国家作为温室气体排放重要贡献者却并未承担量化的减排责任，发达国家应对气候变化的意愿明显降低；其次，传统能源结构短期内不会发生根本性变化，应对气候变化技术还没有取得突破性进展，应对气候变化行动面临着技术、经济和社会各方面的困难，减缓气候变化信心不足；第三，减少温室气体排放对国家能源、工业、交通、建筑等重要部门影响很大，采取强制的减限排措施将削弱国家竞争力，应对气候变化的长期目标与国家发展的短期目标发生了冲突。因此，虽然应对气候变化的要求日益迫切，各国在“约束力”和“灵活性”之间都更倾向于“灵活性”，国际气候体制在缔约方的“推动”下走向松散。

三、《联合国气候变化框架公约》进程的演变

自《公约》诞生至今 20 余年，应对气候变化国际合作进程既有成

功的经验，也有失败的教训。人们也更深刻地意识到应对气候变化是一项全球性的、长期的任务，不能一蹴而就，需要一个循序渐进的过程。气候变化相关的科学、政治、经济和技术的发展也在不同程度上反映了各个阶段的成果，这些成果也为未来应对气候变化国际合作进程奠定了坚实的基础。

《公约》诞生和生效这一阶段是国际气候谈判最高效的时期。1990 年 12 月，联合国大会决定成立一个政府间谈判机构（INC）来拟定《公约》文本；1991 年 2 月开始，历经 15 个月、五次会议之后，INC 就确定了《公约》最终案文；1992 年 6 月，《公约》在联合国环境与发展大会上正式签署；1993 年 12 月 22 日，《公约》生效的条件已经满足；1994 年 3 月 21 日，《公约》正式生效。

《公约》体现了各方极高的政治智慧，达到了求同存异的目的，为今后应对气候变化国际合作进程打下了良好的基础。《公约》取得的最重要的三项成果分别是目标、原则和各方义务。

——《公约》第二条确立了应对气候变化的目标，即“将大气中温室气体的浓度稳定在防止气候系统受到危险的、人为干扰的水平上。这一水平应当在足以使生态系统能够自然地适应气候变化、确保粮食生产免受威胁并使经济发展能够可持续地进行的时间范围内实现……”这段话精炼地概括出了既要减少温室气体排放又要规避气候变化风险这一应对气候变化的终极目标，为后续的减缓和适应行动指明了方向。

——《公约》还明确了应对气候变化国际合作应遵循的原则，包括公平原则、共同但有区别的责任原则、各自能力原则、预防原则、成本有效性原则、考虑特殊国情和需求原则、可持续发展原则和鼓励合作原则，全面考虑到了应对气候变化的各个方面，为各方参与国际合作提供了保障。

——《公约》还根据各国的责任和能力作出了国家分类，即附件一国家、附件二国家和非附件一国家，并明确了各类缔约方应对气候变化的义务：附件一国家应率先开展控制和减少温室气体排放的行动，到 2000 年将排放降低至 1990 年的水平；附件二国家应为非附件一国家提供新的和额外的资金支持，并采取有效措施促进气候友好技术向非附件一国家的转让。

在这一时期，国际社会对于环境与发展的问题高度关注。1987 年，世界环境与发展委员会发布了《我们共同的未来》报告；1988 年 IPCC 发布了《气候变化第一次评估报告》；1992 年在巴西里约热内卢召开的联合国环境和发展大会上，183 个国家代表团、102 位国家元首或政府首脑到会，达成了包括《公约》在内的“环境三公约”。这时期的科学研究基础以及政治力量的推动是开启应对气候变化国际进程最重要的力量。

1995 年 3 月的《公约》第一次缔约方大会上，各方通过了“柏林授权”，决定启动进程强化附件一国家的承诺。为此成立的柏林授权特设工作组（AGBM）在 1995 年 8 月至 1997 年 10 月期间组织召开了八次会议，最终各方在 1997 年底《公约》第三次缔约方大会上达成了《京都议定书》。《议定书》达成后，“布宜诺斯艾利斯行动计划”开始，旨在进一步完善《议定书》实施的技术细节，特别是核算规则、灵活机制和履约机制等。这项工作直至 2001 年才完成，最终各方通过了《马拉喀什协定》，以一揽子的方式解决了《议定书》实施规则的问题。然而，《议定书》生效需要超过 55 个《公约》缔约方核准，直至 2004 年底俄罗斯完成核准工作，《议定书》才最终得以于 2005 年 2 月 16 日生效。同年底，《议定书》第一次缔约方大会在蒙特利尔召开，通过了《马拉喀什协定》的多项决议。根据《议定书》第 3.9 条的规定，各方需在第一承诺期(2008—2012 年)结束前 7 年开始审议后续承诺期的减排目标，《议定书》第一次缔约方大会即开启了这一进程，但进展相当缓慢，直至 2012 年多哈会议才最终完成。

《议定书》作为《公约》进程下第一个具有法律约束力的成果，为后续的应对气候变化国际合作留下了宝贵遗产，包括量化目标、灵活机制和法律形式。

——《议定书》首次在应对气候变化国际合作进程中确定具有法律约束力的量化减排目标，不仅明确了发达国家在第一承诺期减排5.2%的总体目标，还将每个国家确定的减排目标列入附件B。

——考虑到成本有效性，《议定书》还建立了三种灵活机制，即排放权交易、联合履约机制和清洁发展机制，旨在通过经济手段为承担减排义务的缔约方提供更灵活的履约方式。从实施效果来看，这三种灵活机制不但实现了最初的设计目的，清洁发展机制项目的实施还大大提高了发展中国家应对气候变化的意识和信心。

——《京都议定书》包含一个具有法律约束力的国际协议所应具有的所有要素，包括目标和时间表、灵活机制、机构设置、核查规则、生效条件、履约机制等，在形式上，是完备的法律文书，堪称范本，在技术层面，确定了大量实施细则。

但《议定书》也留下了关键的未决问题，即如何进一步加强《公约》下各国承诺的力度。经过激烈的交锋，各方最终虽承认现有承诺不足以实现《公约》最终目标，但发达国家没有提高实现目标的意愿，发展中国家也拒绝承担任何发达国家转嫁的责任，关于目标审批的进程无法达成任何结论，直至今日各方仍未能找到弥补承诺力度不足和目标要求差距的方式。此外，美国拒绝核准《议定书》也促进了国际社会对《公约》进程的反思，从长远看，也有助于国际气候合作的健康发展。

与此前《公约》诞生和生效阶段不同，《议定书》谈判过程中国际社会对应对气候变化问题的困难进行了重新评估，与此前过于乐观的态度不同，各方应对气候变化的态度更加务实。随着新兴经济体崛起，温室气体排放规模不断扩大，发达国家要求发展中国家加强减排行动的诉求也越来越强烈。1997年6月，美国伯德－哈格尔决议以95票全票通过，明确提出美国不应签署只包含发达国家减排承诺的法律文书。克林顿政府在核准问题上采取了拖延态度，2001年，小布什政府毫无意外地宣布美国将拒绝核准《议定书》。其他发达国家也有同样的关切，欧盟也一直蓄势启动新的进程，力图将美国和发展中国家的减排承诺纳入国际气候进程。由于缺少了大国的领导力，应对气候变化国际合作进程陷入低谷。

《议定书》谈判受阻，国际社会在应对气候变化问题上的热情遭受打击，为了进一步推动国际气候合作，各方于2005年第十一次《公约》缔约方大会通过新的授权，启动“应对气候变化的长期合作行动对话”，对话的四个主题包括：推动实现可持续发展目标、适应气候变化、全面实现技术潜力以及充分发挥市场机制的作用。与此同时，《公约》谈判之外也开展了很多关于气候变化的非正式对话，正是在这样的场合，很多国家提出在2007年底的巴厘缔约方大会上建立一个路线图，启动关于2012年后机制安排的谈判进程。在漫长磋商之后，各方确立了巴厘路线图谈判的五大构件（building blocks，共同愿景、减缓、适应、资金和技术）。被各方寄予厚望的哥本哈根会议由于程序问题意外失败，实际上最终案文已经获得了188个国家的赞成，仅仅五个国家反对。《哥本哈根协定》实际上已经基本反映了各方的共识，因此在坎昆会议上，各方顺利地找回共识，达成了包含2℃温控目标、“共区”原则、资金支持目标、技术机制安排等重要内容在内的新协议，《坎昆协议》挽救了《公约》进程，也结束了巴厘路线图进程的谈判。

巴厘路线图的谈判主要需解决三个问题，即发达国家和发展中国家的区分、发达国家目标可比性以及保证发展中国家减缓行动可测量、可报告和可核实，最终达成了一系列积极的成果，其成功之处主要体现在以下三个方面：

——巴厘路线图确立的“双轨制”。关于“双轨”，一种理解是《议定书》第二承诺期谈判和长期合作行动特设工作组谈判的安排，另一种理解是巴黎行动计划下发达国家减缓承诺和发展中国家“国家适当减缓行动”（NAMAs）的安排。客观的说，应对气候变化需要发展中国家采取力所能及的行动，这一理念在巴厘路线图进程中得到进一步明确，发展中国家的积极参与也使美国重新回到国际气候进程中来，而双轨制安排和防火墙建立也确保了发展中国家参与国际气候合作的安全感。

——巴厘路线图的成功之处还体现在五大构件的确立。减缓一直是国际气候进程的核心，但随着对气候变化问题认识的不断加深，适应、资金和技术的重要作用不断凸显。特别是对于发展中国家来说，适应气候变化以及获得开展应对气候变化资金技术支持的需求迫在眉睫。而共同愿景作为《公约》目标的延伸，同减缓、适应、资金和技术紧密联系，极大地丰富了《公约》的原则和思想。

——巴厘路线图开始，“承诺+审评”的自下而上的模式便初步确立。各国在共同愿景的谈判中分歧较大，但最终还是达成了相对灵活全面的2℃温控目标，为自下而上的承诺模式留下了空间。澳大利亚在谈判中提出基于国家计划（national schedule）的提案，奠定了“承诺+审评”的基调，但美国推动的透明度规则的建立，也起到了进一步完善这一模式的作用。

经历《议定书》的低潮之后，巴厘路线图肩负着重振应对气候变化国际合作的任务。在巴厘进程谈判的几年之间，气候变化问题也受到了国际各界的广泛认可，国际社会对应对气候变化问题重要性的认识得到前所未有的提高。哥本哈根会议举办时，气候变化在国际政治领域达到了一个新的高度，各主要大国都以积极的姿态参与应对气候变化的主流进程。但过高的政治预期也可能是导致哥本哈根大会失败的关键因素，出于对大国主导进程的不满，最终由于五个国家的反对，《哥本哈根协定》未能经过各方协商一致得以通过，在这个“南北”分歧为主导的国际多边机制之下，第一次出现了小国集团与超级大国之间的较量。当然，这也是世界政治经济格局的真实反映，新兴经济体的发展必然会导致新的矛盾的产生，各方博弈的结果是最终并未达成一个具有较强法律约束力的成果，松散的机制显然不能满足应对气候变化的要求，因此在巴厘路线图进程尚未完全结束时，新一轮的谈判进程就开始酝酿了。

德班平台在一定程度上是巴厘路线图进程的延续,《坎昆协议》的法律约束力不尽如人意,各方就其法律形式的探讨无果而终,各方承诺也仅达到2℃目标要求的60%。2011年,在南非主席国的推动下,德班平台进程启动,谈判的授权包括达成一个2020年生效的具有法律约束力的新协议,以及提高2020年前的行动力度。德班平台启动时,关于《议定书》第二承诺期的谈判尚未结束,发展中国家对发达国家在此问题上的拖延极为关切,在2012年多哈会议期间,发展中国家特别是基础四国表达了坚定的立场,《议定书》多哈修正案最终得以通过,谈判正式由两轨转为一轨,德班平台谈判全面展开。2013年的华沙会议决定邀请各方开始准备"国家自主贡献",并于2015年巴黎缔约方大会之前提交,在加强2020年前行动方面呼吁发达国家提高减缓目标、发展中国家完善NAMAs,并开启了识别具有减缓潜力政策措施的技术检验进程。2014年,利马会议各方就"国家自主贡献"的范围、所需信息和力度审评进行了激烈的讨论,但各方分歧过大,未达成任何相关的有力成果,缺乏对"国家自主贡献"的基本界定,发展中国家希望将适应、资金、技术和能力建设支持纳入"国家自主贡献"的诉求没能得到反映,而各方已开始陆续提交"国家自主贡献",一定程度上出现了"木已成舟"的局面,给后续谈判提出了更大的挑战。

随着德班平台谈判进程的不断深入,应对气候变化在国际政治舞台上再次升温,各方都试图通过在此问题上发声以提高国际影响力,近两年来,在国际重要的多边双边场合,气候变化都成为重要议题,各国国内低碳发展也呈现良好的势头,对于推动德班平台进程有重要的积极作用。但在过去的几年中,经济危机给发达国家带来困难的同时,新兴经济体仍保持了良好的发展势头。世界政治经济格局的变化使国际气候谈判中关于"两分法"的争论愈演愈烈,在新协议中落实"共区"原则成为谈判的焦点问题。同时,持续增长的温室气体排放量和有限的碳排放空间之间的矛盾加剧,以欧盟为代表的部分发达国家强烈支持建立一个提高减缓力度的审评机制,但各方在是否建立机制、机制的目标和实施规则等问题上分歧依旧显著。剩下的时间已经不多，德班平台谈判面临的任务和挑战依然十分艰巨。

随着气候变化的负面影响逐步显现，特别是极端气候事件频发，气候变化对社会经济和自然生态系统造成的损失和损害越来越大，同时相关的观测证据不断完善，当前以全球变暖为特征的气候变化逐渐被广泛认同。

关于气候变化产生的原因的讨论则更加复杂。由于气候系统具有高度的复杂性和长期性，并且人类系统观测气候变化的历史还非常短，科学界仍然缺乏评估气候变化原因和预测气候变化趋势的可靠手段，对于气候变化的科学性和不确定程度也存在很大的争议。最初，科学界确认了温室效应，并研究得出当代的变暖趋势与人为活动温室气体排放有密切的联系，减缓气候变化应通过控制温室气体，特别是二氧化碳排放来实现。但这种观点受到了很多质疑，人们并不相信全球变暖是人类活动造成的，也认为控制温室气体排放实际上是制约其他国家发展的手段，"气候变化阴谋论"的阴影始终挥之不去。但随着科学研究水平的提升，人为活动排放的温室气体导致以全球变暖为特征的气候变化这一结论得到了更有力的支持。IPCC 历次评估报告关于气候变化归因的表述的变化体现了这一过程。第一次评估报告中，IPCC 指出："人类活动导致的温室气体排放，增加了大气中温室气体浓度，并增强了温室效应，使平均温度上升"，只能够认定人类对气候产生的影响的直接证据还非常有限。IPCC 第二次评估报告指出："自 19 世纪末以来，全球平均地面温度上升了 0.3—0.6℃，这一变化不可能完全是自然产生的"，"各种证据的对比分析表明了人类对全球气候有可辨别的影响"。IPCC 第三次评估报告进一步指出温室气体浓度增加导致温室效应的可能性，指出"最近 50 年观测到的大部分变暖可能（66%—90% 的概率）是由于温室气体浓度的增加。人类活动造成的温室气体和气溶胶排放继续以预期影响气候的方式改变着大气，由于人类活动，大气温室气体浓度及其辐射强迫继续增加"。这一结论在第四次评估报告中得到了扩展："观测到的 20 世纪中叶以来大部分的全球平均温度的升高，很可能（>90% 的概率）是由于观测到人为温室气体浓度增加所导致的。" IPCC 第五次评估报告明确指出："人类对气候系统的影响是明确的。极有可能（95%—100% 的概率）的是，观测到的 1951—2010 年全球平均地表温度升高的一半以上是由温室气体浓度的人为增加和其他人为强迫共同导致的。"

在此基础上，应对气候变化的必要性和紧迫性逐步突显并引起人类的重视。20 世纪 70 年代人们已经开始意识到全球气候变化对当代和后代生存发展的潜在风险，是人类可持续发展面临的共同挑战，到 80 年代末 90 年代初，各方在联合国框架下开启了全球应对气候变化国际合作的讨论和框架公约的谈判。2006 年英国经济学家尼古拉斯 · 斯特恩在调研一年的基础上主持完成并发布了《斯特恩报告》，报告指出不断加剧的温室效应所带来的气候变化是对全球的严重威胁，其后果会对全球经济发展产生影响，其严重程度不亚于世界大战和经济大萧条，亟需全球作出反应。而考虑到减少温室效应的成本与收益，尽早采取应对气候变化行动的收益大于成本，目前的行动会对未来 40 年到 50 年的气候产生有限的影响，但对于更远的未来影响深远。全球各国在共同努力应对气候变化问题上已经达成了趋于一致的看法，低碳发展已成为全球的新潮流。

第二节 《联合国气候变化框架公约》的科学基础、原则和长期愿景

一、科学基础和任务

如前文所述，气候变化问题是由于全球工业革命以来传统的高碳

发展路径和生产、生活方式引发的累积排放造成的，发达国家对此应负不可推卸的责任。全球气候问题不能仅单纯视为环境问题，它首先是发展问题。问题的根源来自发展，解决问题的最终出路也在发展，即可持续发展。要应对全球气候变化，一方面要通过改变发展模式，减缓气候变化，将全球累积碳排放、大气中的温室气体浓度和辐射强迫控制在危险范围内；另一方面要提高适应气候变化能力以抵御气候变化已经造成且可能继续造成的负面影响。而上述减缓、适应气候变化行动均需要相应的技术、资金、能力保障和机制政策安排。

全球气候治理体系（参见图 5-2）旨在解决全球气候变化这一科学问题。建立全球气候治理体系的最终目的在于实现《公约》的目标，即"将大气中温室气体的浓度稳定在防止气候系统受到危险的人为干扰的水

图 5-2 全球气候治理体系的科学基础和任务

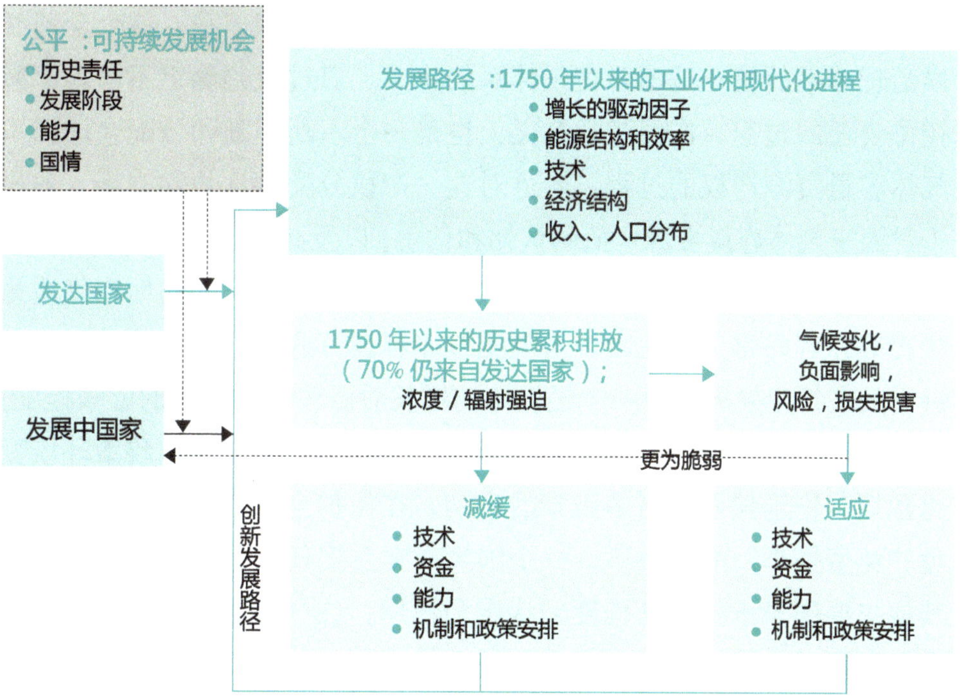

资料来源：傅莎，邹骥．基于"发展路径创新"的后 2020 年国际减排机制设计 [R]. 国家气候战略中心研究报告，2014.4.

平上。确保各缔约方促进可持续发展的权利，认识到所有国家特别是发展中国家需要得到实现可持续的社会和经济发展所需的资源，以统筹兼顾的方式把应付气候变化的行动与社会和经济发展协调起来，以免后者受到不利影响，同时充分考虑到发展中国家实现持续经济增长和消除贫困的正当的优先需要”。(《公约》第2条，前言和3.4条）具体而言，需完成如下两大基本任务：

一是通过全球气候治理体系管理好全球气候这一人类公共财富、解决好温室气体排放全球外部性和跨代外部性问题，将大气中温室气体的浓度稳定在防止气候系统受到危险的人为干扰的水平上，并适应气候变化的不利影响。对于全球气候公共财富的管理，目前还不存在一个凌驾于各国主权之上的纯粹超主权治理模式，无法在国际层面像主权国家内部那样依靠国家强制权力解决国际外部性和全球公共财富管理问题，只能通过跨国谈判和协商进程，通过各国共享的价值观和发展理念等，形成不断演进、强化的全球气候治理体系。这个全球气候治理体系的核心是国际责任体系的定义。通过谈判确定不同国家减控排责任的过程，在经济学意义上也是一个不断明晰和分配全球温室气体容量资源产权的过程。各方对这一产权分配的认识和立场，取决于对温室气体容量资源价值的认知和估价，以及此估价与其他方面（如重要双边外交关系、国家形象、与化石能源消耗相关的局地空气质量状况等）利弊得失的比较。对温室气体容量资源产权价值的估计，还会受到在特定发展阶段一个经济体的增长对温室气体排放的依赖程度、化石能源供求与价格长期变动趋势、相关技术变动趋势、政策环境变化预期、伦理偏好等诸多因素的影响。在保护全球气候公共财富、避免“免费搭车”的背景下，一个国家对自身责任的认定，还取决于对其他主要国家承担责任程度的认定和预期。这些都影响着不同国家对自身承担减排责任多寡、利弊、能力的判断。因而，责任确定的过程也必然是一个不断动态调整和演变的过程。

二是协调好全球气候保护与全球社会经济发展之间的关系。气候变化归根到底是发展路径，全球气候治理在某种意义上也是全球发展治理的有机组成部分。全球气候治理体系应当向各国决策者、投资者和消费者发出明确的引导信号，对重塑世界和各国的经济、改变人类发展进程和创新发展路径，形成国际体制支持。

上述两大任务密切结合，相辅相成，需要一体化地综合完成。其中，“创新发展路径”应视为建立全球气候治理体系的核心概念框架。这有助于将有关全球气候问题的基本争论归结到发展路径上，建立温室气体排放与发展之间的紧密联系，使减排目标和实现目标的资金、技术、体制和政策过程紧密联系；同时，这一概念框架将历史与未来排放轨迹特征与影响排放的驱动因子建立联系，将辩论引向如何通过控制引导深层的影响排放的因素去影响目标排放轨迹，且可以用来揭示处于不同发展阶段的国家其发展议程与排放特征的区别，有利于深化和拓展关于气候变化历史责任的论述，具体体现在处于不同发展阶段的发展中国家与发达国家的区别。这一概念框架着眼于实质的行动和措施，在可持续发展的总体框架下，围绕发展路径的创新和变迁，识别出发达国家与发展中国家各自行动的方向和重点，因此它在政策含义上具有建设性。总之，“发展路径创新”的出发点是处于不同发展阶段的国家都要以积极的姿态行动起来，探寻与本国发展阶段相符的发展路径创新；其着眼点是影响排放轨迹和路径的发展驱动因素，以求在更实质、更实际、更具体的层面上打破全球气候治理谈判的僵局。

二、原则

作为气候变化领域的全面的框架性的有法律约束力的文件，《联合国气候变化框架公约》是全球气候治理体系的重要组成部分，也是设计 2020 年后全球气候治理体系的基石。如前文所述，全球气候体系的最终目的是促进《公约》得到完整、有效和可持续的实施，并加强《公约》下的基于规则的多边机制，以实现《公约》的最终目标。因此，《公约》的核心原则特别是公平和“共区”原则都应在全球气候治理体系的构建过程中得到充分体现。

其中，公平和“共区”原是为全球气候治理体系的核心原则。公平的核心是可持续发展权利和机会的均等。在将碳排放与发展脱钩的历史进程中，平等的可持续发展权利与机会并不必然意味着均等的人均排放和历史累积排放，更不意味着发展中国家必然要重蹈发达国家在历史上的排放轨迹。强调公平并不意味着不行动，事实上，发达国家和发展中国家都有采取强化行动应对气候变化的责任，只是需要在“公平”的框架下采取行动并承担有区别的责任。只有一个“公平”的治理体系才能吸引最大范围的广泛参与，才能有现实存在价值和实施可能。要实现“公平”，一方面要确保发展中国家公平获取可持续发展的均等机会，促使广大发展中国家走上一条可持续的发展道路，在公平获取技术、资金、教育机会和知识的前提下，以气候友好的方式去实现脱贫和改善人民生活质量的发展目标。另一方面要确保发达国家基于其历史责任及在发展阶段、能力和在世界政治经济体系中占支配地位的优势，遵循“共同但有区别责任”的原则，充分履行其减排责任，发挥率先垂范的作用，并向发展中国家提供技术、资金和能力建设支持，帮助促进发展中国家完成发展路径的创新和转型。

同时，正如上文所述，发达国家和发展中国家在所处的发展阶段、面临的发展需求、历史责任和现有能力等方面仍存在巨大差别，利用两分法落实“共区”原则依然具有重要的现实意义。同时，考虑到发展中国家内部差异的客观存在，在未来全球气候治理体系中，可在“两分法”前提下，考虑小岛屿国家和最不发达国家的特殊情况，在具体贡献方式和义务分担上，表现出适当的灵活性。但发展中国家内部的差异并不能取代发达国家和发展中国家的区别。

此外，《公约》确定的其他原则也应成为全球气候治理体系的原则。在全球气候治理体系下，各缔约方应当采取预防措施，预测、防止或尽量减少引起气候变化的原因，并缓解其不利影响。要有风险管理的意识，不能以科学上的不确定性作为理由推迟行动。应付气候变化的政策和措施应当讲求成本效益并考虑不同的社会经济情况，确保以尽可能最低的费用获得全球效益。同时，全球气候治理体系还应帮助建立有利的和开放的国际经济体系促进各方合作和可持续发展，提高各方应对气候变化的能力。

三、长期愿景

气候变化问题是全球范围内的长时间尺度的挑战，因此，在设计全球气候治理体系时需要建立全球长期视野为全球低碳转型和提高气候抵御力提供积极信号。

气候谈判中有关长期愿景/目标的争论由来已久，主要聚焦于如下四个问题：①是否进一步强化温度目标到1.5℃？②是否在温度目标基础上进一步提出长期量化排放目标，如2050年全球在2010年基础上至少减排60%/40%—70%/70%—95%的长期量化减排目标，和全球到本世纪末/2060—2080年期间实现零排放（碳中和）甚至负排放的目标？③是否建立长期减排/适应目标和有关条件即资金、技术、能力建设支持的联系，即是否提出长期资金和技术目标？④是否提出长期适应目标？如是否要提出量化的全球适应目标？是否将温升、减排力度、适应成本和需求及损失和损害挂钩？

考虑到目前科学上关于1.5℃的研究仍有较大局限性和不确定性，还不足以用来支持决策，而且，目前从排放到浓度，从浓度到辐射强迫，从辐射强迫到温升的各个环节都仍存在相当大的科学不确定性，应避免将谈判重点放在确切数字上，使谈判陷入“谴责游戏”的僵局。全球气候治理体系的长期愿景应放在如何实现全球和各国长期发展路径转型上，重点是向全球传达积极信号，表达转型决心。具体而言，全球气候治理体系的“长期愿景”应包含《公约》第二条、温升目标、长期转型愿景、长期减排目标、长期适应目标及与长期减排/适应目标相联系的长期支持目标等一揽子安排：

(1) 重申 2℃目标及《坎昆协议》中有关 1.5℃的表述，强调目前科学上关于 1.5℃的研究仍有较大局限性和不确定性，还不足以用来支持决策。

(2) 强调绿色低碳转型大方向，适时以适当方式提出全球发展路径创新，走绿色低碳可持续发展且具备较高气候抵御力的创新发展道路的长期努力愿景，释放积极信号。强调根据最新科学研究，要实现 2℃温控目标，要求各国实现技术、经济和制度上的全面转型。发达国家和发展中国家都面临创新发展路径，走低碳可持续的具备气候抵御力的创新发展道路的“共同”任务，但考虑双方在历史责任、能力、发展阶段和国情上的差异，发达国家和发展中国家应承担“有区别”的发展路径创新的任务。对发达国家而言，其主要任务是发挥示范作用和市场需求导向作用，加速下降温室气体排放，率先垂范一条创新的低碳可持续发展示范路径，并通过向发展中国家提供资金、技术、能力建设支持及政策、经验交流，将上述示范路径推广和扩散到发展中国家。发达国家的示范作用对于全球创新发展路径有着至关重要的作用。而对发展中国家而言，考虑其后发优势、全球排放空间约束和国内资源环境约束，有必要也有可能改变现有发展模式，通过发展路径的跃迁，开创一条比美欧等发达国家传统发展路径更为低碳的、更早达到更低排放峰值的创新发展路径。但强调受发展中国家自身能力的限制，其实现转型、克服路径依赖的可能性在很大程度上取决于发达国家所给予的支持程度和示范程度。可适时“邀请发达国家缔约方和准备好的发展中国家缔约方适时提出本国创新发展路径、应对气候变化的中长期战略、规划或方案，强调发达国家及相关国际机构应当为发展中国家制定实施上述战略、规划或方案提供资金、技术和能力建设等方面的支持”。

(3) 坚持平衡考虑减缓和适应长期目标，强调不能孤立地讨论长期目标的数字问题，应建立长期目标与相应的条件（技术、资金、能力建设支持等）的不可分割的关联。有关长期减排目标的表述可参考巴厘路线图下“共同愿景”的谈判成果，提及最新科学成果，强调全球大幅减排、尽早实现峰值，实现长期发展路径转型的愿景，但反对量化的减排目标。坚持确定全球长期减排目标要以发达国家对发展中国家提供充足的和量化的资金技术支持的承诺为保障，发展中国家的减排要在可持续发展的框架下进行。

（4）关于长期适应目标的表述以定性为主，重点放在定性描述的全球适应愿景（建立气候抵御力社会）及发达国家的适应支持上。

第三节　全球气候治理的核心要素

一、减缓

减缓一直以来是《公约》下应对气候变化国际合作的核心，在要素设计上应遵循《公约》的原则，特别是公平、“共区”和各自能力原则基于已有的机制，如《京都议定书》和巴厘行动计划的相关安排。为应对气候变化的严峻挑战，应在充分考虑历史责任、能力、发展阶段和发展需求的基础下鼓励各方广泛参与并促进各方提高力度。

（一）长期目标

关于减缓的长期目标，根据 IPCC 最新发布的第五次评估报告的最新研究成果，在相对工业化前温升不超过 2℃的情景下，气候变化的负面影响和风险仍是可以适应和接受的。IPCC 报告进一步确立了 2℃温控目标这一已形成全球政治共识的长期目标的科学合理性。但是同时 IPCC 报告也指出，目前人类对气候系统的认识和模拟仍存在巨大的不确定性和局限，科学上仍很难将温升目标转化成具体确切的排放空间和路径，仍是一个较宽泛的范围。因此，应将 2℃目标的长期排放空间和路径目标作为全球应对变化的意向性、指导性和方向性目标，而非约束性目标，重在帮助缔约方定位和明确努力方向。

全球长期目标并不是孤立的，而是与实现目标的有关条件(context)，如资金、技术、能力等密切联系，互相支撑。只有在资金、技术等资源最优化配置的基础上，才有可能通过减缓行动实现温控目标。

2020 年后的减缓行动和效果是建立在 2020 年前减缓力度的基础之上的，越早采取有力度的减缓行动，实现 2℃温控目标的可能性就越大，未来减缓的成本就越低。提高 2020 年前减缓力度的关键是发达国家落实已有减缓承诺和进一步提高目标，这对于维护政治互信、增强全球应对气候变化的信心和降低未来减排成本和气候变化的风险至关重要。

（二）各方承诺

关于各方的减缓承诺，为遏制温升趋势，各国应切实加强减缓承诺和行动，在《公约》原则的指导下，在反映各国国情、历史责任、能力和发展需求的基础上，为全球减缓气候变化作出实质性的贡献。

发达国家的减缓目标应与其历史责任和能力相符，继续发挥领导力，率先减排，转变行为模式，并起到更好的示范作用。发达国家应继续采用全经济范围量化减排目标，各国之间的减缓承诺应具有可比性，且应主要通过国内行动实现其减缓目标。

发展中国家需结合国内消除贫困和可持续发展等目标，在资金、技术和能力建设等方面的支持下，探索低碳发展的转型路径。发展中国家应以低碳发展的最优实践为标准，通过应用最新技术和调整投资模式避免在基础设施、交通、能源供给和消费等领域造成高碳锁定效应，尽早实践低碳可持续的发展模式。考虑到发展中国家内部在资源禀赋、发展模式和能力上存在明显的差异，采用同样的形式作出减缓承诺并不合适。因此，发展中国家应依据国情和能力在《坎昆协议》NAMAs 的基础上继续在国内采取适当的减缓行动，减缓承诺的形式可以包括偏离 BAU、部门措施、强度指标、低排放发展战略等各种形式。有能力的发展中国家可采用“意向性总量控制目标”辅以“具体转型指标”的形式作出减缓承诺，在可持续发展的前提下最大限度地实现转型和管控温室气体排放。

关于减缓的相关机制安排，《公约》进程下有很多可以借鉴的经验，应体现公平、共同但有区别的责任等《公约》原则，应建立高效合理的机制安排和明晰的规则，鼓励缔约方积极参与并作出有力度的承诺。

发达国家的承诺目标应写入协议附件；为确保减缓贡献的可比性，各国应采用相同的基准年（或提供足够的参考信息）和承诺期，并采用统一的核算标准和方法，包括温室气体的种类、部门、全球潜温趋势等；相关信息应通过通用模板进行汇报。

发展中国家的减缓行动和目标也应作为附件纳入协议。考虑到发展中国家减缓行动的多样性，采用共同的基准年和承诺期并不现实。核算上应尽量采用统一的指标和方法，但由于发展中国家能力的欠缺，在其减缓行动的核算和报告上应保留一定的灵活性；发展中国家减缓行动和目标的相关信息应通过国家信息通报、两年更新报等方式报告。

二、适应

（一）适应气候变化全球治理是公约要求

《公约》向各缔约方提出采取适应气候变化行动的具体要求，并要求发达国家缔约方对发展中国家缔约方适应气候变化行动提供支持。具体而言：

——《公约》第 4.1(e) 提出“（所有缔约方）合作为适应气候变化的影响作好准备；拟订和详细制定关于沿海地区的管理、水资源和农业以及关于受到旱灾和沙漠化及洪水影响的地区特别是非洲的这种地区的保护和恢复的适当的综合性计划”；

——《公约》第 4.1(f) 提出“在它们有关的社会、经济和环境政策及行动中，在可行的范围内将气候变化考虑进去，并采用由本国拟订和确定的适当办法，例如进行影响评估，以期尽量减少它们为了减缓或适应气候变化而进行的项目或采取的措施对经济、公共健康和环境质量产生的不利影响”；

——《公约》第 4.4 条提出“附件二所列的发达国家缔约方和其他发达缔约方还应帮助特别易受气候变化不利影响的发展中国家缔约方支付适应这些不利影响的费用。”

《公约》第 4.4 条充分考虑了发达国家和发展中国家在温室气体排放方面不同的历史责任，以及在气候抵御能力方面的差异。发展中国家受制于资金、技术、能力等各方面因素的影响，面临着更急迫的气候危害、更高的暴露度和脆弱性。发达国家在工业化过程当中累积的历史排放所造成的负面气候变化影响，对于所有发展中国家而言，都是一个额外的适应负担，理应由发达国家为其过去无节制的排放行为买单，承担起相应温升情景下的适应所需的资金、技术和能力建设。

根据最新的 IPCC 第五次评估第二工作组报告《气候变化 2014：影响、适应和脆弱性》，已观测到的气候变化影响和风险变得越来越广泛，无论是以发达国家为主的欧洲、北美洲和大洋洲还是以发展中国家为主的亚洲、非洲和南美洲，都存在着实质性的风险。气候变化已对所有大陆和海洋的自然和人类系统产生了影响：包括海平面上升、沿海洪涝和风暴潮；干旱、降水变化导致的粮食安全；内陆洪水；水资源短缺带来的农业生产力下降对农村生计和收入带来损失的风险；极端事件导致基础设施和关键服务崩溃的系统性风险；提供沿海生计的海洋生态系统和服务丧失的风险；提供陆地生计的陆地生态系统及其服务的丧失的风险；极端高温引发的健康风险，等等。

面对气候变化带来的不利影响，越来越多的国家正积极采取适应行动，提高气候抵御力，减小气候风险。一方面，虽然各国正在依靠国家甚至地方治理体系根据自身国情和能力落实《公约》关于各缔约方采取适应气候变化行动的要求，但是考虑到适应行动涉及领域的广泛性和复杂性，区域合作、信息经验分享的需求日益增长；另一方面，气候变化的影响和所带来的风险具有超越国界和各国发展阶段的特征，且考虑到未来气候风险将随着全球持续温升而不断增强，落实《公约》关于发达国家缔约方向发展中国家适应气候变化行动提供支持的需求将愈加强烈。这两方面的国际合作需求必须通过全球治理体系来实现。

（二）适应气候变化全球治理具有科学基础

按照最新 IPCC 报告研究内容，气候变化对自然系统和人类系统所带来的影响是由三个因素作用下形成的：气候变化危害（hazards）、自然和人类系统的脆弱性（Vulnerability）以及暴露度（Exposure），三者互相作用产生了气候变化风险，如图 5-3 所示。如何降低气候风险，最小限度减少气候变化对自然和人类系统的影响，是我们应对气候变化行动的终极目标。

气候变化危害是指气候变化相关的事件或趋势造成人员伤害和其他健康影响以及对财产、基础设施、生计、服务设施和环境资源造成

图 5-3 气候变化对自然系统和人类系统所带来的影响

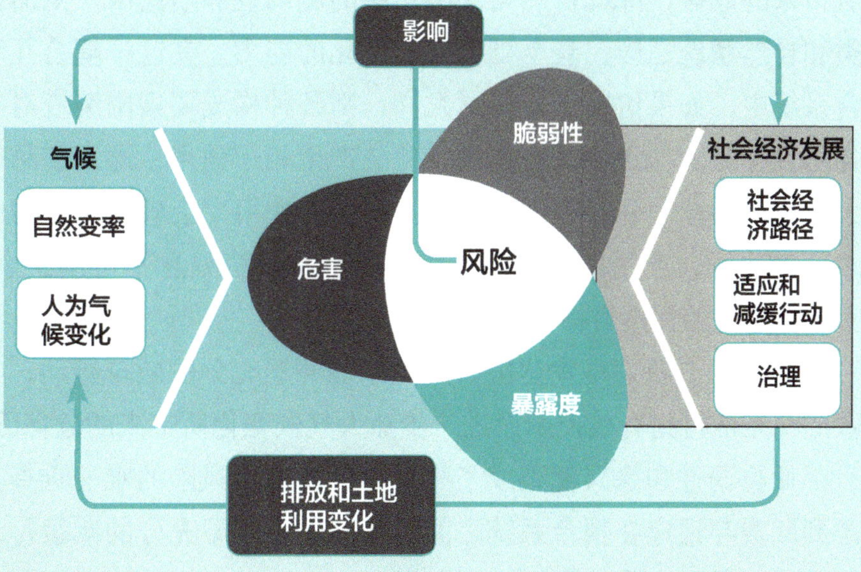

资料来源：IPCC.Climate Change 2014：Impacts, Adaptation and Vulnerability.

的损害和损失。危害因素是气候系统衍生出来的，除去气候系统的自然变率，主要是指人类经济行动增加温室气体排放而造成气候系统变化。为降低危害因素所带来的气候风险，需要开展控制温室气体排放、增加碳汇为主的减缓行动。

气候变化影响 (Impact) = f { 气候变化危害 (Hazards)× 暴露度 (Exposure) × 脆弱性 (Vulnerability) }

脆弱性和暴露度两个因素是指自然与人类系统面对气候变化时的不利因素。脆弱性是指受到不利影响的倾向和趋势，它包含多个元素，如敏感度、易感性、缺乏应对和适应能力等。暴露度是指人类、物种、生计、生态系统、环境服务和各种资源、基础设施或经济、社会或文化资产出现在受严重影响地区的程度。为降低暴露度和脆弱性所导致的气候风险，需要在自然和人类系统中开展大范围的适应气候变化行动。

综合来看三者关系，为达成最大程度降低气候变化对人类和自然系统所带来的影响，可以通过减少温室气体排放控制温升降低气候变化所带来的危害，降低自然与人类系统的脆弱性和暴露度，全面提高人类和自然系统适应、甚至抵御气候变化的能力。在这一耦合互动的复合系统中，如果协同作用得以发挥，则各种构成要素围绕应对气候变化总目标将产生协调放大作用，使之产生相干效用，即系统总体效用大于局部功能之和，否则，所产生的负向作用力会破坏各种构成要素之间的协调，产生反向放大作用，使整个系统处在不协调状态下，大大降低系统的预期目标。

如果缔约方特别是发达国家集团未能开展充分的减排行动，必将导致温升无法得到有效控制而进一步加大气候变化所带来的危害因素，那么降低脆弱性和暴露度则成了弥补高气候变化风险的唯一选择，再考虑到减缓与适应在耦合系统之间的相互作用，不充分的减缓行动则意味着更高程度的适应需求以及适应成本。

然而，与减缓议题下的温室气体排放量、大气中温室气体浓度、碳排放强度等指标不同的是，适应议题一直以来就没有明确统一的量化标准和指标供探讨，也未能与温升目标关联在一起。适应政策和适应技术在多大程度上降低了脆弱性、又在多大程度上提升了适应能力、相应的适应成本又是多少，这些定量估算较少，而且多以区间段的形式给出（例如，对于发展中国家的适应成本，只有世界银行、UNDP等几家机构进行过系统的分析，所得结果在每年40亿—1090亿美元之间），再考虑到数据不足、方法缺陷等因素，结果具有较大不确定性，在一定程度上制约了适应气候变化领域应有的机制安排。

近年来，适应定量方面的研究，特别是集中在成本预估的方法论方面逐步呈现，例如UNEP发布的适应差距报告使用了定量的评估方法对适应成本进行评估，非洲集团则采纳了科学界的研究成果应用于谈判，在其提交的德班平台2015协议适应提案中开创性地提出将适应成本与全球温升挂钩，基于温升情景分析计算适应成本，评估不同温度情境下的气候影响及其成本，将0℃温升与其他温升情景下的成本差值定义为全球适应目标。在此基础上，基于实际温升情况（由减排力度决定）对全球适应目标进行调整，最终决定发达国家所应提供的适应支持。

三、资金

气候资金问题的讨论始于《公约》对发达国家缔约方向发展中国家缔约方提供资金支持的要求。《公约》第4.3条不但明确提出了资金由发达国家提供给发展中国家这一流向，还规定资金应该是“新的”和“额外的”，且针对某些《公约》规定的发展中国家应承担义务所产生的费用提出了由发达国家承担所“招致的全部费用”。《公约》第4.3条还进一步规定了发达国家提供的发展中国家所需资金“包括用于技术转让的资金，以支付经议定的为执行本条第1款所述并经发展中国家缔约方同第11条所述那个或那些国际实体依该条议定的措施的全部增加费用”。该条款甚至明确提出资金应“充足和可以预测”，发达国家缔约方间应适当分摊出资负担。《公约》第 4.7 条进一步明确将发展中国家缔约方应对气候变化行动与发达国家缔约方资金支持挂钩。

《公约》谈判进程中的气候资金问题更多被放在伦理层面和国际责任划分的语境下讨论。具体而言集中反映了“共同但有区别的责任”原则是否和如何适用于气候变化国际合作的争论。因此，谈判的焦点集中在谁出资、出多少、资金来源和性质、谁受资、受资国政策和体制环境以及资金透明度等方面。这些构成了以《公约》为基础的全球气候治理体系中资金问题的主要内容。事实上，正如前面所述，《公约》对这些问题都作了框架性规定，但落实到不断变化的现实情况和各方关切的治理体系构建，则须经历不断谈判和妥协的过程。

关于谁出资，尽管《公约》作了明确规定由发达国家提供资金支持，但随着部分发展中国家经济实力的提升，特别是通过南南合作渠道对外援助的资金规模逐年扩大，发达国家要求“有能力”的发展中国家出资的呼声越来越大。一些发展中国家在气候变化领域也已经采取行动，帮助其他发展中国家减缓和适应气候变化。但从国际责任和义务的角度，发展中国家普遍认为发展中国家之间的相互帮助应与发达国家出资在性质上作出明显的区分。

关于资金量，目前并没有一个完整、准确和权威的发展中国家应对气候变化资金总需求量的评估。IPCC 第五次评估有关结论① 显示，在实现 2℃的情景下全球 2010—2029 年对非化石能源的投资增量需求约为每年 1470 亿美元，对提高能效的投资增量需求约为 3360 亿美元。根据《公约》秘书处② 的估算，全球 2030 年的适应成本将达到 280 亿—

① IPCC, 2014, IPCC 2014: Mitigation of Climate Change, Cambridge University Press, Cambridge, United Kingdom and New York, NY, USA, 2014.

② UNFCCC, 2007: Investment and Financial Flows to Address Climate Change. Climate Change Secretariat, Bonn, Germany.

670亿美元，而世行[1]预测全球2050年的适应成本更高达700亿—1000亿美元。2009年的哥本哈根会议上，发达国家承诺2010—2012年连续3年每年动员100亿美元快速启动资金，并在2020年实现每年动员1000亿美元的长期资金。发达国家是否兑现了快速启动资金的承诺，是否能够实现长期资金的目标，围绕这两个问题的量化分析很多，也没有统一的结论，争议较大，分析的不确定性也很大。气候资金不论从需求还是从发达国家提供的支持两方面都无法也没有必要陷入无休止的数字争论。越来越多的研究和政府间谈判开始关注气候资金如何推动各国走向绿色低碳转型的去碳化道路。

关于资金的来源和性质，按照《公约》关于发达国家向发展中国家提供资金支持的要求，大部分发展中国家认为履行《公约》义务的载体是政府，政府应提供公共资金；而发达国家出于自身资本输出、投资回报的考虑，更强调资金流的概念，将从发达国家流向发展中国家的资金都算作气候资金，突出私营部门资金的作用。另外，即使针对公共部门资金，发达国家和发展中国家对《公约》提出的“新的”和“额外的”要求也存在争议。发展中国家认为气候资金应该有别于官方发展援助，然而发达国家在统计报告时大多将其与气候变化沾边的官方发展援助资金计算在内。

四、技术

气候技术的创新突破、大规模扩散和应用是实现全球应对气候变化目标的核心，也是全球气候治理的重要工作内容。气候技术的开发和转让与一般技术有共性也有区别。与其他技术一样，气候技术要经历研发、示范、商业化应用等环节，部分减缓技术可以且正在通过商

① World Bank, 2010: The Costs to Developing Countries of Adapting to Climate Change: New Methods and Estimates. 2010. World Bank, Washington DC.

业途径按照一般性技术的过程实现着从发达国家向发展中国家的转让。然而，由于气候属于全球公共物品，气候技术的转移面临着市场失灵的风险。一方面，还有相当一部分气候技术由于缺乏盈利能力，在国际技术市场中没有商业价值，需要政府的干预；另一方面，面对那些有盈利能力的气候技术，发展中国家企业面临使用和维修费用高、知识产权过度保护、技术标准垄断、自生能力无法提高等困难。发达国家企业基于现有知识产权体系和自身资金、人才能力、政策和体制优势保持其技术的国际竞争力本无可厚非，但从全球应对气候变化的紧迫性来考虑，需要找到两者的平衡。

《公约》正是出于寻求维护国际技术市场秩序和加快气候技术开发与转让步伐平衡的目的，对发达国家提出了向发展中国家转让气候技术的要求。《公约》第 4.5 条明确提出发达国家“酌情促进、便利和资助向其他缔约方特别是发展中国家缔约方转让或使它们有机会得到无害环境的技术和专有技术，以使它们能够履行本公约的各项规定。在此过程中，发达国家缔约方应支持开发和增强发展中国家缔约方的自生能力和技术。”

五、能力建设

关于发达国家向发展中国家提供能力建设方面的支持，《公约》提出了详细的要求，所涉及的条款包括《公约》第 4 条第 1、第 3、第 4、第 5 和第 7 款及第 5 条、第 6 条和第 11 条第 1 款等，内容包括熟练掌握技术的本地能力、研究观测能力、加强教育和培训以提高公众意识等。

全球气候治理的能力建设活动首先应充分考虑发展中国家的切实需要，考虑到发展中国家的能力需求会随着经济、社会发展以及应对气候变化行动的深入实施发生动态变化，应按照国别不断跟踪评估各国的具体能力建设需求，并反映各国可持续发展战略重点和行动规划，能力建设是一个连续、渐进和互动的过程，在执行中应以发展中国家的优先事项为依据。能力建设要注重实践，通过开展示范项目等实践活动不断学习和了解发展中国家需要进一步发展的具体能力。

能力建设应对特别易受气候变化不利效应影响的发展中国家给予特殊对待。需要照顾最不发达国家和小岛屿发展中国家的特殊情况：① 生态系统脆弱；② 人口压力大、地理位置偏僻；③经济薄弱，收入低，极为贫困，缺乏外国投资；④ 土地退化，荒漠化；⑤ 气象 / 水文服务和水资源管理等各种服务欠发达；⑥缺乏自然灾害管理方面的预警系统；⑦ 缺乏粮食安全。

在全球气候治理下对发展中国家开展的能力建设具体内容可包括：①加强或建立国家气候变化中心；② 国家信息通报；③ 国家气候变化方案；④ 温室气体清单，排放数据库管理以及用来收集、管理和利用活动的数据和排放因素的系统；⑤ 气候脆弱性和适应评估；⑥ 执行适应措施方面的能力建设；⑦缓解办法执行情况的评估；⑧ 研究和系统观测，包括气象、水文和气候方面的服务；⑨ 技术的开发与转让；⑩ 提高决策能力，包括为参与国际谈判提供援助；⑪ 教育、培训和宣传。

第四节　《联合国气候变化框架公约》的组织结构和机制安排

一、组织结构

应对气候变化多边进程在《公约》的指导下经过 20 多年的谈判，机制安排不断细化，围绕《公约》和《京都议定书》对减缓、适应、资金、技术等要素的要求，组织机构，逐步丰富，形成了以《公约》和《议定书》缔约方会议为最高决策机构，其他各机构责任和分工明确、关系清晰的管理体系（见图 5-4）。

图 5-4 《联合国气候变化框架公约》机制机构

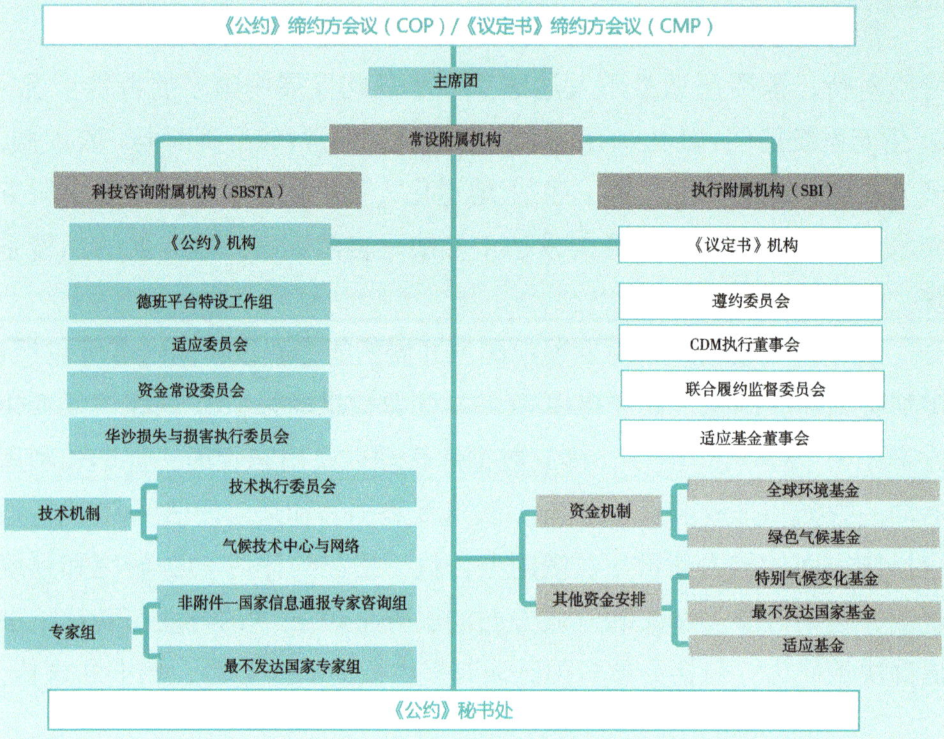

资料来源：《公约》网站，http://unfccc.int/2860.php.

(一)《公约》和《议定书》实施的管理机构

缔约方会议是《公约》的最高决策机构。《公约》缔约方在缔约方会议上审评《公约》及其所有具有法律效力决定的执行情况，并作出包括机制和行政安排在内的提高《公约》有效实施的必要决定。

《公约》的最高决策机构——《公约》缔约方会议还承担着服务于《京都议定书》缔约方会议的任务。《议定书》缔约方应参加由《公约》缔约方会议承担决策功能的《议定书》缔约方会议，非《议定书》缔约方的《公约》缔约方作为观察员参加会议。《议定书》缔约方会议评估《议定书》及其决定的执行进展，并作出提高《议定书》实施的决定。

科技咨询附属机构通过为与《公约》和《议定书》相关的科学和

技术问题及时提供信息和建议支持《公约》和《议定书》缔约方的工作。

执行附属机构通过对《公约》和《议定书》实施情况的评估和审评支持《公约》和《议定书》的工作。

增强行动德班平台特设工作组是《公约》下的一个附属机构，于2011年由第17次缔约方会议授权建立，旨在不晚于2015年在《公约》下达成一项有法律效力的、对2020年国际应对气候变化机制作出安排的协议。工作组的授权还包括为2020年前提高减缓力度识别弥补力度差距的行动选项，确保所有缔约方作出最大的减缓努力。

《公约》和《议定书》缔约方会议主席团为缔约方会议提供意见和指导，包括非缔约方会议期间《公约》及其《京都议定书》下的有关工作、谈判会议的组织、秘书处的运转等。主席团成员由联合国五大区域外加小岛屿发展中国家选举产生。

（二）《公约》下机构

适应委员会是COP全体在适应气候变化负面影响的顾问团。其任务是在《公约》指导下全面推进并加强适应行动的执行。根据第16次缔约方会议有关决议（1/CP.16），适应委员会的工作范围包括：

（1）向缔约方提供技术支持与指导方针；

（2）加强落实信息、知识、经验以及最佳实践分享；

（3）促进国家、地区，国际组织、中心，和网络的协同参与；

（4）向缔约方会议提供参考信息和建议；

（5）考虑缔约方之间关于适应行动的监督、审评、支持等方面可能存在的需求与差距等信息。

资金常设委员会向缔约方会议在气候变化资金、资金机制合理化，金融资源调转以及资金支持透明度等方面提供政策建议。

华沙损失与损害国际机制执行委员会由第19次缔约方会议有关决议授权建立（2/CP.19），主要任务是为华沙损失与损害国际机制的实施提供指导。执行委员会在《公约》下，受《公约》指导，向《公约》负责。

公约下的技术机制由技术执行委员会（TEC）和气候技术中心与

网络（CTCN）两个机构组成，前者是技术机制承担技术机制政策研究功能，后者承担技术机制的执行功能。TEC 主要为缔约方提供有关技术开发和转让的政策建议，具体包括开展对技术开发与转让过程的障碍分析（如知识产权）、向缔约方和利益攸关方提供减缓、适应以及与包括资金机制在内的公约其他机制联系的信息和建议等。CTCN 在缔约方会议指导下开展行动，其决策机构为咨询理事会。理事会由 25 名代表和正副两名主席组成。代表包括 TEC 的正副主席、绿色气候基金成员、适应委员会成员、资金常设委员会成员、CTCN 主任以及公约观察员组织机构代表。CTCN 的主要任务是回应发展中国家缔约方通过国家指定实体提出的技术项目咨询服务请求。

另外，公约下还有两个专家组，包括非附件一缔约方国家信息通报咨询专家组（CGE）和最不发达国家专家组（LEG）。CGE 由第 5 次缔约方会议（8/CP.5）授权成立，其主要功能是为非附件一国家制定国家信息通报提供技术支持和建议。由于发展中国家将于 2015 年开始提交两年更新报（涵盖了国家信息通报的主要内容），第十八次缔约方会议（8/CP.18）授权 CGE 对两年更新报提供技术支持。LEG 是为落实《公约》第 4.9 条而设立的，围绕最不发达国家的特殊情况与特殊需求开展以下七个方面的工作：①为国家适应计划进程提供技术指导方针与建议；②支持国家适应行动的执行；③帮助最不发达国家中女性及其他弱势群体；④支持最不发达国家工作计划的执行；⑤对最不发达国家缔约方提供准备、修改与更新国家适应行动的支持；⑥针对国家适应行动进程以及 LEG 工作进行拓展延伸；⑦在《公约》下全面支持协调适应工作。

（三）《议定书》下机构

遵约委员会由第一次《京都议定书》缔约方会议（27/CMP.1）确定成立。该委员会有两个分支（强制分支与促进分支）和一个全体会议。每一分支由十人组成，各有一主席和一副主席。委员会旨在便利、促进和强制各缔约方遵守《京都议定书》规定的义务。遵约委员会的全体大会每两年召开一次会议，两分支按要求尽可能多的召开会议。遵约委员会每年向CMP进行汇报。每个分支向全体会议汇报其工作成果。如果相应缔约方认为其被拒绝合法诉讼，根据《京都议定书》第3.1条，《议定书》缔约方会议可以考虑对强制分支的决定提出上诉。

CDM执行理事会负责监督CDM的执行。[①] 执行理事会共有十个委员及十个候补委员，一年中平均召开五到六次会议，并每年向CMP进行报告。其支持结构包括：①认证小组，以第三方验证与检验者身份针对CDM项目提出建议；②造林和再造林工作组（与方法论组一同）准备提交新底线的提案建议，并监督CDM造林和再造林项目的方法论；③二氧化碳捕集与封存工作组，与方法论组一同准备提交针对新底线的提案建议，并监督方法论；④方法论组，发展执行理事会关于底线与监督计划的方法论的建议，并为新底线与监督方法论提交提案建议；⑤小型工作组，为已提交的针对CDM小型工作组的新底线与监督方法论的提案准备建议。

联合履约监督委员会负责监督发达国家缔约方间通过项目级的合作获得减排单位(ERU)的全过程，包括减少或移除附件一缔约方在《京都议定书》下承诺的排放量、依据该项目取得的减少或移除成果分发ERU等。委员会有十个委员及十个替补委员，平均一年召开三次会议，并每年向COP进行汇报。

适应基金被用来资助《京都议定书》下发展中国家缔约方的适应计划，重点关注受气候变化负面影响最为严重的国家。适应基金来源于CDM项目（除最不发达国家）取得的减排收益分配和包括自愿捐款在内的一些其他来源。

适应基金董事会有十六名董事，正副两名主席，在《议定书》缔约方会议的指导下运行。世界银行为该基金的临时受托人，全球环境基金（GEF）为董事会的临时秘书处。董事会每年召开三到四次会议，且每年向《议定书》缔约方会议汇报。

① 黄婧.《京都议定书》遵约机制探析[J].西部法学评论，2012（1）.

（四）资金机制

目前《公约》下的资金机制主要包括全球环境基金和绿色气候基金。另外，在《公约》的实施过程中也推动成立了具有特殊用途的专门性基金。

1.《公约》资金机制

GEF是独立运转的金融机构，为包括《公约》在内的多个国际环境多边进程提供资金。GEF每年向《公约》缔约方会议汇报，并会议的指导。GEF已经开展了大范围气候变化项目投资，自成立以来，减缓行动开支共计约40亿美元，通过适应基金窗口支持的适应活动开支约为2亿美元。GEF对《公约》进程中的发展中国家需要承担或要求开展的国家信息通报、技术需求评估等工作也给予了资金支持。

绿色气候基金（GCF）由第十六次缔约方会议授权成立（1/CP.16），第十七次缔约方会议（3/CP.17）通过了其管理办法。GCF由二十四人制董事会与两名主席共同管理。世界银行为资金中间托管者，在基金启用后每三年需进行一次审评。GCF董事会建立了独立秘书处以支持其各类活动。GCF董事会每年向《公约》缔约方会议报告运行情况并接受指导。GCF将资助强化适应行动、减缓行动（包括REDD-plus）、技术开发与转让（包括碳捕集和封存）、能力建设等。GCF还将支持发展中国家缔约方实施气候变化战略规划以及项目层面的行动。GCF将特设私营部门基金，直接或间接为私营部门减缓和适应活动提供资金。

2．专门基金

气候变化特别基金由第七次《公约》缔约方会议（7/CP.7）授权成立，旨在为适应、技术转让、能力建设、能源、交通、工业、农业、森林化、废弃物处理和经济多元化项目提供资金。基金有适应和技术转让两个资金窗口。基金由GEF委托管理，受《公约》缔约方会议的指导。GEF理事会每年以基金理事会的身份召开两次会议，由GEF首席执行官与GEF理事会主席担任主席。附属执行机构（SBI）在基金发展指导方针以及对基金进行评估等方面为《公约》缔约方会议提供支持。

最不发达国家基金由第七次《公约》缔约方会议授权成立，旨在为最不发达国家开展国家适应行动。基金受《公约》缔约方会议指导，由 GEF 委托管理。

另外，适应基金也作为专门基金，专注于向发展中国家适应项目提供资金支持。

二、透明度与信任

报告（reporting）和审评（review）是保障《公约》实施和各缔约方履行承诺的透明度的重要机制，对于建立各方互信有重要的作用。经过 20 多年的发展,《公约》下已经建立了一套相对完整的透明度机制，包括国家信息通报、国家温室气体清单的报告和审评,《京都议定书》下的相关机制包括对附件一缔约方国家系统和登记簿相关信息的报告和审评，随后《坎昆协议》下又建立了发达国家两年报（BR）和国际评估与审评（IAR）机制、发展中国家两年更新报（BUR）和国际磋商与分析（ICA）机制。

（一）国家信息通报的报告和审评

根据《公约》第 4 条 1—2 段和《公约》第 12 条，各缔约方应向缔约方大会提交国家信息通报和国家温室气体排放清单。

1．附件一国家缔约方

对于附件一国家缔约方，国家信息通报内容包括：国家温室气体排放量，气候相关政策和措施，GHG 排放量预测，气候变化脆弱性和适应，为非附件一国家提供的资金技术支持，旨在提升公众意识的行动，以及国家温室气体排放清单。

附件一国家按照报告指南每 4—5 年提交国家信息通报，迄今为止，附件一国家共提交了六次国家信息通报，第六次报告于 2014 年 1 月 1 日截止。自提交日起 1—2 年，其国家信息通报经由国际专家审评小组

(ERT) 审评，小组成员由秘书处从 UNFCCC 专家名册中选出，秘书处在选择小组成员时将考虑附件一国家和非附件一国家人员数量的平衡，并考虑地区平衡。

审评的目的是为了对附件一国家落实公约的情况提供彻底且全面的技术评估，从而确保缔约方大会有足够的信息量审评《公约》的落实情况。审评不会对附件一国家的落实情况作出政治性的论断。

附件一国家信息通报采用专家审评小组到访审评的方式，并将审评结果详细呈现到审评报告草稿中。在正式的审评报告提交到 UNFCCC 网页和缔约方大会前，这份草稿将事先交与缔约方征求意见。此外，除了到访审评这一模式外，国家信息通报审评还采取了集中审评模式。

2．非附件一国家缔约方

对于非附件一国家缔约方，国家信息通报内容包括：国情概述，为落实《公约》而采取的行动和措施，气候变化易损程度和适应，技术转让和获取，科研项目的落实，提高公众意识和能力建设项目的落实，以及资金、技术、能力建设的需求和已获支持。大多数非附件一国家缔约方的第一次国家信息通报已被秘书处整理，从 1999 年至 2005 年，秘书处已对综合报告进行了六次汇编并整合。

根据第 1/CP.16 号决定，考虑到“共区”原则、与发达国家有区别的时间表以及资金支持是否及时到位，非附件一国家信息通报每四年提交一次，其内容与附件一国家的国家信息通报相比更简化。

为了帮助非附件一国家缔约方准备国家信息通报，在第五次缔约方大会上，建立了专家咨询小组（CGE）。一直以来，专家咨询小组对国家信息通报的贡献得到缔约方大会的认可，在第十八次缔约方大会（第 18/CP.18 号决定）上，缔约方认识到专家咨询小组也将在方便为编制和提交第一份两年更新报告提供技术咨询和支持方面发挥重要作用。第十九次缔约方大会（第 19/CP.19 号决定）决定将专家咨询小组的授权延期五年（2014—2018 年），CGE 将为非附件一国家的国家信息通报和两年更新报告提供技术咨询和支持，包括提供行动方案，双边、区域和多边资金和技术援助来源的信息等。并在秘书处协助下，利用专家咨询小组的最新培训材料，最迟于2014年，为国际磋商和分析（ICA）进程下提名的技术专家（见下文）制定适当的培训方案，以改进其技术分析。

（二）国家温室气体清单的报告和审评

《公约》下引入温室气体清单报告和审评机制的目的是对缔约方减排温室气体并增强碳汇的过程进行技术评估。国家温室气体清单包含温室气体排放量的相关信息，例如活动数据、排放因子和排放估算的方法学。附件一缔约方每年需参考《公约附件一缔约方报告指南》，并基于IPCC的国家温室气体清单方法学上报其清单。最新的指南融合了《2006 IPCC国家温室气体清单指南》的方法学，并在第15/CP.17号决定中被采纳。2012年10月至2013年5月期间，缔约方可自愿参考更新后的《公约附件一缔约方报告指南》进行上报，2015年提交后，所有缔约方需强制遵守该指南。

每年的温室气体清单包括一个国家清单报告（NIR）和统一的上报格式（CRF）表。NIR包括定量和定性信息，例如方法学描述，包括排放因子、活动数据、重复计算、排放趋势、主要类别和不确定性分析以及准备清单的相关机构和程序安排（归档、质量保证和质量控制）。各国的清单预估通过CRF表格提交给秘书处，以便于数据的分析和比较。CRF表格包括清单预估中的活动数据、隐含排放因子和排放结果。关于所有温室气体源和汇以及年排放估计值时间序列（起始年为1990或基准年）等信息的全面报告有助于评估方法学和数据来源的连贯性，并评估排放趋势。

国家温室气体清单由专家审评小组（ERTs）每年进行审评，审评应遵循《公约审评指南》(1999年在第6/CP.5号决定中采纳，2002年在第19/CP.8决定中完善)。公约审评指南有助于确保：COP可获得客观、连贯、透明、完整且全面的信息以及关于清单的技术评估；清单与经同意的上报指南的一致性以及清单的质量逐渐提升；且 COP 可获得关于附件一缔约方承诺，即《公约》第 4 条 1（a）和第 12 条 1（a）执行情况的技术评估。

根据第 6/CP.5 号决定，关于附件一缔约方国家温室气体清单的技术审评于 2000 年启动。随着试运行阶段的结束，自 2003 年起，对每个附件一缔约方国家温室气体清单强制实施审评。

温室气体清单的审评主要侧重评估清单估算和准备过程中采用的方法学和数据来源与参考指南的一致性，具体指南包括《更新的 1996 IPCC 温室气体清单指南》、《温室气体清单最佳实践和不确定性管理》、和《LULUCF 最佳实践指南》。具体国家的温室气体清单审评通过到访审评、集中审评和案头审评（desk review）的方式开展。根据《公约》审评指南，对一缔约方的温室气体清单审评每五年至少采用一次到访审评的方式。这种类型的审评确保更加全面的检视缔约方的方法学和假设、活动数据、排放因子、文件记录以及机构和程序安排。由于专家审评小组在不同地点的协调难度以及报告质量等因素，案头审评的方式在近几年鲜少使用。

根据第 19/CP.8 号决定，清单审评流程分为三个阶段进行：初步检查、综合和评估（包括两部分）和个别审评。在每个阶段结束时，被审缔约方有机会针对各阶段产出的报告草稿（分别是现状报告、综合和评估报告一、二及个别审评报告）发表意见。《公约》审评指南中明确给出了准备这些报告以及提出意见的具体时间安排。

个别审评阶段，由专家审评小组负责。审评小组的专家从 UNFCCC 专家名单中选取，专家小组的构成体现地域间平衡性以及附件一和非附件一缔约方专家人数的平衡。通常，每个小组包括一名跨部门专家以及来自各个清单部门（能源、工业、农业、LULUCF 和垃圾处理）的一或两名专家。每组选出两名主要审评专家，分别来自附件一国家和非附件一国家。对于集中审评（四到五个缔约方集中参审），秘书处通常邀请两名跨部门专家和两到三名审评专家负责一个部门。

对于到访审评，每组一般包括六名审评专家，每个部门一名专家和一名跨部门专家。审评的成果由专家审评小组共同准备，并通过年度审评报告的形式发布。报告最终发布到 UNFCCC 网站之前将提供给缔约方针对草稿提出意见。

（三）《京都议定书》下的报告和审评

对于同时是《京都议定书》缔约方的附件一缔约方，为了审评其在《京都议定书》下的履约情况，还需根据《京都议定书》第 7 条下“信息准备指南”提供额外的信息，在年度温室气体清单报告和国家信息通报中提供关于《京都议定书》执行情况的信息。对于国家信息通报来说，额外的信息如下：

(a) 国家系统的描述；

(b) 国家登记簿的描述；

(c) 解释缔约方如何利用《京都议定书》的机制来辅助其国内行动；

(d) 缔约方实施《京都议定书》第二条下政策与手段的信息；

(e) 描述缔约方国内和区域的法律体系建设、生效情况以及管理程序；

(f) 技术转让、能力建设、研究与合作，以及《京都议定书》第 10 条下的其他行动；

(g) 资金来源。

对于温室气体清单，补充信息包括：

(a) 评估温室气体清单方面的进步；

(b)《京都议定书》第 3 条第 3、4 段下 LULUCF 相关的温室气体的源与汇的信息和估算，包括核算上述信息并将其增加至分配数量或从分配数量中减去；

(c) 缔约方持有的和交易的《京都议定书》单位；

(d) 国家系统的变化；

(e) 国家登记簿的变化；

(f) 缔约方实施其气候变化减缓的过程中尽量减少其对发展中国家

社会、环境和经济负面影响的相关信息。

除了《京都议定书》下要求的年度额外信息之外，在第一承诺期内，同时是《京都议定书》缔约方的附件一国家还需提交两份一次性的报告，分别是第3条第2段背景下的可论证的进展报告（demonstrable progress report）和初始报告（initial report），并需单独提交报告以覆盖调整期（true-up period）期间活动的相关信息。调整期报告应于调整期结束时提交，以决定缔约方是否履行了其在《议定书》第3条第1段下的承诺。调整期报告必须包含承诺期内GHG排放的最终信息以及KP分配数量单位的增减情况，包括所有为了遵约而使用的分配数量单位的变动情况。

《京都议定书》下的审评适用于《〈京都议定书〉第8条下的审评指南》，并与《公约》下的审评一同进行，因此同时是《京都议定书》缔约方的附件一国家根据《公约》和《京都议定书》的要求，只需进行一次审评。

《京都议定书》下的审评相对来说更加严格并且有遵约机制作为补充。特别是，如果审评专家组识别出与强制性义务相关的、影响缔约方履约的问题，并且该问题在审评过程中没有得到解决，专家组则需要将其记录在最终报告中，列为“实施问题”。遵约委员包括促进事务组和执行事务组，所有的审评报告将被转给遵约委员会，之后再分配到合适的事务组。

遵约委员会执行事务组会考虑在审评过程中识别出的方法学和与清单要求实施相关的问题。审评过程中识别的“实施问题”如未得到解决则会影响到缔约方在《京都议定书》机制（排放交易、清洁发展机制、联合履约）下的资格认定，资格认定将延期至问题解决。

与《公约》下的审评专家组相比，《京都议定书》下的审评专家在排放核算上有专门的任务。如果审评专家组在审评过程中认定缔约方的清单不完整或者与IPCC方法学不吻合，他们将建议对相关的排放评估进行调整。类似的，对《议定书》第3条第3、4段下的相关活动

(LULUCF）的评估也可以进行调整。此外，审评专家组还负责评估缔约方持有和交易相关配额的信息，如果专家组在某一交易中发现了问题，会建议根据《京都议定书》配额核算规则进行修改。

与《公约》下的机制类似，秘书处对《京都议定书》的审评专家进行培训，内容涵盖国家系统、《议定书》第5条第2段下的调整手段、分配数量核算的程序、国家登记簿和分配数量信息的审评，以及《议定书》第3条第3、4段相关行动的审评。只有通过了考试的专家才能够参加《京都议定书》的年度审评。

（四）发达国家两年报和国际评估与审评

根据第1/CP.16号决定，发达国家（附件一国家）每两年提交一次报告以提供其全经济范围量化减排目标实施的相关信息。两年报的内容应包括：温室气体排放及趋势的信息、全经济范围量化减排目标，实现全经济范围量化减排指标的进展。

——温室气体排放及趋势信息：国家温室气体清单中关于温室气体排放及趋势的信息摘要；国家温室气体清单编制的组织安排信息；

——全经济范围量化减排目标：描述其全经济范围减排目标，包括目标的条件和假设；基准年、气体的种类和部门、全球增温潜势、LULUCF核算规则、利用国际市场机制完成目标的情况等；

——实现全经济范围量化减排指标的进展：减缓行动及其影响的信息（政策措施、按部门的减缓行动、按气体类型的减缓行动，国内履约、监测等机制建设，应对措施的经济和社会影响），市场机制及土地利用碳汇单位的使用情况（基准年的总温室气体排放量、土地部门碳汇信息，使用市场机制的情况），预测（2020年和2030年的温室气体排放预测信息），向发展中国家提供的资金、技术和能力建设支持的信息（附件一国家有向非附件一国家提供资金、技术和能力建设支持的义务，应报告相关信息以说明支持是新的和额外的）。

此外，也鼓励附件一国家在两年报中提供对其目标履约情况及相关科学要求的评估的信息，以及国内履约机制的相关信息。

发达国家两年报应通过国际评估和审评（IAR）机制进行审评。IAR 机制下审评的内容包括：减排目标以及源 / 汇信息，假设、条件及方法学，减排进展，资金、技术和能力建设支持。

IAR 分为两个阶段：首先进行技术审评，与国家信息通报和国家温室气体清单同时进行，最终结果为技术审评报告；第二阶段为多边评估，秘书处将在 SBI 会议期间组织对发达国家的履约进展进行审评，审评基于附件一缔约方的两年报、国家温室气体清单以及清单报告和国家信息通报，会议之前各缔约方可以书面形式提交问题和评论，被询问方须一一回答，最终多边评估的结果将以秘书处准备的深度审评报告、SBI 会议总结、多边评估环节的问题及回应，以及被审评缔约方在 SBI 会议两个月内提交的其他信息。

发达国家应于 2014 年 1 月 1 日前提交第六次国家信息通报，并以附件的形式提交第一次两年报。3 月到 8 月，秘书处组织审评专家完成第一组附件一国家两年报的审评报告；8 月到 10 月，进行第二组附件一国家两年报的审评工作；9 月至 11 月，其他缔约方与被审评的缔约方就两年报及其审评报告交换意见；12 月，在 SBI41 会议上，秘书处将组织针对第一组附件一国家两年报的国际评估与审评；2015 年 3 月，完成第一组附件一国家的 IAR 报告；6 月 SBI42 会议组织第二组附件一国家的 IAR，并于 9 月完成第二组附件一国家的 IAR 报告；若仍有未完成的 IAR 工作，将继续在 2015 年 12 月的 SBI 会议上完成。

（五）发展中国家两年更新报和国际磋商与分析

除上述国家信息通报外，第 16 次缔约方大会决定（第 1/CP.16 号决定），非附件一国家应该根据自身能力和所获资金支持，提交两年更新报告。第 17 次缔约方大会（决定 2/CP.17）进一步对两年更新报告作出决定，要求非附件一国家应该根据自身能力和所获资金支持，于 2014 年 12 月提交第一份两年更新报告，这份报告既可以作为国家信息通报的相关部分概要提交，也可以作为单独的一份更新报告提交。其中，最不发达国家和小岛屿国家可视情决定。有关非附件一国家缔约方两年更新报告的指南详见 2/CP.17 决定附件三。

非附件一国家两年更新报告内容包括：国情信息、国家信息通报的机构安排、国家温室气体排放清单、国内减缓行动及其效果，在资金、技术和能力建设方面的需求及已获支持、在两年更新报告准备和编写上已获的支持以及国内可测量、可报告和可核实（MRV）信息。

针对两年更新报告，第 16 次缔约方大会决定在附属履行机构下开展国际磋商和分析进程（ICA），国际磋商和分析以非侵扰性、非处罚性和尊重国家主权的方式进行，旨在通过技术专家与有关缔约方磋商进行的分析和交换意见，提高发展中国家减缓行动的透明度及其效果。

此外，为发展中国家缔约方进行的第一轮国际磋商和分析将在发展中国家缔约方第一轮两年更新报告提交 6 个月之内开始，并根据第一轮的经验，于 2017 年前对 ICA 的模式和指南进行修订。有关 ICA 的指南详见决定附件四。

第 17 次缔约方大会提出（2/CP.17, Annex IV），ICA 进程包括两个步骤：第一，ICA 下的技术专家小组（TTE）针对非附件一国家提交的两年更新报告提交一份技术分析报告草稿，草稿将提交给非附件一国家缔约方，邀请其在三个月内对草稿提出意见和反馈。在技术专家小组磋商后，最终形成一份摘要报告提交给附属执行机构（SBI）。技术专家小组的技术分析报告应考虑国家温室气体排放清单报告、减缓行动、行动效果及其影响、行动落实的 MRV 进程以及发展中国家所获支持等方面。第二，共享技术分析报告和两年更新报告的信息。附属执行机构会定期召开交流研讨会，方便各缔约方共享两年更新报告和技术分析摘要报告的信息。ICA 最终的成果文件包括上述技术分析报告的摘要报告和信息共享记录。

第 19 次缔约方大会对技术专家小组的构成和模式作出决定 (20/CP.19 Annex)，技术专家小组应由经提名被纳入《公约》专家名册的专家组成，并且，提名的专家要完成咨询小组制定的培训方案后才有资格在技术专家小组任职。此外，大会决定技术专家小组的总体构成中，来自非附件一国家缔约方的专家占大多数。技术专家小组应优先考虑并尽可能做到至少纳入一名专家咨询小组成员，咨询小组成员最多可占技术专家小组三分之一的名额。在确定技术专家小组其他专家时，应优先考虑曾担任专家咨询小组成员的专家。

三、不断提高行动力度的周期机制

随着全球气候治理体系逐渐由“自上而下”向“自下而上”过渡，“国家自主决定”逐渐确立为各缔约方确定应对气候变化贡献的核心原则，如何平衡解决“国家自主决定”与全球应对气候变化要求（包括但不限于实现 2℃全球温控目标）之间可能存在的差距，如何通过周期性的机制安排促使各方不断提高行动力度以满足应对气候变化要求成为各方争论的焦点问题之一。

在现有谈判中，发达国家力推在巴黎协议中设立“减缓循环”的周期性安排，即各方按照统一规则定期更新减排目标，国际社会开展定期审评以推动各方持续提高减排力度。欧盟更是力推对各方减排目标开展实施前审评和实施后审评。综合来看，目前关于不断提高行动力度的周期机制的争论焦点主要包括如下方面 ：

（1）周期性机制是否涵盖所有要素？是仅仅针对减缓还是覆盖减缓、适应、实施手段等各个要素？减缓 / 适应周期与实施手段的联系？

（2）所有缔约方是适用同样的贡献周期安排和审评机制还是应体现区分，并针对不同的承诺类型和承诺要素制定不同的贡献周期和审评机制？

（3）周期性提出贡献目标安排与审评 / 盘点机制的关系？是否要求各方根据审评 / 盘点的结果提出 / 更新目标？贡献进程与其他已有进程如 2020 年前的透明度安排，《公约》下的其他审评进程以及 IPCC 科学评估进程的协调和关系？

(4) 周期性机制是否需要统一各方的贡献周期？ 5 年还是 10 年？如何解决已有贡献周期不一致（2025 年和 2030 年）的问题？

(5) 关于周期性机制中的审评 / 盘点机制：① Whether: 即是否要有事前审评 / 盘点 / 磋商进程和中期 / 事后审评 / 盘点 / 磋商进程？② Why: 即为什么要有事前审评 / 磋商和中期 / 事后审评进程？设立事前审评 / 盘点 / 磋商进程和中期 / 事后审评的目的何在？是致力于澄清目标，提高透明度和增进各方对各自贡献的理解还是为了了解承诺 / 贡献的事实情况，促进遵约？抑或为了填补力度差距提高贡献力度水平？ ③ What: 即事前审评 / 盘点 / 磋商和中期 / 事后审评的对象和内容？审评 / 盘点 / 磋商是仅仅针对减缓还是覆盖减缓、适应、实施手段等各个要素？审评 / 盘点 / 磋商是针对单个缔约方的审评还是针对整体力度（加总贡献）的审评 / 盘点？是力度审评 / 盘点还是公平性审评 / 盘点，抑或实施情况审评 / 盘点？ ④ How: 即事前审评 / 盘点 / 磋商和中期 / 事后审评 / 盘点的方法或形式？审评 / 盘点是基于碳预算自上而下分配方案的评估？是基于特定指标体系如基于原则的参考框架的评估？是基于科学要求或各自责任、能力、国情的评估？还是主要基于澄清和对实施效果的评估？ ⑤ Who: 即由谁来开展事前审评 / 盘点 / 磋商和中期 / 事后审评 / 盘点及审评 / 盘点形式？是各缔约方的自我评估还是缔约方的评估？抑或特定专家组或者第三方机构的审评 / 盘点？还是主要基于交流、咨询、研讨？ ⑥ When：即事前审评 / 盘点 / 磋商和中期 / 事后审评 / 盘点的时间？事前审评 / 盘点是在初始贡献提出后立即审评，还是在 2015 年后进行？抑或设立相应的阈值指标作为开启审评 / 盘点的信号？事后审评是在承诺期结束后开展还是在承诺期中期开展？

针对上述问题，考虑到“不断提高力度的周期性安排”是《公约》下的，应受《公约》原则，特别是公平和“共区”原则的指导，且应符合德班平台的授权，涵盖减缓、适应、资金、技术、能力建设和透明度等全部要素。

同时，考虑到建立不断提高行动力度的周期机制的目的是致力于通过持续的、互动的、强化的和“干中学”的国际合作进程减少各国特别是发展中国家面临的社会、经济、技术等方面的挑战和不确定性，以提高个体和整体应对气候变化的力度，实现《公约》目标，因此，周期性机制安排应区分缔约方的后续贡献和审评 / 盘点安排。

具体而言，“不断提高力度的周期性安排”可考虑包含如下五个阶段（见图 5-5）：提出贡献（2015 年）；澄清贡献和贡献总体力度评估（2015 年）；确定贡献并写入 2015 协议（2015 年）；中期 / 事后盘点：贡献实施情况和审评 / 盘点，缔约方可考虑对目标进行自愿调整（2020 年后）；新一轮贡献目标和遵约（2030 年）。

（1）提出贡献：关于提出贡献，各缔约方应根据相关已有决议提交贡献及贡献相关的信息，贡献的通报应有助于澄清、透明度和对贡献的理解。

（2）澄清贡献和贡献总体力度评估：关于提出目标后的澄清进程设计，重在贡献最终确定前，为缔约方提供机会就贡献内容和力度进行澄清。其重点应放在增进对各国贡献的澄清和理解上，避免在《公约》下就各国个体贡献的公平和力度开展审评，各国可自愿对自身贡献公

图 5-5 不断提高力度的周期性进程安排

资料来源：张晓华，傅莎，陈济，祁悦，等．2015 协议框架和要素设计的基本设想 [R]. 国家气候战略中心政策研究报告，2014.5.27.

平和力度提供看法。进程设计要确保顺其自然，不必过于拘泥于形式，可通过《公约》外双边、小多边、大国间相互交流、公众评议等多种方式灵活开展。这一进程的目的是增进理解和提高透明度，而非强制性提高贡献。事前审评特别是基于技术评估小组和第三方机构的事前审评是没有授权的，不符合“非侵入性”(non-intrusive) 的原则，且容易使得谈判陷入混乱和无休止的争论，不利于达成协议。

对各国整体预期贡献的评估可通过秘书处综合报告或者在《公约》下由相关机构自行根据兴趣开展，仅作为参考信息。

（3）确定和写入贡献：贡献可作为协议的附件或者以单独的附件或者决定的形式确定和写入。贡献的批约和后续调整可借鉴KP的经验，也可更为灵活。

（4）中期 / 事后审评 / 盘点：为确保《公约》目标和全球2℃温控目标的实现，在各方确定2015协议下的贡献水平后，可在贡献执行中期（如2025年）对贡献实施情况进行盘点。

关于对贡献实施情况进行盘点的透明度的规则应建立在巴厘进程中关于透明度的机制安排基础上，同时加强对支持的透明度机制安排，并充分体现发达国家和发展中国家贡献的区别。

此外，可考虑建立战略审评的机制，旨在对各方实现《公约》目标的总体进展和发展路径变迁情况开展盘点，此盘点将涵盖所有要素，旨在基于审评 / 盘点分享优良经验，识别强化国际合作的机会和需求，以确保充分的国际合作。此审评 / 盘点将仅仅是技术审评，其产出也将仅仅作为信息工具，不作为缔约方后续行动的依据。

（5）新一轮贡献目标：在中期 / 事后审评 / 盘点后，各方可视情自愿调整贡献目标或提出下一阶段的贡献目标。

四、决策机制

《公约》下各缔约方谈判采用协商一致的原则。气候变化是影响到所有国家的全球挑战，因此这样的决策机制有利于使各方掌握对等的话语权，充分表达各自关切，但另一方面，协调各方利益诉求难度颇大，这种决策机制也给《公约》进程带来一定的风险。在哥本哈根，事实上当时绝大多数的缔约方都已经接受了《哥本哈根协定》，但出于对大国主导进程的不满，最终由于五个国家的反对，《哥本哈根协定》未能经过各方协商一致得以通过。协商一致的决策机制也被人诟病效率低下，一些学者建议采用世界贸易组织（WTO）体制下投票的方式进行决策，并且应按照温室气体排放规模分配决策权利。

但是气候作为全球公共物品，各方从中获利和受害都没有任何排他性，不论一国是否采取行动都将受到气候变化以及应对气候变化行动的影响，并且这种危害不仅限于经济方面的损失，还是对生存和发展的根本威胁，这与经贸领域的国际协定有根本性的区别，在国际气候进程中不应忽视任何一国的利益和诉求，国际气候谈判中协商一致的决策机制仍是不二之选。

第五节　关于巴黎协议的几点建议

巴黎会议能否取得成功，取决于各方能否在“共同但有区别的责任”原则落实、长期目标、不断提高行动力度等焦点问题上找到妥协。基于对这些问题的深入分析，本书提出了对巴黎协议的几点建议。

一、对几个关键问题的考虑

（一）“共区”原则的落实

“共区”原则是否以及以何种方式在2015年协议中体现，关系到未来全球气候治理是否继续基于《公约》的原则和条款并在《公约》

的框架之内，关系到各缔约方是否按照《公约》设定的责任和义务开展国家合作，也关系到全球气候治理是否能够沿着公正合理的方向发展，也是中国在协议谈判中的核心关切，也是所有发达国家以及部分发展中国家极力寻求变化的领域。德班决定（1/CP17）和多哈决定（2/CP18）明确要求和认可巴黎协议是在《公约》之下，必须在《公约》原则的指导之下开展德班平台的工作。而“共区”原则是《公约》的一项最重要、最基本的原则，必须在新协议中得到体现。

从目前的谈判形势看，各缔约方对 2015 年协议原则性提及“共区”原则已基本达成一致，利马决议借用中美气候变化联合声明的相关表述已找到各方均可接受的“共区”原则案文表述。巴黎会议则要解决如何将“共区”原则体现在各要素中的问题，而解决这一问题的核心是以何种方式体现“共区”原则。

目前在缔约方中至少有四种对“共区”原则的解读，除两分法外，还有自区分、中心区分和毕业制。自区分是各缔约方基于各自提出的国家自主贡献自我区分；中心区分（又有一说为“同心圆”区分）是巴西提出的概念，核心是要求各缔约方责任逐渐向发达国家现有责任趋同；毕业制是动态调整承担减排义务的国家名单，缔约方具备一定能力即可毕业加入承担责任的行列。落实“共区”原则应满足四个条件：一是要能够完全反应历史责任、能力、发展阶段和国情的差异；二是能够推动各方提高力度；三是能够反应动态变化的现实；四是必须政治上可行。自我区分无法反应历史责任，也无法推动各方提高力度；毕业制也同样无法提供各方提高力度的激励机制；毕业制和中心区分都涉及重新分组缔约方，政治上可行性较低。

发达国家与发展中国家的“两分法”是《公约》“共区”原则的基础和前提，它反映了世界上存在着发达国家和发展中国家的基本政治和经济现实。根据《公约》的要求，发达国家应率先大幅度减排，并向发展中国家提供资金、技术和能力建设支持。《公约》实施的 20 多年，在两分法的框架下，“共区”原则始终确保了《公约》的生机和活力，彰显了极大的包容性。事实证明，两分法并没有阻碍广大发展中国家根据自身能力为全球应对气候变化作出更多贡献，而是为发展中国家增强行动提供了政治保障和动力。经过 20 多年的实践，只有两分法已经用事实证明能够满足上述所有条件，应该成为巴黎协议落实“共区”原则的最佳基础。

按照两分法架构，可对协议各要素作如下考虑 ：

——减缓。推动协议明确发达国家 2020 年后承担全经济范围绝对量化减排指标，发展中国家 2020 年后在可持续发展框架下、在获得发达国家资金和技术支持情况下，采取多样性的强化减缓行动。在此基础上，协议可以鼓励“准备好的发展中国家”提出不包含资金和技术需求的行动目标并利用本国资源落实相关行动，同时允许小岛屿国家、最不发达国家和非洲国家在减缓贡献的类型和力度方面享受灵活照顾。

——适应。明确发达国家在历史上的累积排放已经造成了过去、今天和未来短中期内显著的气候变化并造成了负面影响，发展中国家抵御此负面影响的能力十分脆弱。在各缔约方均应加强适应以减少气候变化负面影响时应特别加强发展中国家的适应行动，重点加强早期预警预报能力和适应基础设施的建设，发达国家应对发展中国家开展适应行动提供资金、技术和能力建设的支持。

——资金、技术和能力建设。明确发达国家 2020 年后继续承担向发展中国家提供资金、技术、能力建设支持的国际义务，通过政治宣言或后续进程安排确定发达国家至 2030 年不低于每年 1000 亿美元的资金支持程度，强化现有相关机制安排，反对发展中大国 2020 年后承担出资、技术转让和能力建设等方面的法律义务，欢迎有关发展中国家向其他发展中国家提供帮助的行动，鼓励在气候变化领域南南合作行动对实现《公约》目标发挥补充促进作用，但强调其性质与发达国家向发展中国家提供支持的义务有本质区别。

——透明度。应坚持沿用《公约》下巴厘路线图针对发达国家和发展中国家设立的两套透明度规则，要求在协议中规定 2020 年后透明度规则在《公约》下现有机制安排的基础上进一步得到加强，由缔约方会议进一步明确加强的方向并细化、完善相关规则。

（二）长期目标

气候变化问题作为全球范围内的长时间挑战，需要建立全球长期大尺度视野为全球低碳转型和提高气候抵御力提供积极信号。2015 协议中有关长期目标的谈判所涉及的重点问题包括：①是否进一步强化温度目标到 1.5℃？②是否在温度目标基础上进一步提出长期量化排放目标，如 2050 年减排目标或者零排放（碳中和）甚至负排放目标？③是否建立长期减排 / 适应目标和资金、技术、能力建设支持等有关条件的联系，即是否提出长期资金和技术目标？④是否提出长期适应目标？目前科学上关于 1.5℃的研究仍有较大局限性和不确定性，还不足以用来支持决策。应避免将谈判重点放在确切数字上，使谈判陷入“谴责游戏”的僵局，影响协议达成。

解决长期气候问题的关键是各国创新发展路径，建议将有关长期目标的谈判重点放在如何实现长期发展路径转型上，认为巴黎协议重点是向全球传达积极信号，表达转型决心。巴黎协议应重申 2℃目标，强调绿色低碳转型大方向，适时以适当方式提出全球发展路径创新，走绿色低碳可持续发展且具备较高气候抵御力的创新发展道路的长期努力愿景，释放积极信号。发达国家缔约方和准备好的发展中国家缔约方适时提出本国创新发展路径、应对气候变化的中长期战略、规划或方案，强调发达国家及相关国际机构应当为发展中国家制定实施上述战略、规划或方案提供资金、技术和能力建设等方面的支持。巴黎协议应包含定性描述的全球适应愿景。

（三）不断提高行动力度

自华沙会议提出各缔约方准备各自国家自主贡献以来，以欧盟为代表的发达国家充分利用其解读国家自主贡献概念的机会，在将国家自主贡献仅仅限定在减缓范围并抹杀发达国家与发展中国家区别的前提下，设计出了制定、提交、实施、审评和更新国家自主贡献的循环体系，重点解决各方行动总和与2℃全球温升控制目标要求差距的问题。随着越来越多的缔约方提交国家自主贡献，以自主贡献为基础的自下而上的减排模式基本形成，以避免这种减排模式很可能带来的力度不足问题为理由，欧盟力推通过审评而不断更新行动力度的机制安排，一方面在保持国际气候治理体系效力和持续性方面具有一定的道理，但同时因其抹杀发达国家与发展中国家自主贡献的区别且在自主贡献中不能平衡反映适应、技术、资金和能力建设等其他要素，另一方面有可能在他们建议的循环机制下使未来国际气候进程加速背离《公约》，犯下方向性错误。

考虑到《公约》第4条第1款已有关于所有国家制定、发布、实施和定期更新国家减缓和适应气候变化国家方案的规定，所有缔约方应根据《公约》要求，结合华沙会议和利马会议国家自主贡献的有关决议规定，准备、通报、落实并定期更新国家自主决定的减缓和适应贡献，发达国家按照《公约》要求对发展中国家自主决定的减缓和适应贡献提供资金、技术和能力建设支持。可在协议中要求各方在提交贡献时提供相关信息，以增加贡献的透明度，并由缔约方会议根据利马会议决定进一步细化信息要求，防止强制各方遵循统一的信息要求或模板。

贡献的时间跨度问题可通过缔约方决定予以解决。可建议在巴黎成果中规定，到2020年，提出到2025年贡献的国家提出到2030年的贡献，提出到2030年贡献的国家细化到2025年实施贡献的措施，缔约方会议将于2030年前适时邀请各方准备并提交2030年后自主贡献。

二、核心要素的设计建议

（一）减缓

1．核心协议涉及的问题

巴黎协议在减缓要素上的核心是确保各方行动力度不倒退及持续

提高。由于利马会议决定已提出各方自主贡献要在其现有承诺基础上有所进步，各方可在此基础上寻找合适的表述。

2．缔约方决定涉及的问题

——资金和技术支持“条件”的问题。巴黎会议决定应允许发展中国家保留提出包含资金、技术和能力建设支持条件的减缓贡献的权利，或发展中国家在发达国家进一步的资金、技术和能力建设支持下进一步提高减排力度或作出额外减缓贡献。

——减缓贡献可量化、可比较、可汇总的要求。发达国家提出此要求意在按照统一规则将各方减缓贡献折算成绝对减排量。考虑到发展中国家减缓贡献类型多样，对可比性问题有较大关切，且可量化、可比较和可汇总的要求也与自主决定的安排相违背，巴黎会议决议应突出减缓贡献在推动绿色低碳发展方面的效应，防止通过简单折算各方排放及减排量来比较评价各方贡献。

——市场机制。考虑到市场机制的设计和建设有待讨论的问题多且复杂，可在巴黎协议达成后再讨论市场机制相关问题。

——土地利用。土地利用主要涉及农业和林业问题。欧盟、澳大利亚、新西兰等发达国家要求协议规定各方可通过土地利用变化完成减缓贡献。阿根廷等发展中国家反对将农业纳入减排行业范围，不同意使用土地利用表述，要求使用减少毁林排放的表述。各方分歧较大，从政治现实出发，巴黎会议决定可暂不提及该问题或原则性提及土地利用和减少毁林排放。

（二）适应

——国家自主决定的适应贡献。发达国家在本国开展适应行动的基础上，在技术、资金和能力建设层面为发展中国家特别是小岛屿国家、最不发达国家和非洲国家制定和实施国家适应计划、开展相关项目提供支持。发展中国家在已有的 NAPs 和 NAPAs 基础上，在发达国家的支持下，通过国家适应计划识别需求和障碍，加强行动，提高能力。发展中国家适应气候变化的关键是增加发展中国家预警预测、防灾减灾等方面的能力，发达国家应为此提供充足的资金、技术和能力建设支持。关于适应贡献的审评和定期更新，应侧重经验教训交流分享及对发达国家提供适应支持规模和发展中国家适应行动所需支持满足情况的评估上。

——损失与损害。推动华沙损失损害国际机制的切实运转，根据发展中国家的实际需求，进一步完善机制。

（三）实施方法（资金、技术和能力建设）

——资金。要求发达国家明确2020年后资金的规模以及出资的时间表和路线图，2020年后的资金应在每年1000亿美元的基础上逐年提升，强调发达国家所提供的资金应主要来自公共部门，还应强调绿色气候基金的运行应在《公约》下，接受《公约》指导。

——技术。敦促发达国家积极解决气候技术转让中的资金和知识产权问题，推动气候技术的联合研发，保障气候技术转让项目的资金支持，建立技术转让绩效评估体系，要求发达国家提供可转让的气候技术清单并定期更新。

——能力建设。应要求建立专门关于能力建设的国际机制，制定并实施能力建设活动方案，加强发展中国家减缓和适应气候变化能力建设。

（四）透明度

巴黎协议应建立评估总体行动力度的机制安排，有关具体设计内容如下：

——评估内容。应包括减缓、适应及抵御能力、资金技术支持三个方面，评价各方落实贡献、实施《公约》的总体进展情况。

——整体审评和个体审评。巴黎协议应避免评估个体力度，但可盘点各国行动总体效果。

——评估目的。巴黎会议应明确评估的目的是促进贡献的澄清、透明度和理解，交流创新可持续发展方面的经验、政策措施和最佳实践，识别发展中国家实施面临的困难、障碍和需求，发掘加强国际合作的机会。

（五）法律形式

巴黎会议成果可以核心协议加缔约方决定的形式。在法律约束力的问题上，不应仅仅聚焦在减缓要素，适应、资金、技术、能力建设和透明度等德班会议有关决议确定的协议要素都应具有法律约束力。

【参考文献】

[1] 李玉娥，马欣，何霄嘉．《巴厘行动计划》以来适应气候变化谈判进展及未来需求分析 [J]. 气候变化研究进展，2014，10（2）：135-141.

[2] 祁悦，阿柔娜，刘林蔚，张晓华．《公约》下报告和审评机制综述 [R]. 国家气候战略中心政策研究报告，2014.5.20.

[3] 徐丽丽，张晓华 .2015 协议中关于适应要素各种提案简析 [M]// 李俊锋，邹骥，徐华清，等．气候战略问题研究 2014. 北京：中国环境出版社，2014.

[4] 张晓华，傅莎，陈济，祁悦，等．2015 协议框架和要素设计的基本设想 [R]. 国家气候战略中心政策研究报告，2014.5.27.

[5] 张晓华，祁悦．“预期的国家自主决定的贡献”概念浅析 [M]// 李俊锋，邹骥，徐华清，等．气候战略问题研究．北京：中国环境出版社，2014.

[6] 张晓华，祁悦．应对气候变化国际合作进程的回顾与展望 [J]. 气候战略研究，2015(13、14)（总第 76、77 期），2015.8.24.

[7] 张晓华，祁悦，陈济，等．IPCC 第五次评估第二工作组报告主要结论解读 [M]// 李俊锋，邹骥，徐华清，等．气候战略问题研究 2014. 北京：中国环境出版社，2014.

[8] IPCC, 2014, Climate Change 2014: Impacts, Adaptation and Vulnerability. Cambridge University Press, Cambridge, United Kingdom and New York, NY, USA, 2014.

[9] Spencer, T., Fu Sha, Zou JI, et al. (2015), Regular Review and Rounds of Collective Action and National Contributions the 2015 Climate Agreement: A Proposal, Working Papers, IDDRI, Paris, France. Available at: http://www.iddri.org/Publications/Regular-Review-and-Rounds-of-Collective-Action-and-National-Contributions-the-2015-Climate-Agreement-A-Proposal.

[10] Spencer, T., Zou, Ji, Fu, Sha, Colombier, M., Ribera, T. (2014), The mitigation framework in the 2015 climate change agreement: from targets to pathways, Working Papers n° 07/14, IDDRI, Paris, France, 20 p.

[11] The African Group (2013). Submission by Swaziland on behalf of the African Group on adaptation in the 2015 Agreement [online] Available at http://unfccc.int/bodies/awg/items/6656.php#workstreamone.

[12] UNEP, The Adaptation Gap Report 2014. UNEP: Nairobi. Available at:http://www.unep. org/climatechange/adaptation/gapreport2014/portals/50270/pdf/AGR_FULL_REPORT. pdf .
[13] Zou Ji , Fu Sha. The challenges of the post-COP21 regime: interpreting CBDR in the INDC context[J], International Environmental Agreements: Politics, Law and Economics, Volume 15, No.3, September 2015, 10.1007/s10784-015-9303-8, available: http://link.springer.com/article/10.1007/s10784-015-9303-8.

第六章

全球治理下的气候治理

Chapter Six

Climate Governance under the Global Governance

摘要

伴随着全球化进程的不断深入，经济、安全、贸易、投资等全球特性日益突出，推动世界建立新型治理体系，以满足解决不同的全球性问题的需要。这些专门领域的全球治理体系在全球治理的大格局下，反映全球治理体系的不断演进和发展，受到全球治理格局的制约，同时也互有联系，相互影响。气候治理一方面受到全球治理体系下相关领域多边进程和国际机制的影响，另一方面，随着全球气候治理中的主要国家主体在越来越多的问题上能够寻找到合作的领域和方式，全球气候治理正对全球治理体系产生外溢效益，部分发展中国家期望通过在气候变化领域的合作更多的和更深入的参与全球治理，一些体量较小的发达国家也期望通过在气候领域积极行动，争取提升在全球治理中的地位。本章介绍了第二次世界大战以来全球治理体系的形成及近20年的沿革及与气候治理的关系，并选取了安全、可持续发展、贸易和金融四个专门领域重点论述了相关治理体系与气候治理的联系和影响。

第一节　全球治理与气候治理

一、全球治理体系

二战以后，美国通过联合国、国际货币基金组织、世界银行、关税和贸易总协定以及地区安全和军事防卫联盟等，建立了一整套战后国际秩序。冷战之后，美国巩固和强化了这一优势地位。21 世纪以来，随着经济全球化以及中国、印度等为代表的一批高速发展的新兴经济体的崛起，财富和经济相对实力发生了从西方向东方的具有深远影响的历史性转移。大量的国际趋势研究报告都认为，到 2025—2030 年这一时间段，二战以后所建立的国际体系将发生重大变化，一个全球性的多极体系将占据主导地位。各个国家或者国家集团之间的整体实力对比将发生变化，权力分散和多极化趋势日益明显。伴随着全球格局的多极化，全球治理体系将发生重大但非根本性的变化。

美国为主导的全球治理体系正日益受到快速增长的发展中国家崛起带来的挑战。新兴国家在全球体系中有着日益突出的利益诉求，同时由于实力的增长，也具备维护自身全球利益的能力，因此提出了对现有国际治理体系改革的强烈要求，希望在国际议程设置中获得更大的发言权，以便与自身日益增长的综合实力相匹配。

20 世纪末期，不断的科技创新带来了世界生产力的加速发展，形成了以经济全球化为标志的整体发展趋势。伴随全球化，现存国际治理体系更加开放、更多依赖以大量国际规则为基础的多边协调机制。新兴国家的崛起，一方面对现有全球治理体系提出了挑战和改革的诉求，另一方面由于各个国家的相互依存，新兴国家的进一步增长和发展，又深深依赖于现有的全球治理体系，尤其是世界经济体系中的投资和贸易，因此新兴国家将倾向于在现有体系中寻求更为有利的相对位置，而不是试图颠覆现有体系。中国国家主席习近平在第 70 届联合国大会的一般性辩论中提出，“中国将始终做国际秩序的维护者，坚持走合作发展的道路，继续维护以联合国宪章宗旨和原则为核心的国际秩序和国际体系”。这体现了中国对现有国际治理体系的立场，也代表了广大发展中国家的立场。

全球气候治理体系是在全球治理体系不断演进和发展的过程中建立、完善和运转的，反映了全球治理体系的变迁，受到气候治理格局的制约，与全球治理下的安全、经济、贸易、投资等国际体系以及航空航海和邮政等具体国际合作领域相互影响和联系。

二、全球气候治理的外溢效应①

当前的经济发展无法摆脱对化石燃料的依赖，经济活动所赖以进行的现有技术体系和能源资源禀赋还不能避免排放温室气体。因此在全球变暖对温室气体浓度提出上限要求的情况下，温室气体排放容量正在成为新的稀缺战略资源，事关经济发展战略空间和发展权利。全球气候治理体系关系到国际责任体系建立过程中的权利分配问题，已经成为国际战略的新边疆，并将重新定义长期竞争力，具有广泛而重大的国际贸易、国际投资含义，进而关系到国际战略竞争的格局。

从历史排放轨迹的全貌来看，目前大气中由人类活动导致的温室气体存量绝大部分仍然是由发达国家造成的，发达国家对此负有不可推卸的责任。从 1751 年到 2010 年附件一国家能源相关 CO_2 累积排放占全球总累积排放的比重达到约 70%。如果考虑消费端的排放特别是产业和贸易转移排放带来的影响，则发达国家的历史责任更高。由于

①本节内容部分参考了由国家气候战略中心与中国人民大学公共承担的国家发改委中国低碳发展宏观战略研究项目“中国低碳发展国际环境课题“的研究成果。

全球 2℃温升目标对全球累积排放提出了上限要求，发达国家对历史累积排放的过多占有，必然严重压缩发展中国家未来排放的增长空间。在现有发达国家所开创、主导、扩散和强化高碳发展路径的能源技术体系下，温室气体排放作为能源资源利用不可避免的副产物，必然与发展权益紧密相关。

随着全球范围内的城市化，城市人口增加，新兴国家中等收入人群扩大，消费升级，将导致对全球重要资源，包括能源、矿产、粮食以及水资源等的需求激增，各国围绕海洋、大气、资源、环境等全球公共物品的争夺将更为激烈。目前的全球公共物品治理架构和程序主要由发达国家主导，而新兴国家由于经济快速增长，对全球公共物品的需求更为迫切，而且供应的稳定和安全性问题日益突出，因此希望全球公共物品供应的治理架构和程序能够充分反映自身诉求。新兴国家希望能主动参与到应对全球资源和环境问题的全球解决方案之中，在诸如气候变化等国际谈判中发挥更积极的作用，而不是将全球公共物品的配置权让给发达国家。

当然，全球公共物品治理架构的改革和转型，必然伴随着发达国家与发展中国家以及新兴国家之间的利益冲突。但是，必须意识到，气候变化问题是全人类共同面临的重大挑战。由于气候变化问题影响的深度和广度，任何一个国家都不能独善其身，需要世界各国通力合作，携手应对。因此，虽然存在发达国家与发展中国家之间关于排放责任和排放权益相关的竞争和冲突，但是构建新型国际合作关系，仍然是全球针对这个问题的共识。全球气候治理体系作为变革中的全球治理体系中的一个重要性日益突出的有机组成部分，其历史沿革以及创新改革诉求正深刻地反映了 21 世纪以来全球政治和力量格局的变迁。正如《蒙特利尔议定书》作为首个成功实施的全球环境协议对气候变化公约的借鉴作用，由于气候变化问题的紧迫性以及在这个问题上的全球共同利益，全球为共同应对气候变化而构建新型国际合作关系更具有可行性，这也将对其他全球治理领域形成很好的示范。从气候变化问题可以看出未来全球治理体系的新的格局。这也是中国国家主席习近平在 70 届联大一般性辩论发言中所倡导的“建立合作共赢为核心的新型国际关系，打造人类命运共同体”的重要体现。

由于中国整体实力上升，未来权力分散化和多极化的全球政治格局要求中国在全球治理体系中处于更为关键的位置，拥有更大的话语权和影响力，相应在全球范围内也需要承担更多的责任，因此中国在各种政治、经济和环境等全球议题的讨论中，将面临更大的责任压力。而随着中国地位的上升，中国需要主动承担责任，参与塑造稳定的国际环境。这一角色转变对于中国而言提出了很多新的挑战，从政府、企业、非政府组织到国民，都需要适应这一重大角色转变，并且需要大量的熟悉国际规则、能够参与塑造国际规则的机构和个人。而这一转变带来的机遇是，在权力分散化和多元化的全球格局中，中国将不再仅仅是全球规则的接受者，而将转变为全球规则的重要制定者以及全球秩序的维护者。换言之,中国也很可能成为全球规则的既得利益者。中国可以凭借在各个国际机构中地位的提升，营造有利于自身发展转型和实现战略发展目标的国际环境。

因此，在主动参与创新型全球气候治理体系的构造中，中国事实上也在参与对二战以来所建立的现有全球治理体系的改革以及新的全球治理体系的构造，这正是中国所倡导建立的以合作共赢为核心的新型国际关系的重要组成部分。

作为世界第二大经济体，最大的发展中国家，最大的温室气体排放国，中国在气候变化问题中的角色是多元的。一方面中国是易受气候变化不利影响的国家，积极应对气候变化符合中国自身利益，也有利于中国自身社会经济发展转型。另一方面，中国作为发展中国家，发展任务仍然很繁重，在国际社会应承担与自身国情相适应的义务。最后，中国不仅要尽可能承担更多的减排责任，在支持其他发展中国家应对气候变化方面，也可能要在资金、技术和能力建设方面提供更多的支持。所以，中国既是温室气体排放大国，也是受气候变化不利影响的大国，既是气候变化领域的受援国，也是南南合作的援助国，需要全面平衡地看待自身在气候变化问题上的权益与责任，在全球气候治理体系中的角色与作用。

中国已经从全球气候体制的参与者逐渐转变为建设者和主导者，在全球气候体制的构建中发挥越来越重要的领导作用。中国根据自身国情、发展阶段、可持续发展战略和国际责任担当，确定了到 2030 年的自主行动目标，包括二氧化碳排放 2030 年左右达到峰值并争取尽早达峰，单位国内生产总值二氧化碳排放比 2005 年下降 60%—65%，非化石能源占一次能源消费比重达到 20% 左右。中国的目标兼具减排雄心和公平考虑，给广大的发展中国家甚至很多发达国家作了很好的示范。中国还宣布拿出 200 亿元人民币建立“中国气候变化南南合作基金”，支持其他发展中国家应对气候变化，包括增强其使用绿色气候基金资金的能力。

更重要的是，中国自身践行生态文明，探索创新发展路径，对其他国家形成很好的示范作用。正如国家气候战略中心针对中国自主减排贡献的评论所言：“作为最大的发展中国家，中国被很多发展中国家视为发展样板。中国可通过创新发展路径，垂范一条低碳可持续发展示范路径，为后发的发展中国家提供可资借鉴的示范模型和经验，支持这些发展中国家规避传统高碳路径依赖和锁定效应，尽早走上‘低污染低排放’的高效创新发展道路。同时，中国也能通过自身转型促进全球转型，为重塑全球发展路径作出贡献。”①

由于全球气候保护与全球社会经济发展之间的关系，全球气候治理不能单纯视为环境问题，而应首先视为发展问题，问题的根源以及最终解决出路也都在发展。这就意味着全球气候治理在某种意义上也是全球发展治理的有机组成部分。全球气候治理的结构变迁，有利于

① 傅莎、邹骥、刘林蔚等，解读中国国家自主贡献（INDC）目标 [R]. 国家气候战略中心政策评论文章，2015 年 6 月，http://www.ncsc.org.cn/article/yxcg/zlyj/201506/20150600001484.shtml.

向各国决策者、投资者和消费者发出明确的引导信号，有利于重塑世界和各国的经济，改变人类发展进程和创新发展路径。正是由于全球气候治理具有综合战略意义，中国已经在全球气候治理中逐渐发挥主导作用，树立了负责任、公平正义的良好形象，可以全球气候体制作为重要平台，谋求在更广泛国际事务中提升自身地位和影响力，促进并提高在更广义的全球治理中的地位和话语权。[①] 而中国深度参与全球气候治理，也在思想观念、意识、国内体制机制、人才队伍培养等方面形成铺垫，为进一步拓展其他全球治理领域奠定基础。

2014 年习近平在中央外事工作会议中提出“中国必须有自己特色的大国外交”。气候外交既是构建全球气候治理体系的重要手段，也是有中国特色大国外交的重要组成部分。

气候变化作为全球各国面临的共同挑战，引起了大国的广泛关注。主要国家经历了 2008 年的全球金融危机及之后的普遍衰退与复苏乏力，也将绿色低碳发展作为促进发展转型的重要抓手。随着 2015 年底巴黎气候大会临近，全球各国间密集开展气候外交。气候变化已经成为多边和双边外交的重要议题。

在中国社会经济发展进入新常态背景下，低碳发展可望成为中国经济升级的助推器和切入点，也是中国顺应国际低碳发展和科技革命潮流、提高在国际领域低碳竞争力的必由之路。习近平主席说，应对气候变化是中国可持续发展的内在要求，这不是别人要我们做，而是我们自己要做。因此，中国积极稳妥地设定了国家自主贡献目标，探寻低碳发展路径的创新，在实践中不断摸索行之有效的战略和政策措施，并加强国际合作，形成内外互动的综合战略，使中国国内可持续低碳发展的努力和构造全球应对气候变化的治理有机结合起来。加强气候变化领域的多边与双边合作，特别是与重要的大国之间开展务实

① 邹骥、傅莎、陈济等，基于“发展路径创新”的后 2020 年国际减排机制设计 [R]. 国家气候战略中心研究报告，2014 年 4 月 .

双边合作，包括中美气候变化合作、中欧气候变化合作等，正是这一战略的重要组成部分。通过加强中美、中欧等双边气候变化合作，可望实现下列目标：

第一，通过双边合作，可充分借鉴欧美发达国家在应对气候变化立法、能源与气候变化管理体制、政策制定与协调、低碳技术研发、商业模式创新、价值链整合、公众意识提高等方面的先进经验，从而提高中国自身应对气候变化领域的政策制定能力和管理水平，增强创新能力。

第二，中美、中欧乃至中德、中英、中法等双边关系都是中国特色大国外交的重要组成部分，具有重要的战略意义，也有很大的发展潜力。中国与这些国家在气候变化问题上互补性很强，共同利益多，合作空间大，合作氛围好，将成为发展更加密切的双边政经关系的一个支柱，加强国家之间的互信和相互尊重，并为加强在其他领域的合作做好铺垫。

因此，加强双边气候变化合作，可以作为联合国多边机制的有益补充，有利于中国借鉴欧美发达国家经验实现国内低碳发展转型，有利于双方建立更加紧密的双边政经关系，有利于开展平衡和多元的气候外交，构建有利于中国自身发展的全球气候治理体系，并改善在气候变化问题上的国际舆论环境。

以中美关系为例，作为世界上最大的发达国家和最大的发展中国家，作为守成大国与崛起大国，中美双边关系某种程度上被视为 21 世纪最重要的双边关系，决定了世界的未来。

而 2014 年以来，中美两国将应对气候变化作为两国最重要的合作领域之一。双边对话与合作不断深化，体现了多层次、多渠道、多主体、多频率等特点，基调已由国际气候谈判互为对手转变为务实有效的双边合作，而且合作的深度与广度前所未有。应对气候变化议题已经被纳入中美双边最高议事日程，成为中美建立新型大国关系的重要内容。

由于气候变化问题所具有的长期战略意义，中美两国在气候变化问题上互补性较强，共同利益较多，合作空间和潜力较大，所涉议题敏感程度较低，不易受短期政治形势的影响，稳定性较强，合作氛围较好，因此可能“成为双边关系的一个支柱”，帮助建立两国之间的互信和尊重，成为加强两国在国际政治、经济、金融、安全等领域合作的铺垫，促进中美新型大国关系的构建。

中美两国在气候变化全球治理体系中具有突出地位。中美加强应对气候变化合作，是两国国内气候变化政策的重要延伸，有利于各自减轻在国内采取强有力减排行动和签署国际气候协议的压力。还可营造一个能够彼此充分交流和沟通并且相互信任的政策对话环境，避开国际传媒的聚光灯，避免在国际场合中无谓的争吵和相互指责，促进双方加深了解，理解各自困难，求同存异，寻求务实解决方案。这将影响全球气候变化谈判格局，对其他气候谈判中的利益集团带来示范效应，从而有利于 2015 年全球协议的达成。因此，2014 年《中美气候变化联合声明》中明确提出“双方重申将致力于为 2015 年全球应对这一挑战的成功努力作出重要贡献”。全球气候治理在某种意义上是全球发展治理的有机组成部分。中美通过共同促进全球气候治理体系的创新，树立了榜样，有利于推动全球格局的再平衡和新型全球治理体系的构建。中美共同合作，参与全球气候治理体系的构建，为两国合作，共同解决其他全球性问题树立了一个榜样，具有非常重要的意义。

从中美气候变化合作的成功经验可以看出，加强对全球气候治理体系的参与，不仅有利于中国在更广泛的治理领域提高话语权，也有利于中国开展更加务实的多边与双边外交，成为中国特色大国外交战略的重要组成部分，为塑造有利于中国和平崛起的环境作出贡献。因此，看待全球气候治理，需要以更加宽广的视野，意识到其所具有的外溢效应，以更加积极主动的姿态，参与国际政治、经济、发展、安全、环境等领域规则的重塑。

第二节　全球安全与气候治理

一、非传统安全的概念

非传统安全（non-traditional security，简称NTS）是指由军事、政治和外交冲突带来的传统安全威胁之外的，对主权国家和全球社会整体生存和发展构成威胁的各种因素。1983年，美国著名国际政治经济学家R.乌尔曼在《国际安全》季刊上发表《重新定义安全》一文，明确提出国家安全以及国际安全概念应予扩大，使之包容非军事性的全球问题，如资源、环境、人口问题等。2003年，伦敦政治经济学院B.布赞在《新安全论》中提出五个相互关联的安全领域，包括军事安全、政治安全、经济安全、社会安全和环境安全，形成新的安全观，并对传统的安全管理提出挑战。冷战结束以后，美国成为世界上唯一的军事强国，反而凸显了其所面临的各种非传统的安全威胁。因此，美国也是较早在国家安全战略中突出强调非传统安全的国家。中国则在2011年9月发布的《中国的和平发展》白皮书中将恐怖主义、大规模杀伤性武器扩散、金融危机、严重自然灾害、气候变化、能源资源安全、粮食安全、公共卫生安全等攸关人类生存和经济社会可持续发展的全球性问题归纳为人类面临的共同安全问题，并认为任何国家都不可能单独解决这些问题，国际社会必须携手应对。2013年中共十八届三中全会决定设立国家安全委员会。习近平总书记在2014年国家安全委员会第一次会议上指出，“当前我国国家安全内涵和外延比历史上任何时候都要丰富，时空领域比历史上任何时候都要宽广，内外因素比历史上任何时候都要复杂”，并且要求“既重视传统安全，又重视非传统安全，构建集政治安全、国土安全、军事安全、经济安全、文化安全、社会安全、科技安全、信息安全、生态安全、资源安全、核安全等于一体的国家安全体系”。可见，在传统安全基础上，应对非传统安全也在中国被提上了最高议事日程。

二、气候变化是典型的非传统安全问题

气候变化问题因其影响的深度和广度，被国际社会和主要国家视为典型而又重要的一种非传统安全问题。早在 2003 年，美国五角大楼报告《气候突变的情景及其对美国国家安全的意义》就描述了温度升高一旦超过某一阈值之后的气候突变情景，包括食物短缺、关键区域淡水的供给数量和质量下降、洪水和干旱事件更为频繁以及由于大范围的风暴袭击对能源供应带来的威胁等诸多影响，并探讨了这些影响如何对地域政治环境的平衡构成潜在威胁，以及由于资源紧张而导致的冲突、战斗甚至战争。报告因此提出，“由于这些可怕的潜在后果，气候突变的风险尽管不确定性很大，而且发生的可能性很小，但是完全应该将对其的关注提高到美国国家安全的高度，不应只停留在科学争论上。”

IPCC 第五次评估（AR5）报告阐述了气候变化带来的风险，特别提出温升在 1℃—4℃时一些物理系统或生态系统可能处于较高的突变或巨变的危险之中。IPCC 报告提出，如果全球平均温度较工业化前水平上升达 4℃或更高，全球气候变化风险为高至非常高水平，会包括对独特和受到威胁的系统的严重和广泛影响、大量物种灭绝、对全球和区域粮食安全的大风险以及损害人类正常活动的高温高湿。

中国也是最易受气候变化不利影响的国家之一。《国家应对气候变化规划（2014—2020）》认为“气候变化关系我国经济社会发展全局，对维护我国经济安全、能源安全、生态安全、粮食安全以及人民生命财产安全至关重要”。可见，气候变化对我国粮食安全、水资源安全、生态安全、环境安全、能源安全、重大工程安全、经济安全等传统与非传统安全都将产生严重威胁，并对国家整体安全提出了严峻挑战。

《第三次气候变化国家评估报告》认为，中国在特殊的地理环境背景下，更容易受到气候变化的影响。中国平均气温的升温速度高于全球平均水平，到本世纪末，气温最高或将上升5℃，极端天气将频繁出现，灾害损失呈上升趋势。作为《第三次气候变化国家评估报告》的特别报告，中国工程院还发布了《气候变化对我国重大工程的影响与对策研究》研究报告，从水工程和水安全、道路工程、能源工程和安全、沿海工程安全、生态环境与安全、电网安全六个方面，针对气候变化对重大工程的影响进行系统梳理研究。譬如向西部推进的高速铁路网就将面临诸如沿线大风、强降雨、冰雪以及雷害等干扰的气候环境。对于沿海地区，除了考虑气候变化引致的海平面上升，还需要考虑气候变化可能引发或者加剧的风暴潮、海岸侵蚀和海水入侵等海洋灾害，这些海洋灾害会对沿海基础设施等带来重大影响。此外，中国由于能源资源与电力负荷分布的不均衡，建设了一系列的西电东送输电通道，这些通道大都跨越多地理条件和多气候复杂环境的地域，且输电距离远、容量大，也易受不利气候的影响。

反之，正如2014年《中美应对气候变化联合声明》中所说，“经济证据日益表明现在采取应对气候变化的智慧行动可以推动创新、提高经济增长并带来诸如可持续发展、增强能源安全、改善公共健康和提高生活质量等广泛效益。应对气候变化同时也将增强国家安全和国际安全”。

三、气候变化等非传统安全问题依赖国际合作

作为典型的非传统安全问题，气候变化问题具有全球和代际外部性特征，具备明显的跨国性和全球性特征，必需依靠国际合作加以解决。非传统安全问题治理难度大、时间尺度长、综合性强，需要创新思维。非传统安全问题体现明显的社会性和主体多元性特征，无论是受影响者还是采取应对措施的行为主体，都不限于主权国家，需要由非国家

行为体如个人、组织或集团等合作，共同采取行动。非传统安全存在很大的不确定性，譬如气候变化存在超过某个风险阈值之后发生突变的风险，需要全球合作针对主要的敏感生态系统以及临界点构建全球观测与预警网络，并采取预防性措施；非传统安全与传统安全之间还可能存在相互转化，非传统安全问题转化为依靠传统安全的军事手段来解决，甚至演化为武装冲突或局部战争。美国五角大楼报告《气候突变的情景及其对美国国家安全的意义》基于气候变化对全球和局地承载能力的影响，认为如果发生全球性恐慌蔓延，会导致各国采取防守和进攻两种基本策略，极端情况下会发生全球性军事冲突这一场景。防守策略是指拥有能源的国家可能在其周边建立起实质上的堡垒，以保护自身的资源；进攻策略则是能源资源稀缺的国家，特别是那些与邻国具有仇视历史的国家，可能发动战争以获取食物、洁净水源或能源。原本不会结盟的国家可能会首先考虑结盟这一自卫手段，而结盟的目的仅仅是为了生存，而不再是宗教、意识形态或民族尊严。

总之，由于气候变化等非传统安全问题对全人类形成挑战，应对过程中，需要树立高度重视非传统安全的新的安全观，将其和传统安全问题一道纳入全球以及各国关于安全的最高议事日程中，并寻求全球的通力合作。

四、构建多层次多主体的非传统安全国际治理体系

作为全球性安全合作组织中的权威，联合国已经成立 70 周年。在全球化日益发展、国家安全问题日趋复杂和多样化、非传统安全问题日益突出的背景下，联合国在国际事务中仍然具有不可替代的作用。中国外交部发表的《中国关于联合国成立 70 周年的立场文件》中赞赏“70 年来，作为最具代表性和权威性的政府间国际组织，联合国依靠和平与安全、发展、人权三大支柱，为促进人类进步事业发挥着不可替代的作用，并取得了巨大成就”，并建议“安理会作为国际集体安全机

制的核心，在维护国际和平与安全上负有首要责任。加强国际集体安全机制，推动政治解决地区热点问题，让安理会更好地履行《联合国宪章》赋予的职责，这具有十分重要的意义”。事实上，2007 年联合国安理会首次就“能源、安全与气候”议题展开辩论。而全球为应对气候变化而开展的多边谈判，一直坚持以联合国作为主渠道，并与其他渠道形成互补，共同构建多主体、多层次的全球气候治理体系。习近平在第 70 届联合国大会一般性辩论中提出“各国相互依存、休戚与共”，呼吁“继承和弘扬联合国宪章的宗旨和原则，构建以合作共赢为核心的新型国际关系，打造人类命运共同体”。这样一种新型国际关系，对于应对气候变化这类非传统安全问题，是必要而又急需的。

但是，由于气候变化等非传统安全问题的多层次、多主体等特征，应对这类问题的全球治理体系也将是一个复杂的、多主体且相互联系的网络。政府、非政府组织、跨国公司、政府间和地区联盟以及国际非政府组织都将发挥各自作用。一些非正式的国家集团，譬如二十国集团以及新兴国家主导的金砖国家或者可能创建的其他新兴国家集团在国际治理体系将日益重要，与联合国机制形成有益的互补。而且随着经济全球化的深化，全球产业链的布局和劳动的专业化分工的进一步加强，市场将延伸到世界的各个角落，国家经济关系将演变为一种新的模式，庞大的跨国公司、国际经济机构以及大型的地区贸易体系也将在气候变化等非传统安全问题上发挥更大的影响力。

最后，随着互联网的应用深化，知识和信息在全球范围内传播的深度和广度的加强，关注各类具体问题、更加专业化且拥有全球性网络的非政府组织的影响力也将加强，成为多边体制变化的一个重要组成部分。新的信息传播技术与信息传播方式，一方面可能加剧气候变化等非传统安全问题的威胁，但是更重要的是提供了解决问题的机遇。因此，解决气候变化等非传统安全的全球治理体系，需要适应全球新的信息传播方式。譬如针对气候变化对电网安全带来的威胁，中国工程院的建议是，“建立广义智能能源网，深度融合气象信息、电源信息与电网运行维护信息和技术，形成电网的气象灾害预测、电网安全预警一体化防护系统，同时完善应急管理机制”。

第三节　全球可持续发展进程中的气候变化问题

一、全球可持续发展议程和目标

2015 年 9 月 25 日，联合国可持续发展首脑会议在纽约联合国总部召开。193 名会员国共同达成了载入史册的成果文件《改变我们的世界——2030 年可持续发展议程》。文件包括了 17 个可持续发展目标以及 169 个关联目标，以接替 2000 年提出的八项千年发展目标。

在过去 15 年里，面对重重艰险，世界各国为实现千年发展目标作出了前所未有的努力。为刺激与激励世界各国采取行动，推动千年发展目标，联合国秘书长在 2002 年启动了“千年项目”；2005 年，综合报告《投资于发展：实现千年发展的切实计划》提交，为就实现目标的国家一级进程以及国际系统支持国家一级进程的实施及安排提供了建议，并为实现目标的成本和利得进行评估；同年，时任秘书长科菲·安南根据《计划》发布了指导性报告《大自由》；2008 年，在千年发展目标高级别会议上，各方各界积极响应，在筹集了约 160 亿美元资金的基础上纷纷为实现目标做出了新承诺；2010 年，联合国千年发展目标首脑会议的成果文件帮助各国进一步确立了行动议程、政策和战略；2013 年，秘书长潘基文在联大期间发布了报告《人人过上有尊严的生活》，并在高级别论坛上调动了超过 2500 亿美元的资金承诺。2015 年是实现千年发展目标的终止年，空前的努力使各国在脱贫、普及教育、消除性别歧视、保护环境、改善饮用水源等方面取得了可喜进展。

可持续发展目标不仅是对过去的承接与差距弥补，更将借力于千年发展目标所提供的跳板，迈向发展新高度。正如秘书长潘基文所言：“只要具备有针对性的干预措施、合理的战略、充足的资源和政治意愿，即使最贫穷的国家也能取得前所未有的巨大进步。”

其中，目标 13“采取紧急行动应对气候变化及其影响”以《联合国气候变化框架公约》为基础，从灾害应对、战略制定、科研教育、资金动员以及能力建设等方面为应对气候变化设立了两类、五个关联目标。《议程》确认了《联合国气候变化框架公约》的主要地位，指出所有国家都承诺努力达成一项宏大的全球气候协定，决心果断处理气候变化和环境退化带来的威胁。

二、气候变化与可持续发展

可持续发展的内涵是实现经济、社会以及环境这三大要素共同和谐发展。气候变化对人类实现可持续发展带来严重威胁，采取应对气候变化行动总体而言将推动经济、社会和环境发展，促进更全面地实现可持续发展。具体而言，两者存在以下关系：

第一，气候变化的不利影响广泛存在，其威胁将制约可能的发展路径。

全球二氧化碳浓度上升、均温升高、降水变化、极端气候事件增多以及海平面上升等气候变化问题已耳熟能详，而从纵向来看，变化的气候不仅干扰了气候系统本身，更在一定程度上加剧了自然系统与生物系统会面临的风险，而这些风险反过来将使人类和受管理的系统更加脆弱。许多地区的水文系统因为降水变化或冰川消融正发生改变，进而影响了与人类直接相关的水资源量和水质；部分陆地和海洋的生物物种的地理分布、季节性活动、迁徙模式和丰度等都发生了改变，进而损害了生物多样性，干扰了整个生态系统的结构、功能以及服务性。与此同时，气候变化带来的粮食问题更增加了减贫的负担，粮食的减产与短缺不仅增加了贫困的人口营养不良风险，也将催化一系列健康问题，给国家及区域的生计和居住环境造成压力。对大多数经济部门而言，温升 2℃左右可能导致全球年经济损失占其收入的 0.2%—2.0%，若温升达到或超过 4℃，全球所面临的风险将达到高乃至非常高的水平。

第二，实现可持续发展与应对气候变化既存在协同效应，也要注意平衡以减少相关负面影响和风险。

应对气候变化行动将为世界实现可持续发展带来一系列协同效应。积极应对气候变化有利于保障全球卫生，减少因营养不良、疟疾、腹泻和气温过高的死亡人数，减少因健康造成的直接或间接经济损失，尤其对卫生基础设施薄弱的地区意义深远；有利于提高对自然灾害风险的抵御能力，特别是帮助最不发达国家加强基础设施建设，减少因灾害产生的次生问题；有利于能源安全，有助于降低经济发展对化石能源的依赖，对改善能源结构、提高能源效率以及减少能源浪费起到了催化作用；有利于推动新技术的发展，推动科技创新，摈弃粗放型生产模式，提高资源利用率，减少贫困，增加就业，从而转变发展模式，实现经济发展。然而，某些为应对气候变化所采取的减缓措施可能会削弱为可持续发展与公平所付出的努力。应对气候变化不仅是涉及各国利益的政治问题，也受资金与技术的制约。从经济发展的角度来说，气候变化本身在本世纪并不足以对全球经济发展造成特大影响，相反，恰恰是减缓成本过高造成了部分地区的经济发展压力。因此，考虑到发展的不均衡，为实现可持续发展，应对气候变化虽任重道远，却也困难重重。

第三，有效应对气候变化的能力与实现可持续发展的能力高度重合。

气候变化在发展中产生，应对气候变化是实现可持续发展蓝图中重要的一环。因此，为整体提高可持续发展的能力，有必要实施应对气候变化相关的能力建设。应对气候变化的能力主要从三个层面进行。从个人能力建设上，需要加强应对气候变化知识教育，提升公众对气候变化的认识；从制度角度来说，需要将应对气候变化主流化进各级组织与机构的人力管理、资金安排及文化传承中，也需要进一步鼓励组织部门间的相互合作；从系统大环境来说，需要将气候变化的政策制定放在可持续发展的框架中进行，通过合理的监管、评估和问责，确保应对气候变化行动合理进行。

第四节　全球贸易与气候治理

从 1947 年关贸总协定 (GATT) 到 1994 年世界贸易组织 (WTO) 成立之后的一段时间内，全球贸易治理体系都是以多边谈判为主。然而近 20 年来，全球贸易体制正在逐渐发生转变，从多边协商逐渐过渡到多元化治理体系。越来越多的国家都在 WTO 体系外努力寻求更灵活、更优惠的贸易条件和机会，通过双边协商或者多个国家以及区域内协商达成优惠贸易协定 (PTA) 来进一步改善贸易环境。根据 WTO2011 年世界贸易报告，全球优惠贸易协定数量在 1990 年到 2010 年间增长了两倍以上，从 70 个左右协议攀升到近 300 个。[①] 从贸易规模上看，2008 年发生在优惠贸易协定中的贸易规模（包括欧盟内贸易）已经占全球贸易总量的 50%。[②] 全球贸易治理多元化发展的另一个重要特征是区域经济合作势头高涨，例如近年来在亚洲地区由东盟主导的，中国、日本、韩国、澳大利亚、新西兰和印度参与的区域全面经济伙伴关系 (RCEP)，以及在发达国家方面由美国带头组织的跨大西洋贸易与投资伙伴协议 (TTIP) 和跨太平洋伙伴关系协议 (TPP) 等。

随着贸易体制的演变，全球贸易规模、流向和结构也随之改变，

① WTO. World Trade Report 2011 [EB/OL].https://www.wto.org/english/res_e/booksp_e/anrep_e/world_trade_report11_e.pdf, 2015-10-08.

② Bhagwati Jagdish, Krishna Pravin, Panagariya Arvind. The World Trade System:Trends and Challenges [EB/OL]. https://www.sais-jhu.edu/sites/default/files/JB-PK-SIPA-SAIS% 20Draft.pdf, 2015-10-08.

并对各国国内生产活动和碳排放造成深刻影响。贸易与气候变化之间的紧密联系可以通过贸易隐含碳（即隐含在进出口商品生产过程中的碳排放）反映出来。本书将在下节基于可得数据，介绍近几年全球贸易规模、结构变化与贸易隐含碳流向关系，对全球碳排放影响和含义，以及对未来全球贸易治理体系与气候治理体系关系协调发展的期望。

一、全球贸易发展趋势

2000 年到 2012 年全球出口贸易规模迅速扩大（图 6-1），贸易额从 6.5 万亿美元线性增长到 18.3 万亿美元，2008 年金融危机爆发，全球出口规模有所下滑，2009 年出口总量下降到 12.5 万亿美元，比 2007 年末出口量减少 3.6 万亿美元，但随着全球经济复苏，出口总值继续攀升，2011 年出口总额突破 18 万亿美元。

与此同时，全球出口结构发生变化，整体呈现发达国家出口比重下

图 6-1 2000—2012年全球出口总量和国家及地区分布

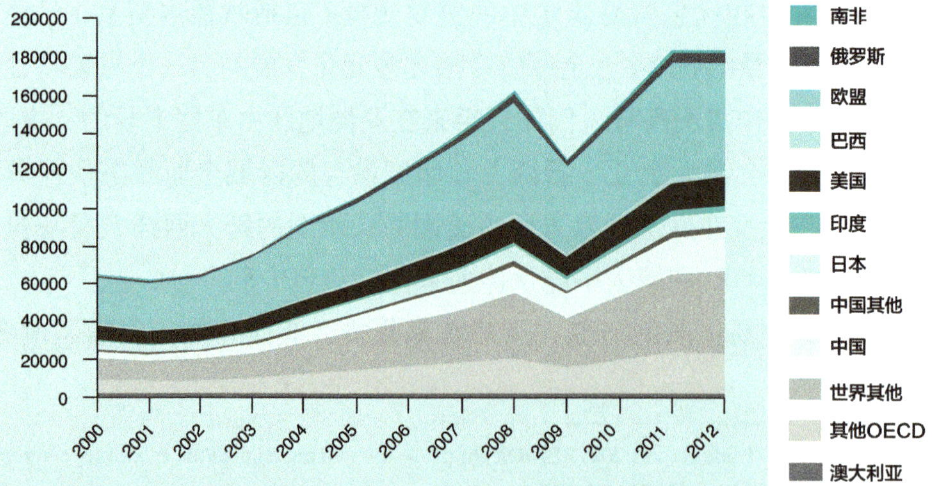

数据来源：WTO database [EB/OL].https://www.wto.org/english/res_e/statis_e/merch_trade_stat_e.htm, 2015-9-10.

图 6-2 2000—2012年全球贸易国家及地区分布

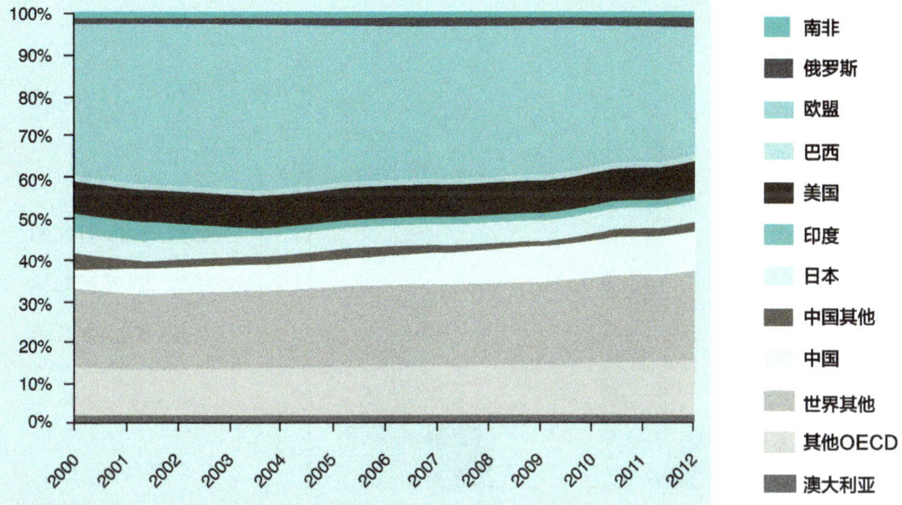

数据来源：WTO database [EB/OL].https://www.wto.org/english/res_e/statis_e/merch_trade_stat_e.htm, 2015-9-10.

降和发展中国家出口规模上升的趋势（如图 6-2）。2000 年到 2012 年，其他 OECD 国家、欧盟、美国、日本等发达国家的出口比重从 72% 下降到 57%，以中国、世界其他发展中国家为代表的发展中国家出口比重则相应上升，占全球出口的比重由 2000 年的 28% 上升为 2012 年的 43%。

二、全球贸易隐含碳概况

（一）全球贸易隐含碳随贸易增长而增长

伴随着贸易的发展，隐含在贸易商品生产过程中的碳排放即贸易隐含碳也在增长。2007 年，全球贸易隐含碳占全球碳排放的比重已经达到近 20%（图 6-3）。

（二）全球贸易隐含碳主要来自发展中国家，去向发达国家，中国是最大的隐含碳出口国

随着贸易流逐渐从发达国家向发展中国家转移，来自发展中国家的

图 6-3 2004 年和 2007 年全球贸易隐含碳变化及占全球碳排放比重

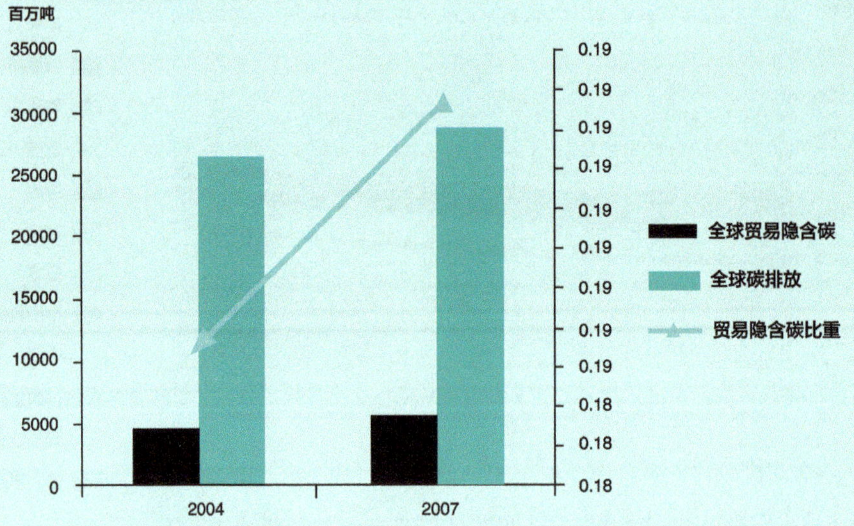

数据来源：全球贸易隐含碳来自 GTAP8 计算结果，全球碳排放数据来自 IEA, CO_2 Emissions from Fuel Combustion 2009[EB/OL]. http://www.oecd-ilibrary.org/energy/co2-emissions-from-fuel-combustion-2009_co2_fuel-2009-en-fr, 2015-9-17.

出口隐含碳迅速增长，同时由于发展中国家国内生产技术较为落后，单位产出的能耗与排放高于发达国家，发展中国家已经成为世界最大的出口隐含碳来源。如表 6-1，2007 年发展中国家出口隐含碳总量为 35.4 亿吨，占全球贸易隐含碳的 64%。[①]2004—2007 年，发展中国家出口隐含碳增量占全球贸易隐含碳增量的 75%，是全球贸易隐含碳增长的主要贡献方。从进口隐含碳分布来看，发达国家是主要的进口国，2007 年进口隐含碳占全球贸易隐含碳的 66%。

从图 6-4 的贸易隐含碳国别分布来看，中国是最大的贸易隐含碳出口国，2007 年中国出口隐含碳占全球贸易隐含碳的 28%。贸易隐含碳进口国都分布在美国、欧盟、日本以及其他 OECD 国家和地区。

① 刘俊伶，王克，邹骥．全球贸易隐含碳变化及其影响分析 [J]. 气候变化研究进展，2015,11(1):54-60.

表 6-1 2004 年和 2007 年南北 * 阵营之间出口隐含碳与出口贸易额

（单位：Mt，亿美元）

年份	出口国	隐含碳 /Mt CO_2 南	北	合计	占全球比例	贸易额 /10^8 美元 南	北	合计	占全球比例
2004	南	869.92	2169.33	3039.25	62%	7509.83	19337.13	26846.96	38%
	北	615.27	1232.62	1847.89	38%	15327.81	28988.30	44316.12	62%
	合计	1485.20	3401.94	4887.14	100%	22837.64	48325.43	71163.08	100%
2007	南	1146.65	2390.78	3537.43	64%	13645.83	30309.12	43954.95	42%
	北	759.70	1253.50	2013.20	36%	23487.04	36466.56	59953.59	58%
	合计	1906.35	3644.28	5550.63	100%	37132.87	66775.68	103908.54	100%

资料来源：刘俊伶，王克，邹骥．全球贸易隐含碳变化及其影响分析 [J]．气候变化研究进展，2015,11(1):54–60.

* 文献基于 GTAP8 数据库，将全球划分为发达国家和地区与发展中国家和地区。发达国家和地区（也称作“北”）包括美国、日本、澳大利亚、欧盟和其他 OECD 国家，发展中国家和地区（也称作“南”）包括中国大陆、巴西、俄罗斯、印度、南非、中国其他地区（香港和台湾）和世界其他国家和地区。

图 6-4 2007年各个国家和地区进出口隐含碳及占全球贸易隐含碳比重

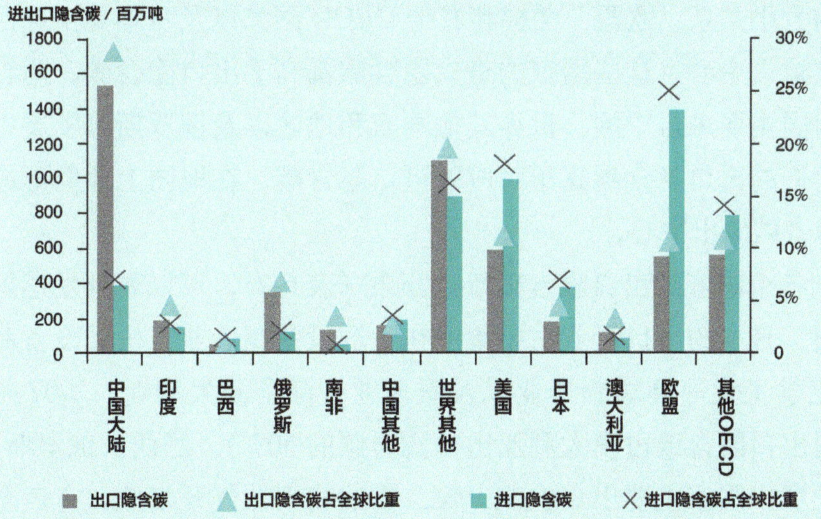

资料来源：刘俊伶,王克,邹骥.全球贸易隐含碳变化及其影响分析[J].气候变化研究进展,2015,11(1):54–60.

图 6-5 2007 年全球贸易隐含碳流向图

（单位：百万吨）

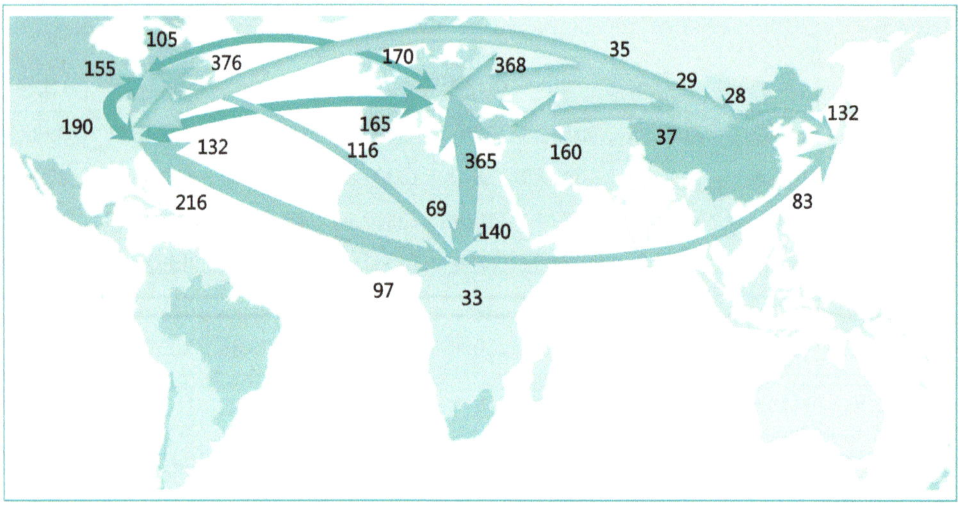

资料来源：刘俊伶. 基于多区域投入产出模型的全球贸易内涵碳排放分析 [D]. 北京：中国人民大学，2012.

图 6-5 为全球贸易隐含碳流向图，线条的粗细代表贸易隐含碳的规模，线越粗代表隐含碳越多，箭头的大小表示两国相互出口商品中隐含碳的大小关系。欧盟、美国和其他 OECD 国家以及日本是全球四大隐含碳输入中心，这些国家的进口隐含碳都高于出口隐含碳，隐含碳的输出国主要来自中国、世界其他国家和地区以及俄罗斯等发展中国家，他们的出口隐含碳高于对应的进口隐含碳，在地图上他们形成了几大隐含碳输出中心。

根据各个国家的出口隐含碳行业分布（表 6-2），可以看到发达国家如美国、日本的出口行业主要集中在运输设备制造业、机械设备制造业和化学工业等高端制造业以及服务业（除了澳大利亚，2007 年有色金属出口隐含碳占澳大利亚出口隐含碳的 50%）。巴西、俄罗斯、南非等发展中国家主要出口农业、煤、石油和天然气开采业、有色金属行业产品，中国以出口机械设备、化工产品和纺织服装为主。出口行业分布总体呈现发达国家位于价值链上端，发展中国家位于价值链

表 6-2 2004 年和 2007 年主要国家和地区出口隐含碳行业分布

（单位：百万吨）

	美国		日本		澳大利亚		中国大陆		印度		巴西		俄罗斯		南非	
	2004	2007	2004	2007	2004	2007	2004	2007	2004	2007	2004	2007	2004	2007	2004	2007
农业	20.4	19.9	0.2	0.2	3.6	2.3	8.3	7.7	4.8	5.1	5.4	4.3	4.5	5.5	3.8	3.0
煤气石油开采	2.7	2.7	0.0	0.0	8.2	9.4	15.2	7.0	0.0	0.0	1.2	2.3	65.2	67.8	3.5	3.2
采矿	3.0	4.2	0.2	0.2	9.0	12.5	4.6	3.9	8.5	12.2	6.7	9.4	1.9	8.4	37.6	38.8
食品和烟草制造	16.6	18.3	0.5	0.5	6.7	6.4	22.1	23.4	7.4	9.0	7.5	6.0	6.4	5.6	2.5	2.2
纺织服装	8.9	6.5	1.0	0.9	0.8	0.6	184.5	226.0	17.7	17.9	1.4	0.9	2.2	0.7	1.1	0.8
木制品	3.4	3.3	0.1	0.1	0.3	0.2	25.2	37.8	0.7	0.7	1.0	0.6	14.1	5.0	1.2	0.8
造纸印刷	15.4	15.0	1.1	1.3	0.6	0.6	11.1	18.6	1.8	1.8	1.7	1.1	2.2	6.6	1.4	1.4
石油产品	25.2	36.5	1.2	2.1	1.6	1.8	35.1	37.6	4.5	9.9	3.2	3.3	33.1	44.2	3.1	2.6
化学工业	112.3	111.7	26.2	33.2	3.4	3.5	121.1	166.7	20.7	25.8	3.8	3.5	60.6	56.3	7.0	6.7
非金属矿物制品	8.8	9.7	4.3	5.0	0.4	0.4	47.9	64.1	4.6	5.5	2.4	2.0	2.6	2.8	2.0	1.2
钢铁	11.5	15.5	16.0	25.2	2.9	3.1	43.0	97.3	12.7	17.7	8.9	7.1	61.0	51.2	16.4	15.1
有色金属	17.4	27.7	2.7	5.0	53.9	61.2	26.0	36.0	8.2	17.5	6.1	5.9	56.8	32.6	26.0	27.0
机械设备制造业	106.6	93.9	58.6	63.2	3.7	3.4	453.9	614.2	12.8	17.2	4.7	4.3	15.8	8.4	7.4	8.3
运输设备制造	51.2	56.7	23.4	29.8	1.2	1.1	37.5	46.7	3.6	4.8	5.0	3.9	5.4	7.5	3.4	3.3
其他工业	4.6	4.2	1.6	1.6	0.3	0.3	34.2	53.0	16.7	15.0	0.1	0.1	1.3	1.2	1.1	1.4
电力	10.1	8.6	0.0	0.0	0.0	0.0	8.1	12.2	0.0	0.3	0.0	0.1	8.2	4.5	8.7	9.3
燃气生产和供应	0.3	1.1	0.0	0.0	0.1	0.3	4.2	4.7	0.0	0.0	0.1	0.2	5.1	14.9	0.0	0.0
服务业	40.4	48.2	4.3	5.6	3.3	3.8	29.4	36.8	11.9	21.9	1.0	1.0	5.8	8.0	2.2	2.8
建筑业	0.8	1.2	1.1	1.7	0.0	0.0	3.5	8.2	0.5	0.6	0.0	0.0	1.9	2.2	0.0	0.0
运输服务业	91.2	105.1	9.2	8.9	12.6	10.6	24.8	29.2	7.0	9.8	6.4	5.8	28.4	23.7	6.4	6.8

资料来源：刘俊伶，王克，邹骥．全球贸易隐含碳变化及其影响分析 [J]．气候变化研究进展，2015,11(1):54−60.

中下端的局面。即使对于发展中国家与发达国家间的行业内贸易，发展中国家也主要出口低附加值产品而发达国家出口高附加值产品。从各主要出口行业看，发达国家单位出口产品完全碳排放系数（即单位产品生产过程中的能源直接消耗排放和投入材料的生产过程中的间接能源消耗排放）都低于发展中国家水平（见表 6-3）。从平均水平来看，发展中国家整体约为发达国家整体平均水平的 3 倍。中国大陆平均完全碳排放系数最高，约为平均完全碳排放系数最低的欧盟的 6 倍。

表 6-3 2007 年发达国家与发展中国家主要出口隐含碳行业完全碳排放系数比较

（单位：t/ 万美元）

行业	美国	日本	欧盟	澳大利亚	中国大陆	印度	巴西	俄罗斯	南非
化工	6.19	3.67	2.96	5.02	19.24	13.86	3.07	24.34	12.14
机械设备	2.76	2.05	1.64	3.67	11.92	11.06	2.43	9.18	8.44
运输设备	2.87	1.67	1.74	2.00	11.85	8.48	2.03	11.69	5.14
煤气油开采	3.83	0.73	2.27	3.91	20.25	5.45	3.82	6.16	8.83
采矿	5.51	4.88	3.43	4.77	13.99	12.09	4.77	20.02	51.61
纺织服装	3.50	1.07	1.61	4.11	12.12	7.02	1.47	7.44	5.66
钢铁	8.68	8.14	5.87	10.60	27.46	24.70	8.36	24.79	20.16
有色金属	8.15	2.65	4.12	20.63	23.09	23.96	9.36	16.37	15.38
平均	5.19	3.11	2.96	6.84	17.49	13.33	4.41	15.00	15.92

数据来源：基于 GTAP8 数据库计算得到。

（三）贸易促进了发展中国家国内碳排放增长，避免了发达国家国内部分碳排放

由于发达国家进口隐含碳大于出口隐含碳，避免了国内部分生产和二氧化碳排放，是贸易隐含碳净进口国。2007 年欧盟、日本进口隐含碳占国内和区域内碳排放总量的 35% 和 31%（见图 6-6）。发展中国家由于出口迅速扩张推动国内碳排放的快速增长。从图上可以看到，中国、印度、南非等发展中国家出口隐含碳都高于进口隐含碳，其中中国大陆和南非 2007 年出口隐含碳占国内碳排放总量的 25% 和 39%。

图 6-6 2007 年各国进出口隐含碳及占国内碳排放比重

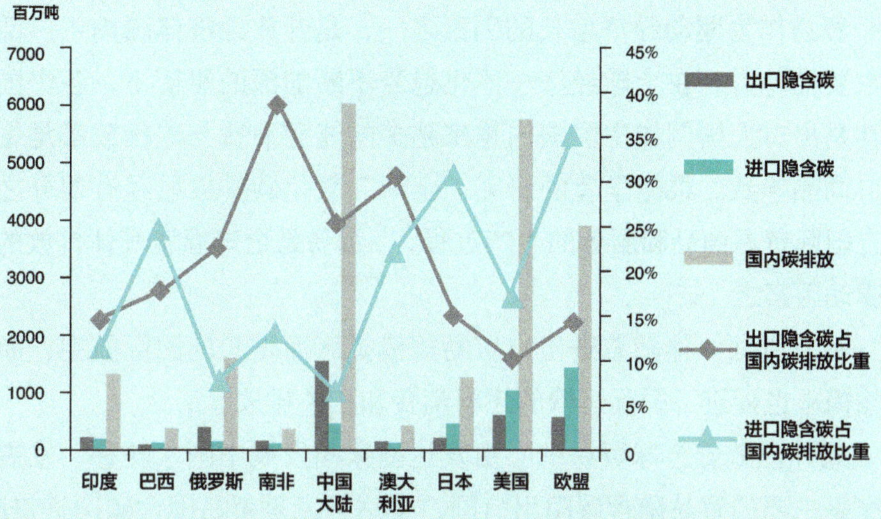

数据来源：进出口隐含碳基于 GTAP8 数据库计算得到，各国国内碳排放数据来自 IEA, CO_2 Emissions from Fuel Combustion 2009[EB/OL].http://www.oecd-ilibrary.org/energy/co2-emissions-from-fuel-combustion-2009_co2_fuel-2009-en-fr, 2015-9-17.

可见，随着国际贸易发展和世界经济一体化程度加深，一个国家的生产和消费活动能够在地域上发生分离，国内生产排放并不完全用于自身消费，部分能够出口，而通过进口也会避免国内部分排放。从目前国际贸易与贸易隐含碳的分布来看，发达国家进口隐含碳大于出口隐含碳，为贸易隐含碳净进口国，发展中国家则为净出口国。发达国家通过贸易进口避免了部分国内排放，而发展中国家在增加出口的同时促进了国内碳排放。

三、贸易与气候变化的联系

随着世界经济一体化的发展，一个国家的排放不仅由国内经济活动和消费决定，还受到国际经济和贸易的影响，与此同时国家与国家之间的贸易关系也逐渐受到气候变化政策的影响。贸易与气候变化的联系表现在以下几个方面：

（一）贸易是全球温室气体排放的主要驱动因素之一

贸易作为驱动经济增长的因素之一，是开放经济体国内生产活动的重要推动力。在全球经济一体化趋势不断加深的驱使下，各个国家尤其是出口大国例如中国国内越来越多的能源消费与碳排放都是生产出口商品导致。根据学者的研究，全球二氧化碳排放已经有四分之一来自国际贸易商品和服务的生产过程，① 贸易是全球温室气体排放的主要驱动因素之一。

另一方面，伴随着进出口贸易规模扩张而增长的国际航空、航海运输需求也促进了国际运输的燃料消费和相关排放。

此外，随着全球贸易中心由发达国家向发展中国家转移，发展中国家是主要的贸易隐含碳净出口国，发达国家是贸易隐含碳净进口国，但由于发展中国家国内生产技术低于发达国家水平，生产活动由发达国家向发展中国家转移会带来全球碳排放的净增长。②

（二）气候变化减缓政策与贸易政策相互影响

各国制定的气候变化政策主要以提高碳价的方式实施，通过降低化石能源补贴、鼓励非化石能源发展、提高能效标准和门槛、建立碳市场等使碳的价格得以体现，并引导生产者与消费者向低碳的生产与消费活动转变。在这一过程中因为改变了原有要素的价格结构，导致各个商品的相对价格发生改变，使得碳密集型商品价格增长大于其他商品。以碳排放密集型商品为主的出口国家产品受碳价的影响会失去比较优势，而以清洁低碳能源为原料和使用清洁生产技术的国家的出口产品则获得市场竞争力，逐渐发展成为新的出口国。气候变化政策的实施会改变各个国家的相对竞争力和全球贸易流向。

---

① Peters, G.P., & Hertwich,E.G..Trading Kyoto [J]. Natural Reports Climate Change,2008, Vol. 2: 40-41.

② 刘俊伶，王克，邹骥．全球贸易隐含碳变化及其影响分析 [J]. 气候变化研究进展 ,2015,11(1):54-60.

反之，贸易政策也会影响全球应对气候变化的行动。对绿色低碳技术的出口管制、出口禁令会阻碍先进技术向发展中国家转移，影响全球控制温室气体排放的进度。此外，进口国对可再生能源生产技术如太阳能、风能零部件蓄意发起的反补贴、反倾销政策也大大降低了气候友好技术在全球范围扩散的速度。例如自 2010 年以来，美国光伏制造联盟以遭受实质损害为由发起对来自中国的晶体硅太阳能光伏电池进行反倾销、反补贴调查的申请，并最终向中国出口企业征收了巨额的惩罚性关税。

四、进一步协调全球贸易体系与气候体系未来关系的展望

在全球化背景下，应对气候变化挑战要求必须协调好气候体系与贸易体系的关系，一方面调解两个体系之间的冲突，一方面推动两个体制的协调发展。

在气候变化治理体系中，要继续遵守 UNFCCC 第 3.5 条规定："为对付气候变化而采取的措施，包括单方面措施，不应当成为国际贸易上的任意或无理的歧视手段或者隐蔽的限制"①，避免以气候政策为名义实施的贸易"保护主义"行为。

在全球贸易体系中，要加快制定、协商与气候变化相关的规则。当前贸易治理体系还没有对关于气候变化相关事宜的全面安排，只在个别的区域贸易协议中进行初步尝试。2011 年 11 月亚太经合组织 (APEC) 通过环境有益商品协议谈判 (Environmental Goods Agreement-EGA)，其中涵盖了风能、太阳能、热泵等与气候变化相关的产品，该协议决定到 2015 年底将环境有益商品的关税降低到 5% 或者更低，并

① UNFCCC.《United Nations Framework Convention on Climate Change》.[EB/OL]. http://unfccc.int/files/essential_background/convention/background/application/pdf/unfccc_chinese.pdf, 2015-9-18.

于2012年制定了54个环境有益商品关税降低目标名单，以促进环境有益商品的自由贸易①。此外，在EGA提供的54个产品清单基础上WTO也在努力制定一份新的范围更广的环境产品清单，供各成员国谈判，以推动关税的逐步降低，但仍然面临环境产品定义的挑战。②除了定义环境产品的内涵，还应当进一步识别与气候相关的产品和技术，形成特定的气候有益产品清单，制定促进该类产品自由贸易的规则和降低关税的时间表。在此基础上再进一步促进达成气候有益服务相关的国际贸易协定。

第五节 全球金融与气候治理

近几年来，气候变化问题逐渐在多边金融体系中主流化。考虑到气候变化可能为投融资带来的挑战和机遇，业界对气候变化的关注度不断提升。虽然气候变化依然存在短期内对投资负面影响的不确定性，金融业者若想获得长久的发展，不可能与气候变化问题完全绝缘。

一、多边金融体系与气候治理的相互关系

在金融分析与预测活动中，气候物理风险被提及的频率逐渐提高，因为气候风险可能会潜移默化地改变资源的需求模式，产品与服务的供求关系、资产的效能表现以及实体的创新动力，从而对投资活动产

① USTR. Environmental Goods Agreement.[EB/OL]. https://ustr.gov/trade-agreements/other-initiatives/environmental-goods-agreement, 2015-10-07.

② ICTSD. 环境产品协定的谈判人员审议产品清单 . [EB/OL]. http://cn.ictsd.org/bridges-news/%E6%A1%A5/news/%E7%8E%AF%E5%A2%83%E4%BA%A7%E5%93%81%E5%8D%8F%E5%AE%9A%E7%9A%84%E8%B0%88%E5%88%A4%E4%BA%BA%E5%91%98%E5%AE%A1%E8%AE%AE%E4%BA%A7%E5%93%81%E6%B8%85%E5%8D%95, 2015-10-07.

生重大影响。同时，包括赤道银行在内的许多金融部门加强了对社会效益的重视程度。不将气候风险考虑在内的投资战略有可能与设立的环境与社会影响目标相冲突，增加不必要的成本，造成社会声誉受损，导致业绩下滑。此外，各级政府逐渐加强并完善了应对气候变化的法律法规，相关政策导向的改变也可能对投资造成影响。

另一方面，为应对气候变化而采取的减缓行动可能为投融资领域带来新机遇。2008 金融危机后，低碳经济被认为是能够带动世界经济实现新一轮增长的主要力量。部分欧洲及拉美国家都宣布了气候中和或 100% 清洁能源覆盖率的战略目标，在政策指引和公众气候意识的推动下，可持续能源、能源效率等领域被认为可能引发第四次工业革命，引发世界经济结构的新调整。

与此同时，应对气候变化行动也迫切需要金融体系所提供的资金支持与资源动力。作为五大构件之一，气候融资对适应及减缓气候变化起到了至关重要的作用。据 IPCC 统计，在 2011 年到 2012 年间，用于应对气候变化的资金流大约为 3430 亿—3850 亿美元。其间，流向发展中国家的公共气候资金总额约为每年 350 亿—490 亿美元，而流向发展中国家的国际私营机构气候资金约为每年 100 亿—720 亿美元。私营机构提供的减缓资金总量的份额约占全球平均水平的三分之二到四分之三，对应对气候变化起到了核心作用。气候治理在金融体系主流化后将获得更多资金与资源以支持减缓和适应行动，金融体系在政策引导与法律约束下投资气候治理行动并获得经济与社会效益，从而将获取更多资金与资源以满足进一步的气候治理需求，帮助实现经济增长与环境保护的协同收益。

二、多边金融体系在全球气候治理中的实践

在哥本哈根会议中首次提出了每年动员 1000 亿美元资金以及设立 300 亿快速启动资金的承诺。德班大会后设立了绿色气候基金 (GCF)，

为在公约下筹集更多发达国家公共资金以大规模动员气候资金用于推动各国低碳转型和提高气候抵御力提供了平台。为实现将均温控制在2℃的长久目标，多边金融机构，尤其是国际政策性金融机构积极开展了全球气候治理的相关工作，为金融体系作出了表率。

从2011年至2014年，多边开发银行（包括非洲开发银行、亚洲开发银行、欧洲重建和发展银行、欧洲投资银行、美洲开发银行和世界银行集团）共同为应对气候变化融资超过1000亿美元。其中，在2014年，世界银行集团在贷款、赠款、担保金、股权融资以及技术支援等方面为来自超过77个国家的224个气候项目融集了119亿美元资金，同时将气候变化囊括进了其新颁布的环境与社会安保政策中；亚洲开发银行通过融资和技术创新引导地区走向绿色道路。从2011年到2014年，亚行共实现了130亿美元的气候融资。到2020年，亚行的气候融资金额将翻倍至60亿美元。到该年代末，亚行用于应对气候变化的支出将增至其总体融资的30%左右；新开业的金砖国家开发银行以及正在筹备的亚洲基础建设投资银行等新兴多边开发银行也将气候变化作为重点创新领域。

三、中国气候变化主流化及绿色金融前景展望

作为全球第一大碳排放国，中国的碳排放量已超过全球碳排放份额的26%。作为负责任的大国，中国宣布大力推进生态文明建设，推动绿色低碳、气候适应型和可持续发展，加快制度创新，强化政策行动。构建绿色金融体系符合绿色产业与可持续发展的需求，但是推动绿色金融体系的建立依然面临诸多挑战。以碳税、排放标准以及技术指标为主导的传统气候政策不能全面体现金融体系所面临的气候风险，为推动绿色金融良好发展，未来气候政策的主流化应当更多考虑资金提供方的需求；金融部门也应当将气候变化问题纳入风险与稳定性评估

框架；保险部门应在资产组合配置与设计时顾及气候与天气风险；各级财政部门应就气候风险给予更多技术性指导；货币与银行监管机构应当提供更充分的资金准备以应对气候风险。除此以外，应加强机构、政策、基础设施以及相关金融法律环境的建设，并挖掘绿色金融刺激机制与方法学，推动绿色金融的发展。

【参考文献】

[1] 陈迎．对 IPCC 第五次评估报告中可持续发展与公平相关问题的解读 [J]. 气候变化研究进展，2014. 10(5), 348-354.
[2] 刘俊伶．基于多区域投入产出模型的全球贸易内涵碳排放分析 [D]. 北京：中国人民大学，2012．
[3] 刘俊伶，王克，邹骥．全球贸易隐含碳变化及其影响分析 [J]. 气候变化研究进展，2015,11(1):54-60.
[4] 马骏．论构建中国绿色金融体系．[C]. 金融论坛，2015.
[5] 普华永道．探索适合中国的绿色金融激励机制执行摘要．2015.
[6] 行动承诺：实现千年发展目标 [C]. 千年发展目标高级别会议相关倡议和承诺汇编 (2008). http://www.un.org/chinese/millenniumgoals/2008highlevel/Commitments%20compilation%20CHINESE.pdf.
[7] 新华社．中美气候变化联合声明 [N][D]. 2014.
[8] 中国人民银行，UNEP. 构建中国绿色金融体系报告摘要．2015.https://www.cbd.int/financial/privatesector/china-Green%20Task%20Force%20Summary.pdf.
[9] ADB (2008) Strategy 2020: The Long-Term Strategic Framework of the Asian Development Bank 2008-2020. ADB, Manila.
[10] ADB (2015). 亚行对亚太地区的气候融资到 2020 年将翻倍至 60 亿美元．
[11] Annan, K. (2005). In larger freedom: towards security, development and human rights for all. Report of the Security General of the United Nations for decision by Heads of State and Government in September.
[12] Arndt, C., & Thurlow, J. (2015). Climate uncertainty and economic development: Evaluating the case of Mozambique to 2050. Climatic Change, 130(1), 63-75.
[13] Bhagwati Jagdish, Krishna Pravin, Panagariya Arvind. The World Trade System: Trends and Challenges [EB/OL].
[14] Climate Technology Centres (2010). A New way to Foster Mitigation and Adaptation In The Developing World http://www.infodev.org/infodev-files/resource/InfodevDocuments_1015.pdf
[15] Dwirahmadi, F., Rutherford, S., Ulrich, W., & Chu, C. (2013). Linking disaster risk reduction and climate change adaptation. Climate Adaptation Futures, 362-370.

[16] Gardener, N. J. L., Gardener, W. H., Gardener, R. M., & Loyd, D. A. CLIMATE CHANGE OPERATIONS PLANNING–FOR ENERGY, REPUTATION AND SAFETY.

[17] General, U. S. (2013). A life of dignity for all: accelerating progress towards the Millennium Development Goals and advancing the United Nations development agenda beyond 2015. New York: UN, http://www. un. org/millenniumgoals/pdf/A% 20Life% 20of% 20Dignity% 20for% 20All. pdf.

[18] Hajat, S., Vardoulakis, S., Heaviside, C., & Eggen, B. (2014). Climate change effects on human health: projections of temperature-related mortality for the UK during the 2020s, 2050s and 2080s. Journal of epidemiology and community health, 68(7), 641-648.

[19] Hoffmann, W. (2014). The Economic Competitiveness of Renewable Energy: Pathways to 100% Global Coverage. John Wiley & Sons.

[20] http://consultations.worldbank.org/consultation/review-and-update-world-banksafeguard- policies

[21] http://www.worldbank.org/en/topic/climatechange/projects.

[22] http://www.adb.org/zh/news/adb-double-annual-climate-financing-6-billion-asiapacific-2020.

[23] https://www.sais-jhu.edu/sites/default/files/JB-PK-AP-SIPA-SAIS%20Draft.pdf, 2015-10-08.

[24] IEA.CO_2 Emissions from Fuel Combustion 2009[EB/OL]. http://www.oecd-ilibrary.org/energy/co2-emissions-from-fuel-combustion-2009_co2_fuel-2009-en-fr, 2015-9-17.

[25] Investing in development: a practical plan to achieve the Millennium Development Goals. Earthscan; Millennium Project, 2005.

[26] IPCC (2014). Climate Change 2014: Synthesis Report of the Fifth Assessment Report of the Intergovernmental Panel on Climate Change [Core Writing Team, R.K. Pachauri and L.A. Meyer (eds.)]. Geneva, Switzerland, 151 pp.

[27] IPCC (2014). Climate Change 2014: Impacts, Adaptation, and Vulnerability - Working Group Ⅱ Contribution to the Fifth Assessment Report. Cambridge University Press, Cambridge, UK and New York, NY USA.

[28] IPCC (2014). Climate change 2014: mitigation of climate change. Contribution of Working Group Ⅲ to the Fifth Assessment Report of the Intergovernmental Panel on Climate Change, 511-597.

[29] IPCC (2014). Climate change 2014: mitigation of climate change. Contribution of Working Group Ⅲ to the Fifth Assessment Report of the Intergovernmental Panel on Climate Change, Ch16

[30] Jackson, T. (2009). Prosperity without growth?: The transition to a sustainable economy.

[31] Marotzke, J., & Forster, P. M. (2015). Forcing, feedback and internal variability in global temperature trends. Nature, 517(7536), 565-570.

[32] Mendelsohn (2009). Robert. Climate change and economic growth. World Bank, 2009.

[33] Necessary, A. (2015). Mainstreaming Climate Change into Financial Governance.
[34] Nordhaus, W. D. (2008). A question of balance: economic modeling of global warming.
[35] Palmer, E. (2015). INTRODUCTION: The Sustainable Development Goals Forum. Journal of Global Ethics, 11(1), 3-9.
[36] Peters, G.P., &Hertwich,E.G..Trading Kyoto [J]. Natural Reports Climate Change,2008, Vol. 2: 40-41.
[37] Stenek, V., Amado, J. C., & Connell, R. (2010). Climate Risk and Financial Institutions: Challenges and Opportunities. International Finance Corporation.
[38] Sullivan, R. (2014). Climate Change: Implications for Investors and Financial Institutions. Available at SSRN 2469894.
[39] United Nations (2015). Transforming our world: the 2030 Agenda for Sustainable Development. New York, USA. Available at https://sustainabledevelopment.un.org/content/documents/7891Transforming%20Our%20World.pdf.
[40] United Nations (2015). Millennium Development Goals Report 2015. United Nations, New York. Available at: http://www.un.org/millenniumgoals/2015_MDG_Report/pdf/MDG%202015%20rev%20(July%201).pdf.
[41] UNFCCC.United Nations Framework Convention on Climate Change[EB/OL]. http://unfccc.int/files/essential_background/convention/background/application/pdf/unfccc_chinese.pdf, 2015-9-18.
[42] WEC. 2015. The Global Risks Landscape 2015. World Ec. http://reports.weforum.org/global-risks-2015.
[43] World Bank (2015). 2014 Joint Report On Multilateral Development Banks' ClimateFinance.
[44] WSJ (2015). U.S., China Build on Plan to Cut Emissions. http://www.wsj.com/articles/u-s-china-build-on-climate-accord-1442342194.
[45] WTO database [EB/OL]. https://www.wto.org/english/res_e/statis_e/merch_trade_stat_e.htm, 2015-9-10.
[46] WTO. World Trade Report 2011 [EB/OL].https://www.wto.org/english/res_e/booksp_e/anrep_e/world_trade_report11_e.pdf, 2015-10-08.
[47] Zenawi, M., & Stoltenberg, J. (2010). Report of the Secretary General's High-Level Advisory Group on Climate Change Financing. United Nations, 5(11).

第七章

全球气候治理中的国家主体

Chapter Seven

State Stakeholders in Global Climate Governance

摘要

国家是全球气候治理体系的主体，随着气候治理体系的不断发展以及治理体系中主权国家国情、发展阶段、历史累积排放责任以及能力的动态变化，各国在全球气候治理中的地位、相互关系和参与治理的方式和目的都将随之调整。目前以《公约》为基础的气候治理体系基本保持发达国家和发展中国家的区分，要求发达国家采取全经济范围大规模减缓行动，并向发展中国家提供资金、技术和能力建设的支持；发展中国家根据自身国情、发展阶段和现实能力采取多样的减缓行动。发达国家通过多年的努力，其总体减排行动尽管离科学要求尚有较大差距，但也已经取得了一定进展，部分发达国家已经开始出现经济增长与碳排放脱钩的现象；发展中国家的发展权利不可侵犯，但也应该为全球应对气候变化做出力所能及的贡献，只有尽量避免走发达国家的老路，创新发展路径才能实现这一目标。发达国家应为创新发展路径做出表率，同时应为发展中国家的路径创新提供支持。本章对全球气候治理下的主要发达国家和发展中国家主体进行了梳理和分析。

第一节　全球气候治理体系主体间的关系

全球气候治理体系的主体是主权国家，随着气候治理体系的不断发展以及治理体系中主权国家国情、发展阶段、历史累计排放责任以及能力的动态变化，各国在全球气候治理中的地位、相互关系和参与治理的方式和目的都将发生变化。目前以公约为基础的气候治理体系主要体现了发达国家和发展中国家的区分。

一般而言，发达国家一方面要保持自身支配和主导地位，限制发展中国家的发展空间，另一方面又要发挥其能源、技术、资金等优势，扩大自身利益。发展中国家有争取发展空间、探索低碳发展这一共同诉求和权利。但在具体问题上，根据不同发展阶段的不同关切，可分为最不发达国家和新兴经济体国家（表 7-1）。其中，前者仍处于社会经

表 7-1　发达国家和发展中国家在气候变化全球治理中的不同关切分析

	最不发达国家	新兴经济体	发达国家
主要利益与现今主要目标	脱贫 降低气候变化的脆弱性	扶贫 基础设施 建立发展社会保障 经济升级	保持支配和主导地位 保持现有的生活方式 寻找新的经济增长引擎
加入国际气候进程的主要动力	获得更多的资金补偿	升级产业与技术 结构调整 环境共生效益	扩大碳市场 既得利益（技术优势） 主导和支配国际进程
最具挑战性的部门	与适应相关的部门	重化工业 可能具有锁定效应的基础设施部门	现有的交通体系 现有的建筑体系 消费模式和消费活动
在气候谈判中的最高关切	在国际决策中被排除在外	因限制能源使用和排放空间而阻碍工业化与城镇化	失去国际进程中的优势和主导权

资料来源：邹骥．在 COP16 边会上的发言，2010.

济发展初期，仍面临严重的贫困问题，且应对气候变化能力较弱。对此类最不发达国家，在气候谈判中获得更多帮助和资源应是其第一要务，而争夺全球治理领导权可能尚不符合其本阶段利益。而后者已经完成了初期发展，面临建立社会保障、落实经济升级、实现产业转型等高层次社会经济发展需要，且全球化浪潮后其国内发展与世界经济联系紧密。对于新兴经济体来说，如何升级产业和技术、改造重化工业，实现环境和发展的共赢，是其能否继续保持发展势头的一大关键。但与此同时，新兴发展中国家由于正处在工业化和城镇化的过程中，碳排放量大，增长速度快，发达国家和部分发展中国家对新兴发展中国家提高减排力度的期望越来越高，新兴发展中国家减排压力逐步增大。

总体而言，发达国家和发展中国家的基本格局没有发生根本性变化。发达国家仍然应该坚持履行《公约》下义务，一方面加强自身减排力度，一方面落实对发展中国家的资金和技术支持。尤其是对于最不发达国家，出于自身责任，都应对其消除贫困、社会发展和应对气候变化的工作努力予以支持，以增强其应对气候变化的能力。对于新兴经济体，发达国家可发挥自身在低碳技术方面的优势，通过技术转让、联合技术研发、技术创新等方式与相关发展中国家开展合作，实现双方共赢互利。

另一方面，全球气候治理也不断推动各国积极采取行动，部分发达国家取得了显著成效，甚至已经开始出现经济发展和碳排放脱钩的迹象，发达国家应该为创新发展路径做出表率。发展中国家的发展权利不可侵犯，但也应该为全球应对气候变化做出力所能及的贡献，只有尽量避免走发达国家的老路，创新发展路径才能实现这一目标。发展中国家的路径创新也应该获得发达国家的支持。基于上述考虑，本书对全球气候治理下的主要发达国家和发展中国家主体进行了梳理和分析。

第二节　美 国

一、美国在国际气候机制中的定位

截止到 2007 年，美国一直是第一大温室气体排放国，人均温室气体排放远高于世界平均水平，在气候变化问题上负有重要的历史责任。美国也是世界第一大经济体，拥有全球领先的技术创新能力，掌握最先进的能源及相关管理技术，资金和人力资源相对充沛，应对气候变化的能力较强。美国在包括气候变化等问题在内的全球治理体制下有至关重要的地位和话语权，在全球气候治理的进程中已经并将继续发挥不可替代的重要作用。

（一）《公约》下的义务和责任

《联合国气候变化框架公约》规定了发达国家和发展中国家在应对气候变化问题上不同的义务，作为发达国家，美国有义务带头依据《公约》的目标，改变温室气体人为排放的趋势；制定国家政策和采取相应的措施，通过限制人为的温室气体排放以及保护和增强温室气体汇和库，减缓气候变化，同时，作为《公约》附件二国家，美国还应提供新的和额外的资金，支付发展中国家为提供国家信息通报所需的全部费用，帮助特别易受气候变化不利影响的发展中国家缔约方支付适应这些不利影响的费用，促进和资助向发展中国家转让无害环境的技术，支持发展中国家增强自身的技术开发能力。

（二）美国在国际气候合作中的基本立场

美国对于“自上而下”减排机制一直以来都持不支持的态度。自 20 世纪 80 年代末各国开始关注气候变化问题，并计划在 1992 年里约环境大会上递交气候变化公约后，各国就开始围绕应对气候变化国际合作展开谈判。当时的欧共体、加拿大和其他一些国家支持到 2000 年把温室气体排放稳定在 1990 年的水平，日本号召工业化国家做出最大努力，而美国则拒绝任何有约束力的限制排放措施，当时的布什总统立场强硬，一再强调“美国人的生活方式是不能谈判的”。

由于《公约》仅定性地提出了稳定大气温室气体浓度的目标，很多国家认为这并不足以使各国采取足够的行动，因此计划发起新一轮关于加强发达国家承诺的会谈，当时欧盟与印度领导的、不包括石油输出国在内的发展中国家集团形成了联盟，媒体和公众也给各谈判代表团施加了压力，美国、加拿大等国最终不得已做出了妥协，各方通过了“柏林授权”。随后，各方开始了《京都议定书》的谈判，并最终于 1997 年达成了共识，确立了发达国家的减排目标，并建立了促进履约的灵活机制。但是，2001 年 3 月，美国就宣布退出已经签署的《京都议定书》，这引起了国际社会的强烈批评，也对日本、加拿大等发达国家产生了消极的影响，《京都议定书》的实际效果也大打折扣，国际气候进程陷入低谷。

奥巴马上台后，在气候变化等问题方面较上届政府政策发生根本性改变，对内通过推动相关立法来应对气候变化；对外积极参与国际谈判，同时拉中、印等发展中大国承担类似的减排责任。美国注重从国家战略安全高度看待气候变化问题，将经济安全和能源安全统筹考虑，在国内将提高能效、发展可再生能源作为刺激经济、增加就业的重要动力之一，培养新的经济增长点，抢占未来经济制高点。但美国受金融危机影响严重，减排将降低其竞争力。虽然目前美国对 2050 年长期减排目标态度积极，但近期不太可能确定激进的国内中长期减排目标，美国气候变化政策的调整将是渐进的。

总体上，美国国会有关气候变化的提案基本采取了减排目标和排放贸易体系的思路，在 2030 年之前的近期目标较为宽松，远期目标分歧较大。

在联邦政府层面，奥巴马政府更重视气候变化问题，将气候变化作为执政重点之一。2009 年 6 月 26 日，美国众议院通过了旨在降低美国温室气体排放、减少美国对外国石油依赖的《美国清洁能源安全法案》，这是美国政府首次向国际社会作出温室气体减排承诺，标志着美国迈出了应对气候变化的重要一步。

奥巴马总统 2009 年上台时，人们对中美两国在气候变化上的双边合作十分看好。同年哥本哈根谈判前，奥巴马访问中国，离开北京前签署了一系列技术交流协议，涉及碳捕获和封存、清洁能源车和清洁能源合作研究等领域。

人们盼望这些初步举动能够为哥本哈根的更深层次合作以及帮助中国拟订一系列更高目标奠定基础，但事与愿违，众所周知的是，由于各种原因哥本哈根谈判严重恶化了两国关系。此后中美关系在许多安全和经济议题上进入了一段紧张时期。然而尽管有分歧和摩擦，且分歧和摩擦将继续存在，但两国还在继续寻找共同利益并在气候变化和能源领域寻求合作。2014 年 11 月中美两国元首在北京发表了历史性的《中美气候变化联合声明》。此后短短的一年内，美国当地时间 2015 年 9 月 25 日，习近平主席同奥巴马总统举行会谈，双方发表了《中美元首气候变化联合声明》。

《中美元首气候变化联合声明》重申坚信气候变化是人类面临的最重大挑战之一，两国在应对这一挑战中将发挥重要作用；重申坚定推进落实国内气候政策、加强双边协调与合作并推动可持续发展和向绿色、低碳、气候适应型经济转型的决心；重申 2014 年 11 月 12 日发表的《中美气候变化联合声明》，坚定决心携手并与其他国家一道努力，达成一项富有雄心、圆满成功的巴黎成果，在考虑 2℃ 以内全球温度目标的同时，推进落实《公约》目标；重申致力于达成富有雄心的 2015 年协议，考虑到不同国情，体现共同但有区别的责任和各自能力原则。此次《声明》向世界再次作出了中美将携手合作、坚定推动巴黎气候谈判的政治承诺。这实际上是对国际社会的一种“再保证”。作为世界最大的发展中国家和最大的发达国家，中美两国元首再次向世界庄严宣誓其决心，向世界传递了强烈的积极信号，进一步提振了国际社会对巴黎气候大会的信心，也有助于为巴黎气候谈判营造良好的政治和舆论氛围。

二、美国应对气候变化的目标和行动

（一）《公约》下的承诺

2012 年美国化石燃料燃烧排放的二氧化碳总量为 50.7 亿吨，是全球第二大排放国，占全球总排放量的 16%。人均二氧化碳排放为 16.15 吨，显著高于世界平均 4.51 吨的水平。

美国曾在《京都议定书》下承诺在 2008—2012 年期间相比 1990 年排放水平减排温室气体 7%，但又在 2001 宣布拒绝核准《京都议定书》，因此该目标并未最终生效。在《坎昆协议》下，美国承诺到 2020 年全部温室气体排放量相对于 2005 年减排 17%，这一目标相当于比 1990 年水平减少约 1.6%。到 2012 年，美国温室气体排放比 2005 年下降约 9.6%，依靠现有的政策措施，美国实现 17% 下降目标有一定的难度。

2015 年 3 月 31 日，美国向《公约》秘书处提交了国家自主贡献（INDC）目标，明确了到 2025 年在 2005 年的基础上温室气体排放总量下降 26%—28%，并将尽最大努力达到 28%。这相当于比 1990 年下降约 12%—14%。美国在国家自主贡献文件中指出其已经采取了必要的政策措施保证实现 2020 年减排 17% 的目标，为实现 2025 年减排 26%—28% 的目标，还需要大幅加速温室气体的减排进程，年均减排速率将翻番，从 2005—2020 年的 1.2% 增加到 2020—2025 年的 2.3%—2.8%。美方案同时强调其 2025 年减排目标力度和实现 2050 年减排 80% 目标的路径相一致。

自 2009 年起，应对气候变化作为奥巴马总统的优先事项得到了大力推进，在机动车燃油经济性标准、建筑节能标准和氢氟碳化物（HFCs）替代品等领域取得了相当的进展。

（二）主要政策措施

1.《总统气候变化行动计划》

奥巴马总统发布的一份应对气候变化国家行动计划是迄今为止美国总统发布的最为全面的气候变化应对计划。如能全面并迅速落实该计划，奥巴马政府将能真正开始落实气候工作。

该计划的目标是全面减少温室气体排放，并保护美国免受日益严重的气候影响。通过制定并切实落实清晰的国家战略，美国政府不但能够保护本国人民，而且能够提振国际社会应对气候变化的雄心。这一目标有可能实现，但前提是多个经济部门必须联合开展进取行动。其中，减排的最大机遇存在于四个领域：电厂、能源效率、氢氟化合物和甲烷，这些领域都已明确列入该项气候变化应对计划。

同时，该计划的内容还涉及应对气候影响和鼓励加强参与国际进程。这些措施将帮助美国在气候变化问题上重振国际形象和领导力。虽然许多细节还有待完善，但总体而言该计划将有利于美国迈向更加安全的未来。

《总统气候变化行动计划》中具体内容包括：

(1) 减少发电厂的碳排放。为了达到这项目标，奥巴马向环保署签发了总统备忘录，要求环保署短期内完成对新建以及现存发电厂碳

表 7-2 《总统气候变化行动计划》的主要内容

《总统气候变化行动计划》 2009 年奥巴马承诺，2013 年再次承诺	
基准年	2005
目标年	2020
减排目标	减排 17% 的全经济范围减排目标
气体种类	CO_2,CH4, N_2O,HFCs, PFCs, SF5, NF3
涵盖的行业	所有 IPCC 考虑的排放源和排放行业，以完整的年度库存衡量（例如能源、交通、工业生产过程、农业、LULUCF、废弃物等）
LULUCF	对 LULUCF 的排放及对排放清除的测量，使用 net-net 的方法，并以 2005 年为基准年。美国考虑识别自然界对排放和排放清除的干扰产生的影响
其他	符合美国法律的规定

排放标准的制定 (2013—2016 年)。

(2) 发展新能源。该计划呼吁美国到 2020 年将可再生能源产量翻一番，并向可再生能源发展开放公共土地，加速清洁能源许可证的发放，力争到 2020 年将公共土地上的可再生能源发电装机容量增加 10 吉瓦，这将满足 260 万个美国家庭的用电需求。该政策为即将出台的电厂排放标准提供了有力补充，并为电厂达到排放标准提供了便利。联邦政府拥有占美国 28% 的国土，有选择地对清洁能源项目开放一些公共土地将有利于缓解电厂和项目开发商在选址方面的担忧。

(3) 激励对清洁能源的长期投资。奥巴马要求能源部颁布一项“联邦登记通知”，发布一项贷款请求的草案，在 1703 号贷款担保项目下，为新一代化石能源项目提供 80 亿的贷款资金。

(4)提高能源效率。提高能效是实现减排最为经济有效的方法之一。能效较高的设备能耗较低，可为消费者节约成本。已有数十项产品符合新的和升级能效标准，其中有些还在等待审批。此外还可通过多种途径在民用、商用和工业领域轻松实现减排目标。最新分析表明，目前共有 6 项标准正在等待审批，而审批每拖延一个月，都会给消费者带来 2 亿美元的储蓄损失并向大气中多排放 300 万吨二氧化碳。对家用电器和公共建筑采取新的或现行能效标准之后，预计可以使 2030 年二氧化碳减排量达到 30 亿吨，如此大规模的减排几乎相当于美国燃煤电厂两年的排放总量。此外，在奥巴马总统的第二任期内，美国政府将再次与领军企业以及其他利益相关方合作，制定 2018 年后重型交通工具的燃油经济性标准。

(5) 确立四年一次的能源评估制度。要求美国政府停止支持在海外用公共融资去建燃煤电厂。

2.《清洁电力计划》

2015 年 8 月 3 日，美国总统奥巴马颁布了《清洁电力计划》，这是美国在电力行业温室气体控制领域的一个重要里程碑。美国电力行业的温室气体排放约占总排放的 40%，该计划是美国在气候变化领域的重要举措，体现了美国应对气候变化的决心。

《清洁电力计划》要求美国的电力行业到2030年在2005年的基础上减少32%的温室气体排放，并从2022年开始强制减排，比之前草案中规定的要晚两年。为了给电力企业留出更充裕的缓冲时间，美国环保署放宽了前几年的减排量要求，此后则逐步提升。这一做法可以让该计划更容易实现。实际上如果从全局看，美国的电力行业已经在2005年基础上减少了15%的碳排放，这意味着减排任务已经完成近半。其中一些州的减排程度超出了人们的预期，这足以证明电力行业是可以作出快速应对的。从2005年到2012年，美国有16个州的电力行业温室气体减排量超过25%，还有9个州的排放量甚至降低了40%以上。相比之下，《清洁电力计划》的要求——到2030年，在2014年基础上减少大概20%（2005年基础上的32%），就显得唾手可得。

《清洁电力计划》为电力行业碳减排项目的投资提供了强大的政策信号和激励机制，能确保电力行业在降低碳排放的道路上继续前行。美国电力行业在2010年到2030年期间计划投资20亿美元用于发电、传输和配电等基础设施的更新换代，这些设备的寿命通常达到几十年。严格、明确的碳排放标准为电力行业提供了清晰的投资信号，并降低了这些投资的预期风险。

从技术层面实现电力行业碳减排完全可行，且现有的技术在成本上已经颇具竞争力。从20世纪90年代开始，随着发电经济性的发展和美国各个州内制定的相关政策，可再生能源和天然气电厂已经在新建电厂中独占鳌头。美国超过半数的州设立了能效目标，节省了可观的电费支出。燃煤和燃油电厂能效提升的技术也已十分成熟。

根据《清洁空气法》，美国环保署的排放标准要能反映已被证明的“最佳减排系统”，同时考虑成本、能源需求和其他因素。遵从《清洁空气法》的这一要求，《清洁电力计划》考虑了电力行业的技术可行性、电力可靠性及区域和技术的差异，为化石电厂的碳排放设立了最低排放标准。标准是基于以下三个模块制定的：

模块1：提高现有煤电厂的效率，降低碳排放；

模块 2：在 2022—2030 年期间，逐渐用天然气替代煤和油发电；

模块 3：逐渐用零排放的可再生能源发电替代化石燃料发电。

这些模块将在 2022—2030 年期间逐渐被采用。同时，环保署考虑了电力跨区域传输的特点，以区域为基础，根据地区的资源条件评估了这些模块的潜力，并在此基础上，选取最宽松的区域标准设定国家标准。环保署为燃煤燃油电厂设定的 2030 年的二氧化碳绩效排放标准为 1305 磅 / 兆瓦时（约 590 克 / 千瓦时），天然气燃机电厂的绩效排放标准为 771 磅 / 兆瓦时（约 350 克 / 千瓦时）。

考虑到能效项目的价值，环保署允许各州使用其减排量进行履约。《清洁电力计划》赋予各州极大的灵活性，在实现排放标准的前提下，允许各州制定自己的州计划。各州计划可以满足或超越国家的标准，并兼顾各州的特点。

州的排放限值可以有四种灵活选择：①直接采用国家对电厂要求的排放标准；②可以对国家的排放标准进行加权平均，不分技术类型，应用到所有化石燃料发电机组，形成唯一的混合标准；③州可以采用环保署为电厂设定的年度总量排放目标；④州可以考虑电力行业的发展，兼顾现有和新建电厂的排放总量。

《清洁电力计划》对美国来说是重要的一步，它是各州和电力公司渴求更清洁、低碳的经济发展的坚实基础。美国的电力行业有足够的能力，确保在电力系统供电稳定和电价平稳的同时，用最经济有效的手段实现并超过《清洁电力计划》规定的碳排放目标。

3．多元化能源战略[①]

2014 年 5 月 30 日，美国能源部发布了题为《多元化能源战略——通向可持续经济增长之路》的报告，该报告总结了奥巴马团队执政 5 年多来能源政策实施的效果，提出了奥巴马剩余任期内的能源政策纲

①张晓华，徐丽丽，胡晓．美国能源部《多元化能源战略——通向可持续经济增长之路》评述，2014 年 6 月 4 日．

领。尤其是该报告提出的“支持经济增长和增加就业，加强能源安全和保障能源供应，发展低碳能源技术为未来清洁能源奠定基础”这三个美国国家能源战略核心要素，在国际社会产生了重要的影响。

多元化能源战略的基本内容主要有四个方面，一是美国能源战略的基本要素，二是能源变革和经济复苏的关系，三是能源变革与保证能源安全问题，四是能源发展的低碳路径。

报告认为，2005 年以来，美国的能源行业正在经历着根本性的变革，其特点是：

第一，长达数十年的能源消费趋势正在被逆转。自 2005 年以来美国石油消费和煤炭消费量在不断下降，其缺口是由天然气和可再生能源补充的。2005 年以来，美国天然气消费量增长了 18%，可再生能源的利用量增加了 5 倍。

第二，能源行业为美国摆脱经济危机，实现经济复苏提供了关键支撑，国内能源供应的提升对 GDP 增长和就业贡献巨大。2012 年和 2013 年，仅原油和天然气生产两项对实际 GDP 增长的贡献率就超过 0.2 个百分点，两个行业在 2010—2013 年期间新增了 13.3 万个就业岗位。太阳能和风能产业也新增了上万个工作岗位。

第三，能源变革降低了美国能源对外依存度，生产增加、消费减少降低了进口，从而大幅度减少了美国的贸易逆差。尤其是随着石油净进口费用和原油对外依存度的降低，美国经济对国际石油供给和价格波动冲击的敏感性开始下降。

第四，美国已成为世界领先的石油和天然气生产国。2013 年美国石油、天然气和其他液体燃料的总产量已超过沙特阿拉伯和俄罗斯。在天然气生产方面美国也已处于世界领先地位。

报告还认为，当前的能源变革是能源行业技术进步推动的，政府只是起到了顺水推舟的作用。报告指出：“早在政府提出战略之前，能源行业就已出现一些发展趋势。总统的多元化能源战略通过对环境负责的方式支持并加速了上述发展趋势。”总统的新的能源战略主要包括以下几点：

一是国家能源战略把天然气作为一种过渡性的燃料，并分步执行，以确保天然气的适度发展。相较于许多其他能源，天然气更加清洁，但是开发利用天然气也需要严格的环境保护措施以减少或避免其环境损害。

二是多元化能源战略支持可再生能源、核能和其他零碳排放能源的研究、开发和应用，并支持在能效领域的投资。政府机构，包括军事设施要率先使用可再生能源和节能建筑。

三是通过推广低碳技术为未来清洁能源奠定基础。报告认为，自2005年以来，美国减少的碳排放总量比起地球上任何其他国家都要多，主要得益于能源转型和能源变革。报告认为由于温室气体排放的外部性，开发和应用低碳技术应该成为国家能源长期战略，政府应该为低碳技术提供市场保障。

4. 其他主要气候能源政策

除此之外，美国应对气候变化领域最主要的政策行动还包括：

（1）清洁空气法案。美国环保署（EPA）于2014年6月提出指导现有电厂运行和电厂新建的规定。要求电力行业到2030年在2005年的基础上减排30%，可以通过改善公共卫生、减少碳污染而获得550亿—930亿美元。

（2）发动机和机动车标准。按照最新的机动车燃油经济性标准，美国市场上各车企2017—2025年款新车的燃油经济性平均值应当达到54.5英里/加仑，约合百公里4.3升油耗，比当前水平几乎提高一倍。

（3）能源效率标准。美国能源部预计这个能效标准到2030年可实现累计减碳30亿吨。该能效标准可以帮助全国消费者每年节约数十亿美元的电费。

总的来说，2012年美国一次能源消费总量相对2007年下降了8.8%，这一下降一方面是经济低迷造成的，另一方面则归功于美国一系列的能源政策大幅度提高了能效效率。其中，作用最明显的是车辆燃油经济性新标准的制定和建筑能效规范等政策，美国能源效率方面的政策行动及预期实施效果详见表 7-3。

美国各类能效政策在其实施过程中存在着不同的驱动力，其中政府对提高能效的政策共识对建立联邦政府与州政府之间统一协调的能源政策会有很大的帮助，但与此同时仍然存在着一些阻碍，例如目前较低的天然气价格不利于对能源的节约利用，提高能效很可能带来的能源销量下降会引起能源供应商的不满，等等（具体见表 7-4）。

近几年美国在提高能效方面已经取得了重大进展，预计在 2020 年美国将成为最节能的国家之一。能源效率改进重点包括机动车能效、家电和设备能效、建筑能效提高。

表 7-3 美国能源效率方面的政策行动及预期实施效果

相关行业	政 策	2011 年节能量（TWh）	预计 2020 年节能量（TWh）
轻型和重型车辆燃油经济性标准	美国环保署（EPA）/ 国家公路交通安全管理局（NHTSA）联合制定规则 2012-16-2017-25	N/A	962
家电及设备标准计划	1987 年、1988 年国家家电节能行动 (NAECA)；1992 年能源政策法案 (EPAct1992)；2005 年能源政策法案 (EPAct2005)；2007 年能源独立与安全法案 (EISA)	400 其中： 电力：242 天然气：156	695 其中： 照目前的政策不变：610 新政策贡献：85
财政支持的能效计划	州级层面立法和监管控制，确立能源效率标准和节能义务	117.3 其中： 电力：81 天然气：36	至少 210 至多 255
建筑规范	州级层面住宅和商业建筑规范	64 其中： 电：37 天然气：26 油：0.5	244 其中： 电：89 天然气：69 油：1
能源服务公司	EISA，第 432 节； 美国复苏与再投资法案 (ARRA)	270	770
合计		851.3	2926

资料来源：IEA，美国能源政策，2014.

表 7-4 能效政策实施过程中存在的驱动力和阻碍

驱动力	阻碍
1. 联邦政府与州政府的政策共识	1. 较低的天然气价格
2. 新的管道能效技术	2. 公共预算减少的压力
3. 越来越严格的州级能效政策要求	3. 能源供应商对销售量下降的反应
4. 能源的协调供应提高了能源利用效率	4. 不缺乏劳动力（劳动力充分）
5. 新能源市场更注重高能效	5. 受制于监测能力
6. 新的联邦立法：Shaheen-Portman Act and the Sensible Accounting to Value (SAVE) Act.	6. 能效提高的边际成本越来越高

政府在为能效技术发展提供资金支持的同时，还不断鼓励私营部门对提高能效领域进行投资。2009 年的经济刺激计划为现有的能源项目提供了额外资金资助。这些项目的直接和间接融资总额超过 300 亿美元，是 2008 年的资金水平的 5 倍。财税政策和能源服务公司的发展也有助于能效监管政策的实施。除此之外，税收优惠政策通过刺激对节能家电或设备的消费，淘汰低效设备，提高总体能效水平。

（4）全经济范围减少其他温室气体的措施。美国环保署和其他机构正通过采取多种不同的措施来减少垃圾填埋、煤炭、石油开采、天然气开采运输过程中以及农业部门的甲烷排放量。国务院在《蒙特利尔议定书》下，努力削减全球工业氢氟烃的排放。美国环境保护署制定了新替代政策（SNAP）以削减国内氢氟烃的排放。此外，私营部门承诺到 2025 年为全球减排氢氟烃 7 亿吨。

（三）美国气候援助的基本情况

1．基本政策目标

美国正在充分利用双边、多边融资以及出口信贷等多种方式，调动民间融资，并进行关于气候变化负面影响适应性建设的战略性投资。具体表现为：减少森林砍伐和森林退化带来的排放，支持低碳发展战略，过渡到一个可持续的、清洁能源主导的经济形式。2014 年美国宣布向绿色气候基金注资 30 亿美元。

通过一系列显著政策指令，气候变化已成为美国外交和发展目标的一大重点。2010 年总统指令政策为全球发展所确定的全球气候变化倡议（GCCI）提供了一个平台，使美国将气候变化因素纳入其对外援助行动之一。2012 年美国国际开发署（USAID）提出了关于气候变化与发展的原则、目标与战略。

此外，海外私人投资委员会（OPIC）调整了政策，将更多的国际私人投资注入气候友好活动中。作为美国政府的金融发展机构，OPIC 调动民间资本以应对发展的挑战，这样做有助于实现美国发展和外交政策目标。

用长远的眼光对经济和环境的可持续发展进行展望，美国正在努力通过受援国自主计划、着眼于可衡量的结果等方式，确保其能力建设和投资支持具有高效性和创新性。

2．资金规模和渠道

2010—2012 年美国共提供援助资金 75 亿美元，其中政府提供的官方援助 47 亿美元，发展融资 19 亿美元，出口信贷 7.5 亿美元。通过多边渠道援助 12 亿美元，其中气候投资基金（CIF）下 20 亿美元（已落实 11.4 亿美元），GEF 下 2010—2014 年累计援助 5.6 亿美元。目前美国也是多边适应基金最大的援助国，向最不发达国家基金和气候变化特别基金注资。

用长远的眼光对经济和环境的可持续发展进行展望，美国正在努力通过受援国自主计划、着眼于可衡量的结果等方式，确保其能力建设和投资支持具有高效性和创新性。美国的气候援助绝大多数是官方发展援助 ODA，很少一部分通过贷款的方式。

3．实施机构

美国双边气候援助的主要实施机构包括美国国际开发署（USAID）、外交部（DOS）、千年挑战集团（MCC）。

美国国际开发署是承担和实施美国大部分对外非军事援助的联邦政府机构，作为一个独立的联邦机构，依照美国国务院的外交政策进行相关援助。由 USAID 的经济增长、教育与环境局下设的全球气候变化办公室主要负责。作为美国双边和多边发展援助的首要实施机构，USAID 对 50 多个发展中国家提供资金和技术等相关气候援助，重点关注适应、清洁能源与可持续景观。

2004 年初，美国总统布什在墨西哥的蒙特雷发出“世界发展新协调”的倡议，将发达国家更多的援助与发展中国家更大的责任相结合，并为此提议建立了一种机制，既“千年挑战账户”，或称作“利于千年发展的账户”。在此机制下，大量的发展援助基金将提供给那些实施良政、促进人类发展及鼓励经济自由化的国家。在两党议员的强烈支持下，美国国会批准建立了初始资金为 10 亿美元的专项援外资金。2006 年起，布什总统承诺将该基金增至 50 亿美元，使目前美国援外资金总量增加了近 50%。基金由一家新的国有公司——千年挑战公司（MCC）负责，确保账户实施效果。MCC 总经理由总统任命、国会批准、受董事会的监督（董事会组成：国务卿、财长、美国贸易代表、MCC 总经理以及总统任命的国会批准的另外四人，国务卿为董事长）。

在利用多边渠道开展气候变化援助方面，财政部和外交部扮演最重要的角色，大部分资金流向国际发展协会、境界技术基金、非洲发展基金、全球环境基金和气候投资基金（CIF）等。

4．优先领域和区域

美国气候援助的优先领域包括适应、清洁能源、可持续景观。适应领域的援助多关注食品和饮用水安全、近海管理和公共卫生，主要提供给最脆弱的国家，包括最不发达国家、小岛屿发展中国家和非洲等；可持续景观多指帮助发展中国家减缓和停止森林破坏，减少温室气体排放，相关气候援助主要提供给世界性的热带雨林地区国家，如中非、亚马逊、南亚地区等；清洁能源相关的气候援助主要提供给新兴经济体，即未来的排放大国。

5．气候援助的主要活动和形式

美国为发展中国家提供气候援助的形式主要包括：提供信息、帮助各国实施改革和相关项目、开展试点、开展合作研究和带动私人投资。

例如，美国相关部门与美国航天局（NASA）合作为相关国家提供更全面的卫星数据和信息，提高农业、水资源、能源、健康、森林、土地利用与规划等方面的适应性。

美国开展的“非洲基础设施”（AIP）项目也将气候援助用于提高清洁能源利用的基础设施项目，帮助撒哈拉以南的非洲开展清洁能源能力建设，帮助相关非洲国家政府和机构规划和实施有关吸引清洁能源私人投资的制度与法律法规的改革。AIP 项目还提供具体的技术援助和咨询服务，以支持政府的评估和清洁能源项目的谈判和实施。

美国还在发展国家资助林业投资方案，加强森林管理和制度的有效性，并采取措施减少森林砍伐。美国资助的森林碳伙伴基金（FCPF）帮助系统开展项目试点，推动林业减排。

同时美国也积极推动相关的国际倡议，使气候行动与合作更具规模，例如美国是全球甲烷倡议（GMI）的领导捐助国，致力于减少全球范围内甲烷的减排和后期资源化处理。

美国也十分重视私营部门在气候变化援助中的贡献，正在努力通过金融发展机构、多边发展银行、主要的多边气候变化基金和出口信贷机构等工具，促使公共投资带动发展中国家在减缓和适应气候变化方面的私人投资。首先，私人投资者管理的资源较公共资源来说较低级，而这些资源往往可以更快、更有效地进行分配；其次，由于气候问题的规模不同，只有公共资源的投入不足以充分解决气候变化问题；此外，私人投资更有效的杠杆作用可以使国家将现有公共资源应用到私人部门不能直接接触的领域和部门；最后，与缓解气候变化相关的投资可以从中获得经济收益。例如美国支持建立了非洲清洁能源基金（ADEF），旨在通过动员清洁能源项目的私人投资，解决非洲撒哈拉以南地区的迫切能源需求。通过外交部提供的 0.2 亿美元资金，美国贸易与发展机构提供的项目计划以及 OPIC 提供的金融工具和风险规避工具，ADEF 将带来上亿美元的私人投资。ADEF 在非常有限的公共援助基础上动员更多的资金以实现气候项目的实施提供了示范。

6．技术转让活动

以促进气候友好型技术和实践的发展为首要目标，美国通过双边和多边的方式和机制，与发展中国家和经济转型国家之间进行技术发展和转让活动。

美国技术转让的原则是支持促进技术转移行动的政策法规、制度和框架的建立与发展。例如，美国帮助相关国家进行电器能效标准的能力建设活动，帮助其制定可再生能源政策以及智能电网的监管制度等。同时，美国与其他国家共同努力，改变市场结构，加快技术流动。在多边层面，美国为UNFCCC技术执行委员会和气候技术中心网络作出了贡献。

7．能力建设活动

美国的对外援助中包含基础设施建设、信息共享等能力建设活动，尤其强调提高发展中国家气候变化适应能力。

三．对美国应对气候变化目标和行动的解读和评价

（一）减缓目标

美国二氧化碳排放于2005年达到峰值，达峰时国内其生产总值约13.1万亿美元，人均GDP约4.4万美元，二氧化碳排放量约为60亿吨，人均排放量约20吨。工业化至今，美国累积二氧化碳排放量约为3600亿吨，相当于中国同期累积排放量的3倍，即使排放已经达峰，美国对目前已经产生的气候变暖和未来将会发生的气候变化依然负有重要的责任。

2001年，美国退出了《京都议定书》第一承诺期，其最初承诺的减排目标也低于当时为欧盟、日本等设置的减排目标。而美国2020年减排目标选择以2005年为基年（排放峰值年），实际上放宽了对2020年排放量的约束，与其他发达国家相比减排力度较弱。美国进而承诺到2025年比2005年减排26%—28%，虽然其宣称2020年到2025年的减排速率比2020年前加倍，但实际减排量以及目标年的绝对排放总量和人均量仍然较高。

即使美国承诺的减缓目标并不具备较强的力度，其实施进展仍不乐观。根据美国第一次双年报技术专家审评报告的相关结论，依靠现有的政策措施，美国并不能实现其 2020 年减排 17% 的目标，仍需要进一步加强行动。

（二）气候支持的充分性

《联合国气候变化框架公约》下规定了发达国家向发展中国家提供新的、额外的资金支持的义务，但是各方对于如何定义“新的、额外的资金”有显著的分歧。发展中国家指出只有超出发达国家既定官方发展援助数额（GNI 的 0.7%）的资金才能够算做“新的、额外的”气候援助。作为世界第一经济大国，战后美国为全球提供了大量的官方发展援助，目前其每年的援助金额仍高于援助规模第二大和第三大国家援助额之和，但是美国在向《公约》下资金机制提供支持方面表现差强人意。2014 年，美国承诺向绿色气候基金（GCF）注资 30 亿美元，并在 2015 年中美元首气候变化联合声明中重申此承诺，后续的落实情况以及提高资金支持规模都有待观望。

（三）国内政策

从《总统气候变化行动计划》可以看出：奥巴马总统就全面推进应对气候变化国际与国内行动向国际社会释放出了强烈的政治信号，得到了国际社会的广泛关注。《总统气候变化行动计划》作为美国未来气候变化行动的纲领性文件，其实施及其所反映出的美在气候变化问题上的未来动向必将对全球应对气候变化的努力产生较大的影响。作为世界第一大国，美国显然时时不忘巩固其领导地位，这也是其吸引国内民众支持的常用手段之一。但美国对领导力的理解并非以自身更有力度的减排行动来率先垂范，而是如何迫使中、印、巴等发展中大国采取更为实质性的减排行动。因此，无论是《公约》外小多边合作、《公约》下新协定还是行业减排行动，其核心都是指向中、印、巴等发展中大国。

另外值得注意的一点是，在该计划中美国重提了环境产品和服务的自由贸易问题，这是 WTO 于 2001 年启动的多哈发展回合谈判下关于贸易与环境部分的重要组成部分。此次将其打包到气候变化，除了符合美善于掌握制定国际规则话语权的一贯做法，很大程度上也处于有助推动美国经济实现绿色增长的考虑。中国政府目前也积极通过战略规划和产业优惠政策扶持环保产业发展，自由贸易的过快推动可能将在一定程度上阻碍我国相关产业的壮大与发展，对中国实现低碳发展产生负面影响，值得进一步关注。

作为美联邦政府迄今为止最为全面和系统的应对气候变化方案，从形式上讲，该计划具有极大的象征意义，对国际应对气候变化进程将会产生很大影响。就实质性内容而言，计划提出的措施，在“绿色新政”、“清洁能源法案”内都有类似的内容，属于“新瓶装旧酒”，缺乏新的、重大关键性的制度措施，只是对于已有燃煤电厂的举措可算其中的亮点，但真正的落实效果存在较大变数，尚待进一步观察分析。

总的来说，美国在制定应对气候变化政策时都是从国内的实际国情及需求出发，在研究确定了可行性之后才颁布实施。有些气候政策看似缺乏雄心，但基本都能保证实施完成。

虽然美国近几年在应对气候变化方面已经取得显著的进展，但仍存在着一些问题。

(1) 美国仍然缺少气候与能源中长期战略与目标，这会导致政策的长期性、持续性、一致性和确定性难以保证。

(2) 联邦层面的能源与气候变化政策缺乏一个统一的框架。一直以来有许多国内外的批评者认为，美国缺乏一个统一的国家级能源政策。为此，在过去六年中，美国制定了一系列能源战略政策，一定程度上解决了对化石燃料的依赖，有利于建立更独立的可持续的能源系统。其中需要重点关注的有“总统的能源未来蓝图”和“各类能源战略”。此外，2009 年制定的《美国复苏和再投资法案》(ARRA) 扩大了对能源基础设施的投资，对能源行业产生了重大积极的影响。通过 ARRA, 美国能源部 (DOE) 投资逾 310 亿美元，大范围地支持清洁能源技术发展。

(3) 联邦层面，能源、环境、经济和安全政策缺乏协调性，职权分散，国会、白宫和州政府以及联邦政府下的行政和立法部门之间协调难度很大。

(4) 页岩气发展可能带来的甲烷泄漏及环境污染风险，以及减排效果的长期可持续性风险。美国页岩气资源的快速发展伴随着更严格的空气质量监管有助于减少排放量用天然气代替煤。排放减少的趋势是否能持续还取决于未来煤炭和天然气的相对价格。如果未来煤炭的价格比天然气更低，就会存在排放回弹的风险。虽然新的排放标准能在一定程度上抑制煤炭的使用，但这个风险仍然存在。此外，另一方面的担心是，如果不控制页岩气生产和分配过程中的甲烷排放，对气候变化也会造成重大影响。

(5) 可再生能源支持政策等存在不确定性。以联邦可再生能源生产税收减免政策（PTC）为例，PTC 是对符合条件的可再生能源发电量进行通胀调整后对每度电予以税收减免。PTC 的实施规则自制定以来进行过多次调整，会在不同时刻偏向于不同的资源类型和设施类型（见表 7-5）。

PTC 调整可能对可再生能源的投资造成影响，因为能源基础设施建设的周期往往比较长。例如某些投资者在风电有税收优惠时开始建

表 7-5 联邦可再生能源生产税收减免政策（PTC）的适用范围

能源类型	起始年	额度
风能	2013.12.31	USD 0.023/kWh
闭环生物质能	2013.12.31	USD 0.023/kWh
开环生物质能	2013.12.31	USD 0.011/kWh
地热能	2013.12.31	USD 0.023/kWh
堆填区沼气	2013.12.31	USD 0.011/kWh
城市固废	2013.12.31	USD 0.011/kWh
水电	2013.12.31	USD 0.011/kWh
海洋能及流体动能	2013.12.31	USD 0.011/kWh

资料来源：IEA，美国能源政策，2014。

设风电项目，但是如果项目还没完工，风电的税收优惠政策被取消，无意将影响投资收益。

2013 年初美国 PTC 延期一年，但是未来是否还能延期存在不确定性。这种政策的不确定性易削弱投资者的信心，造成可再生能源装机量年际之间的波动。

第二节　欧盟

一、欧盟的全球气候治理理念

欧盟是全球气候治理的主要倡导者，是推动气候变化谈判进程的重要政治力量。欧盟支持通过以《公约》为基础的多边进程构建全球气候治理，因此，欧盟参加公约谈判的立场是其全球气候治理理念的充分反映。

欧盟认为全球气候治理应确保最广泛国家的参与和采取有力度的减缓行动，各国的减缓行动应具有国际法律约束力。欧盟认可《公约》原则是全球气候治理体系的基础,各国可以根据《公约》原则有所区分，但是原则必须以动态的方式适用，以反映不同缔约方不断变动的责任和能力，责任最大和能力最强的缔约方均应在全经济领域作出最大程度的减排承诺。欧盟认为全球气候治理体系应建立在清晰的、统一的规则基础上，特别是对温室气体排放的测量、报告、核查以及核算等规则。欧盟是碳排放交易市场制度的倡导者和实践者，积极推动建立国际碳市场，认为通过市场给碳排放定价是全球气候治理的关键内容。

欧盟认为《公约》缔约方在政治上确定的将全球温升控制在 2℃以内的政治共识应该根据有关科学结论量化并据此予以落实。各国一方面要根据国情、责任、发展阶段和现实能力提出应对气候变化的贡献，还应评估贡献是否足以实现 2℃目标，并据此决定是否提高承诺力度。

二、欧盟气候政策与行动

欧盟在《公约》实施以来的 20 多年中，在应对气候变化领域采取了积极努力，取得了一定成效（参见表 7-6）。1990 年至 2012 年期间，欧盟 15 国的温室气体排放总量（不包括 LULUCF）下降了 15.1%（6.4 亿吨 CO_2 当量），欧盟 28 国的温室气体排放总量（不包括 LULUCF 和

表 7-6 欧盟成员国温室气体排放情况

（百万吨 CO_2 当量）
（包含六种温室气体 CO_2, CH4, N_2O, PFCs, HFCs, SF7，不包含土地利用变化）

国家	1990	1995	2000	2005	2006	2007	2008	2009	2010	2011	2012
奥地利	78	80	80	93	90	87	87	80	85	83	80
比利时	143	150	146	142	138	133	136	123	131	120	117
丹麦	69	76	69	64	72	67	64	61	61	57	52
芬兰	70	71	69	69	80	78	70	66	74	67	61
法国	557	553	561	559	547	538	533	509	516	490	490
德国	1248	1118	1040	994	1002	977	980	913	946	929	939
希腊	105	110	127	135	132	135	131	124	118	115	111
爱尔兰	55	59	68	70	69	68	68	62	62	58	59
意大利	519	530	551	574	563	555	541	490	499	487	460
卢森堡	13	10	10	13	13	12	12	12	12	12	12
荷兰	212	223	213	209	206	204	203	198	209	195	192
葡萄牙	61	71	84	88	83	80	78	75	71	69	69
西班牙	284	322	380	431	424	432	398	360	347	346	341
瑞典	73	74	69	67	67	65	63	59	65	61	58
英国	775	723	690	675	672	662	643	590	606	563	581
EU-15	4262	4171	4156	4183	4157	4095	4007	3722	3803	3650	3619
保加利亚	109	76	59	64	65	68	67	58	60	66	61
克罗地亚	32	24	27	31	31	33	31	29	29	29	26
塞浦路斯	6	8	9	10	10	10	11	10	10	10	9
捷克	196	152	146	146	147	147	142	134	137	135	131
爱沙尼亚	41	20	18	18	18	21	20	16	20	20	19
匈牙利	98	78	78	78	77	76	73	67	68	66	62

续表 7-6

国家	1990	1995	2000	2005	2006	2007	2008	2009	2010	2011	2012
拉脱维亚	26	13	11	11	12	12	11	11	12	11	11
立陶宛	49	22	23	23	24	26	25	20	21	22	22
马耳他	2	2	3	3	3	3	3	3	3	3	3
波兰	466	441	399	414	414	415	406	388	407	406	399
罗马尼亚	248	175	141	145	145	143	140	120	116	122	119
斯洛伐克	73	53	50	50	50	48	49	45	45	45	43
斯洛文尼亚	18	19	20	21	21	21	21	19	19	19	19
EU28	5626	5253	5122	5178	5173	5119	5006	4642	4751	4603	4544

资料来源：欧盟提交《公约》秘书处的国家排放清单。

国际航空）下降了 19.2%（10.8 亿吨 CO_2 当量）。欧盟 28 国人均排放量降幅达 24%，从 1990 年的 11.8 吨降至 2011 年的 9 吨。

总体来说，欧盟温室气体排放主要由德国和英国主导，2012 年两国排放量占欧盟 28 国总排放量的约三分之一。在不使用土地部门碳汇和《京都议定书》灵活机制的情况下，德、英两国 2012 年已经实现本国温室气体相比 1990 年减排 5.0 亿吨 CO_2 当量。德国的减排主要是依靠电厂和供暖设备能效提高以及统一后五个新联邦州的经济重组来实现的。英国温室气体的减排主要是依靠能源市场自由化、发电行业燃料由石油和煤炭向天然气转型，以及已二酸生产过程中 N_2O 的减排措施来实现的。

2012 年，法国和意大利分别是欧盟第三和第四大排放国，排放占比分别为 11% 和 10%。2012 年意大利的温室气体排放比 1990 年水平低 11%。1990 年后意大利的温室气体增量主要来自于道路交通、电力、供暖以及石油加工行业。2004 年后意大利的排放量开始呈现下降趋势，2009 年和 2012 年跌幅显著，主要是经济危机和工业输出量下降造成的。法国 2012 年温室气体排放量比 1990 年下降 12%，跌幅主要依靠已二酸生产过程中 N_2O 排放的控制，但道路交通相关的 CO_2 排放和 HFCs 排放在 1990 年至 2012 年期间有明显增长。

波兰和西班牙排放水平在欧盟 28 国分别列居第五和第六，2012

年的占比分别是 9% 和 7%。1990 年至 2012 年期间，西班牙排放量增长 20%，主要来自于道路交通、电力和供暖以及人居和服务业。同期，波兰的温室气体排放量下降 14%，主要是由于低能效重工业的减产以及 20 世纪 80 年代末 90 年代初的经济重组，但交通相关排放仍呈现增长趋势。

（一）德国

德国政府于 2010 年和 2011 年先后出台了《能源方案》和《能源转型一揽子方案》，明确了退出核能和加速发展可再生能源的战略目标。德国政府还设定了能源转型和应对气候变化的量化目标。德国计划到 2020 年与 1990 年水平相比温室气体排放下降 40%、2030 年下降 55%，2040 年下降 70%，2050 年下降 80%—95%；可再生能源消费占终端能源消费的比重 2020 年、2030 年、2040 年和 2050 年分别实现 18%、30%、45% 和 60%，占总电力消费的份额分别实现至少 35%、50%、65% 和 80%；一次能源消耗与 2008 年相比到 2020 年和 2050 年分别下降 20% 和 50%，电力消耗分别下降 10% 和 25%，交通领域终端能源消耗分别减少 10% 和 40%；电动汽车到 2020 年达到 100 万辆，到 2030 年达到 600 万辆。

德国的主要政策措施包括制定节能和提高能效的法律法规、可再生能源固定电价以及能源税和碳排放权交易等。通过法律法规强力促进节能和能效提高。在能效方面，2009 年修订《建筑节能条例》进一步提高新老建筑的节能标准，通过《能耗标示法》和《耗能产品生态设计要求条例》，通过热电联产法，并于 2004 年和 2009 年分别出台了《乘用车强制性能效标识条例》和《新机动车税制条例》。

在可再生能源方面，德国于 1990 年出台《电力接入法》，奠定推动可再生能源发展的政策基础；2000 年实施《可再生能源法》，引入可再生能源发电强制入网和优先购买以及固定入网电价和电价负担均摊等措施。此外，为保证可再生能源的持续发展，德国正在推动电力体系和市场的全面改革，提升电力体系灵活性，整合电力和热力系统。

此外，德国还利用能源税和碳排放交易传递能源价格信号。能源税的征收和管理主要依据2006年颁布并生效的《能源税法》。德国是欧盟碳排放权交易机制的积极参与者，国内建立了完整的认证和核算机构保障交易体系的运行。

（二）英国

英国早在2003年即颁布能源白皮书《我们未来的能源——创建低碳经济》，率先提出了“低碳经济”概念，并提出四项低碳转型的目标：①到2050年温室气体减排60%，并且在2020年取得显著进展；②确保能源安全；③增加市场竞争力，提高生产力；④确保居民享用足够可负担的能源。在此基础上提出面向21世纪的“低碳经济”以及可再生能源发展战略，提出到2020年可再生能源发电量达到20%的目标。

英国政府将可再生能源视为低碳经济发展战略保障体系的核心，通过制定《低碳转型行动计划》和《可再生能源路线图》提出了一系列发展可再生能源的目标。

英国的应对气候变化政策主要着力于节能和提高能效、发展可再生能源以及碳定价机制等。

在节能方面，英国于1995年颁布了《家庭节能法》。2000年公布《气候计划》，制定了一系列的政策和措施以提高能效。设立节能专项基金，包括碳基金和节能基金。2003年能源白皮书提出“最廉价、最清洁和最安全的实现能源政策目标的手段是使用更少的能源”。

在可再生能源方面，英国将发展可再生能源视为国家可持续发展战略的核心，借此推动电力市场改革。20世纪90年代以来，英国构建了以化石能源税和“非化石燃料义务”为核心的政策框架。1989年起对终端电力消费者征收化石能源税，税率为2.2%，并将其用于可再生能源发电的补贴。1990年推行“非化石燃料义务”法规，为可再生能源提供担保。1999年通过《可再生能源义务法》，要求电力供应商在其供应给消费者的电力中必须有一部分可再生能源。2002年实施《可再生能源义务法令》，2011年又发布《规划我们的电力未来：关于发展安全、价格适宜和低碳电力的白皮书》，标志着电力市场改革开始。

在碳定价方面，2008 年英国通过《气候变化法》，以传递长期低碳转型信号并建立了“碳预算管理”制度,为每五年设定全国排放上限。这是全球首次将二氧化碳进行预算式管理和控制。此外，英国还通过 EU-ETS 和气候变化税来推动低碳转型。

（三）法国

2000 年法国颁布了《控制温室效应国家计划》，提出在 2008—2012 年期间，将《京都议定书》规定的六种温室气体的排放总量限制在 1990 年的 1.4 亿吨以下。2005 年法国发布了《法国适应气候变化战略》，将气候变化风险的科学评估与实施适应行动计划有机地结合起来，设立了四个总体目标，包括优先考虑公共安全与健康以及保护人员和物品、考虑社会各方面问题、在风险到来之前缓和不平等现象、降低成本并使收益最大化和保护自然环境。

为实现应对气候变化的目标，法国政府改组并成立专门机构，加强对气候变化的管理，2007 年组建了生态、能源、可持续发展和国土整治部。法国政府还通过《国家预防温室效应计划》、《环境能源和可持续发展协作行动》、《可再生能源发展计划》等一系列战略规划鼓励可再生能源与核电发展。

三、对欧盟应对气候变化行动的评价

尽管欧盟采取了积极的政策与行动，也取得了显著的成效，但从碳排放与经济脱钩的指标看，进展并不明显。从消费端排放来看，经济与排放脱钩的趋势并不显著。1990—2011 年期间，欧盟 -28 消费端 CO_2 排放量的跌幅仅为- 2.68%，且其间消费端 CO_2 排放趋势存在较大波动，无法直接得出经济与排放脱钩的结论。

从 2005 年以来温室气体排放与 GDP 的时间序列来看（见图 7-1），在一般经济条件下，欧盟的排放与经济呈现出一定的脱钩状态，但在

图 7-1 EU-28温室气体排放（相对基年减排百分比）与GDP（相对2005年变化百分比）变化趋势

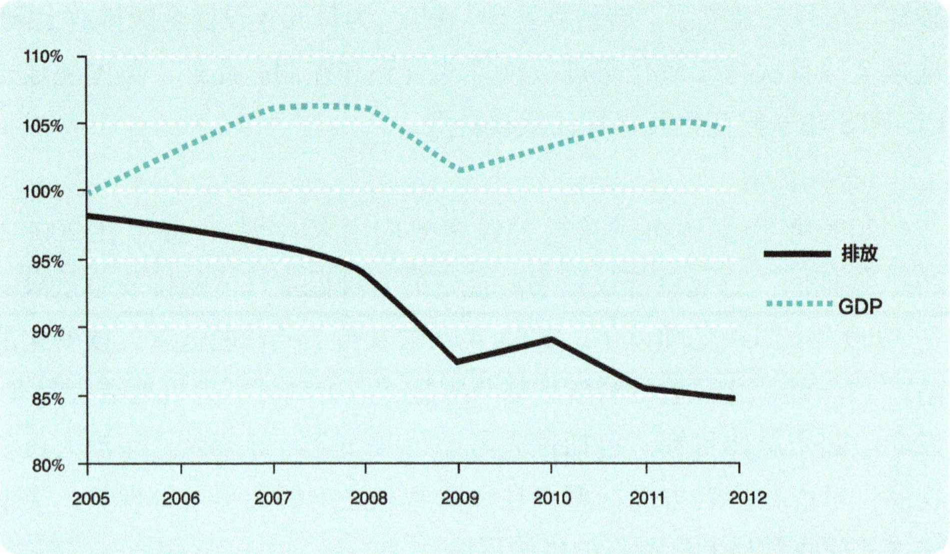

经济危机等突发状况发生时（2008—2010 年），排放水平与 GDP 波动曲线的走势和变化幅度一致性较高。这也进一步证实了欧盟排放与经济脱钩的结论仍有待商榷。

主要排放成员国格局一成不变，欧盟内部分歧显著。根据欧盟 2012 年提交的 NIR 清单，排放前六位的成员国分别是德国、英国、法国、意大利、波兰和西班牙（见图 7-2），六国 2012 年排放总量约占欧盟 -28 总排放量的 71%，比 1990 年水平降低 17%。其中，仅波兰不是欧盟 -15 成员，虽然波兰 2004 年才正式加入欧盟，但其煤炭高依存度和排放水平决定了波兰在欧盟气候与能源政策制定中充当“阻碍者”的角色。

EU-ETS“招牌”地位备受质疑。欧盟碳排放交易体系（EU-ETS）是当今全球最大、运行时间最长的碳排放交易市场。然而近年来，欧盟碳市场面临的最主要问题是配额过剩和供需失衡以及相应的碳价下跌。这一现状与欧盟减排目标缺乏力度以及经济危机的影响不无关系。尽管如此，EU-ETS 仍是欧盟应对气候变化政策措施的核心。近几年，欧

图 7-2 欧盟排放前六位的成员国温室气体排放量（截至2012年）

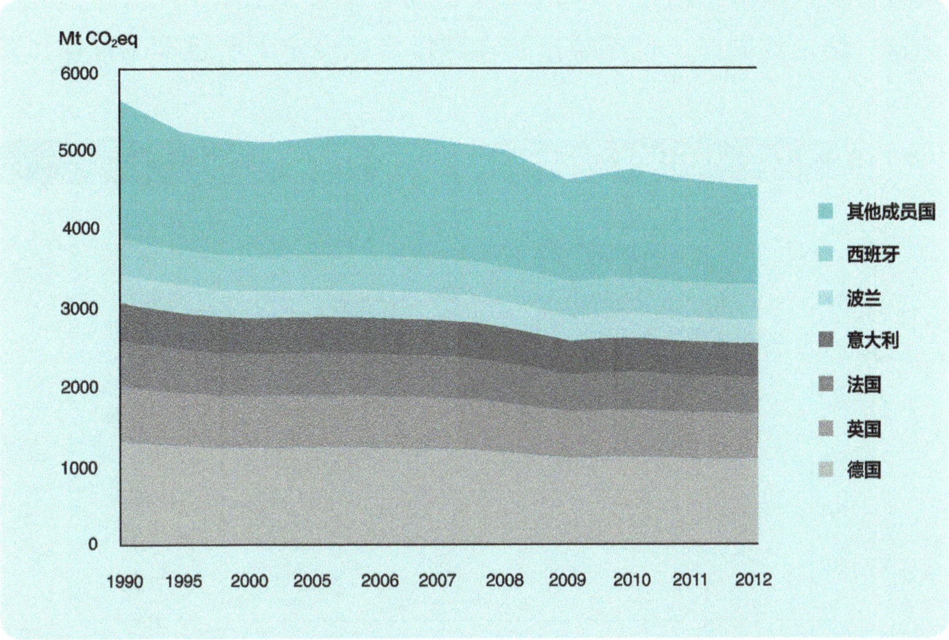

盟已考虑采取相关措施改进 EU-ETS 并提高市场活跃度，例如“折量拍卖”等短期解决方案和 2030 框架中提出的建立“市场稳定储备”等 EU-ETS 结构化改革措施。但缺乏力度的减排目标以及复杂的民主化决策程序从根本上限制了新政在提高碳市场竞争力、刺激碳市场活力上的成效。

此外，由于 EU-ETS 实施导致的“碳泄漏”现象不容忽视。据欧委会统计，欧盟 164 个生产行业及分支均存在碳泄漏的风险，这些行业的产量占欧洲制造业总量的 77%，排放量约为 EU-ETS 排放总量的四分之一。应对气候变化、减排温室气体是全球性的行动，依靠转移高排放企业营造出来的减排假象不利于可持续发展，更加剧了发展中国家的减排压力。

2020 年欧盟履约毫无难度可言，目标风险系数低。欧盟强调其将超额完成《京都议定书》第二承诺期的目标。且在现有政策情景下，

2020年将实现22%的减排量，2030年减排24%（见图7-3）；在额外政策情景下减排将达到26%，2030年减排30%（见图7-4）。欧盟在宣扬这一结论的同时，也反映出其在减排目标设定上所贯彻的规避风险

图7-3 “现有措施”情景下欧盟-28各部门温室气体排放预测

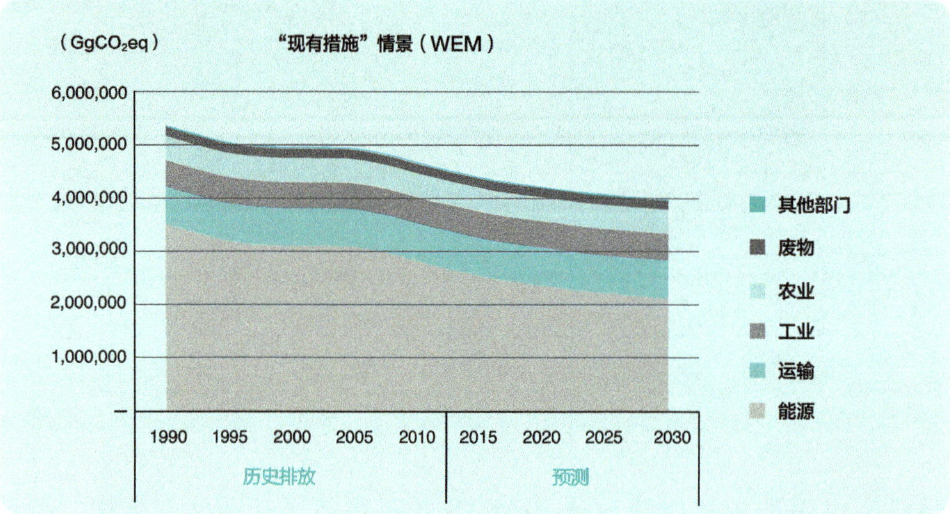

图7-4 “额外措施”情景下欧盟-28各部门温室气体排放预测

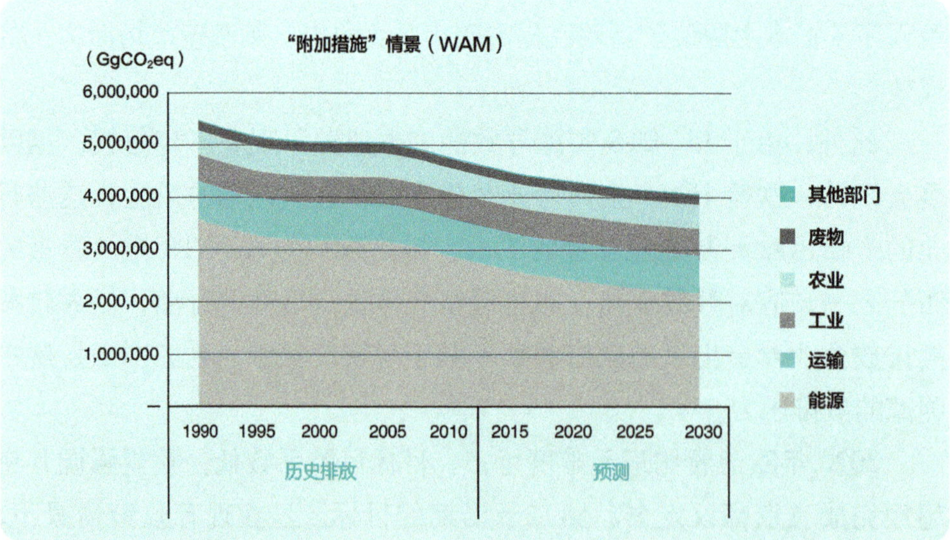

的策略和态度。另鉴于欧盟的决策制度，新目标的通过需要各成员国协调一致，在经过利益相关方博弈后，目标的力度通常会大打折扣。2030框架中40%的减排目标公布之初，外界普遍认为减排目标过于保守，许多环保组织认为要实现2℃全球温控目标，2030年欧盟应达到50%以上的减排量。中期目标的力度将直接影响长期目标实现的难度和成本，欧盟作为气候变化的“领导者”，这种不敢于承担风险的态度不仅使其垂范效应大打折扣，也不利于激励其他主要排放国制定有力度的减排承诺。

四、中欧气候变化合作

2005年，中欧共同发表《中国和欧盟气候变化联合宣言》以来，双方提出了增进气候领域经济、能源、技术等发展的远景目标。为落实合作目标，中国与欧盟于2006年制定了气候变化合作滚动工作计划。在此计划的推动下，双方于2010年和2013年共同发布了《中欧气候变化对话与合作联合声明》和《中欧合作2020战略规划》，并于2015年6月共同签署了《中欧气候变化联合声明》，将中欧气候变化合作推向更深层面。

（一）对话与交流

自2002年第五次中欧峰会以来，中国与欧盟及其主要成员国领导人对气候变化问题给予了高度关注。为促进交流，加深相互了解与分享经验，国家发改委与欧盟气候行动委员会于2010年共同建立了中欧部长级气候变化对话机制及部长级气候变化热线，这为中欧广泛开展首脑级、部长级、高管级和项目级别等多层次的气候对话与合作机制打下了坚实基础。中国政府与欧盟三驾马车间政府决策者、立法委员和多边机构就应对气候变化的高频联系与政策经验交流，为双方落实深化项目开展注入了源源不断的新动力。

（二）项目合作与能力建设

气候变化根本上来源于化石燃料的消费，而目前社会发展对传统能源的依赖性是造成气候变化问题复杂而不可控的主要原因。推动能源革新与产业结构调整势在必行。中国与欧盟及其主要成员国在发展绿色经济方面有相同的诉求，就推动绿色经济与低碳发展达成了共识。中欧在气候科学研究、资金整合与财政援助，技术转让、清洁发展机制、适应气候变化和能力建设等方面展开了多项合作。2009 年中德两国签署了《中德关于应对气候变化合作的谅解备忘录》；2010 年 11 月，中法两国签署了《中华人民共和国国家发展和改革委员会和法兰西共和国生态、能源、可持续发展和海洋部关于加强应对气候变化合作的协议》，正式启动中法应对气候变化双边磋商机制。2014 年 6 月 17 日，中英政府发表了《中英气候变化联合声明》。近年来，互联网技术、大数据理念以及新能源产业的进步有目共睹。科技进步使得绿色发展更加成本有效与风险可控。中欧间广泛的项目合作和技术交流不仅能控制减少排放，更能推动社会经济可持续发展。

第三节　基础四国[①]

“基础四国”机制于 2009 年 11 月 28 日联合国气候变化谈判哥本哈根会议召开前夕，在中国的倡议和推动下形成[②]。之后，“基础四国”定期轮流主持召开气候变化部长级会议，就气候变化相关重点议题、发展中国家的关切进行讨论和立场协调。通过这一有效的协调机制，

①本节主要来自张晓华、胡晓和祁悦发表在 2014 年第 9 期《当代世界》的“气候变化国际谈判中‘基础四国’机制的作用和影响”一文。

② Saibal Dasgupta. “Copenhagen conference: India, China plan joint exit”. The Times of India. Nov 28, 2009. <http://timesofindia.indiatimes.com/india/Copenhagen-conference-India-China-plan-joint-exit/articleshow/5279771.cms?referral=PM> 2014.8.25.

四国在历次气候变化国际谈判协调会和缔约方大会期间以“基础四国”名义进行统一发声，已对谈判进程产生了重大影响，成为“七十七国集团＋中国”[①]中一股不容小觑的代表性力量。

气候变化谈判中“基础四国”机制的形成标志着面对发达国家主导国际体系的现状，发展中大国开始有意识地团结、协调并坚持自身立场，以维护广大发展中国家利益。

一、“基础四国”在全球气候治理中的地位

“基础四国”均属于新兴发展中国家，在全球治理的很多领域都处在相似的地位。冷战结束后，四国经济发展迅速，在国际社会备受瞩目，中国、印度和巴西经济总量都已跻身世界十强。然而，由于四国人口基数大，各项人均经济发展指数仍与发达国家存在显著差距。例如，2009 年，四国中人均 GDP 最高的巴西，其水平也仅相当于发达国家的 1/5 左右，而印度的人均 GDP 仅相当于发达国家水平的 3%；四国人均 GDP 与哥伦比亚、卡塔尔、新加坡等在联合国系统中收入较高的发展中国家相比也有相当大的差距。此外，[②]四国仍有相当比例的人口处于本国划定的贫困线以下，其中印度的贫困人口更是高达 3 亿。[③]这些相似性使得“基础四国”逐渐成为国际社会中一个新的“阶层”。

作为区域大国，四国都希望成为各区域发展中国家的代表。随着经济的飞速发展，这些国家在地区事务中发挥更大影响力的诉求日趋强烈，希望能够作为区域领袖代表其他发展中国家发声，在国际上争取话语权。但是这些诉求都面临着发达国家传统势力的挤压，这使“基

①“七十七国集团”是上世纪 60 年代发展中国家形成的一个旨在反对超级大国控制和剥削的国际集团，目前成员国数已达 133 个。中国虽然不是成员国，但是一贯支持该集团立场，并以“七十七国集团＋中国”这一机制与其共同发声。

② IMF, World Economic Outlook, < http://www.econstats.com/weo/V011.htm> 2014.8.25.

③世界银行 . 世界发展指标 . 2014.7.22. <http://data.worldbank.org.cn/data-catalog/world-development-indicators> 2014.8.25.

础四国”地域间隔遥远并未成为一种障碍，反而进一步增加了各国对各自身份的认同感，成为强化“基础四国”机制的重要因素。

四国在气候变化问题上都面临来自发达国家的压力。发达国家完成工业化之后，其温室气体排放基本处于稳定并呈逐步下降的趋势。但随着经济的发展，四国的排放增长呈加速趋势，面临排放空间约束的挑战日益严峻。尽管四国能源构成和排放源不尽相同，但未来排放总量将不断上升，都将面对来自发达国家的共同压力。目前的国际体系是在西方发达国家的主导下形成的，作为现代国际体系的“后来者”，“基础四国”等发展中国家的利益不能被充分考虑。这一因素使“基础四国”在气候变化问题上自然走到了一起。

这些相似性和客观因素使“基础四国”在气候变化问题上对于维护现有《公约》体制和广大发展中国家团结的立场更为坚定，也因此和力图在气候变化问题上向发展中国家转嫁责任的发达国家形成了某种程度的对立。“基础四国”在2009年年底一成立便成为“七十七国集团＋中国”的支柱力量，其共同的立场主要体现在以下几个方面：在《联合国气候变化框架公约》的体制内，坚持公平和“共同但有区别的责任”原则，严格遵循联合国的议事规则和相关谈判授权；坚持发达国家应率先采取行动，切实履行《京都议定书》和巴厘路线图的承诺，承担绝对量化减排义务，发展中国家在可持续发展的框架内开展适合本国国情的减缓行动；[①] 发达国家要切实落实为发展中国家提供资金、技术和能力建设支持的承诺，支持发展中国家开展符合本国可持续发展需求的减缓和适应行动；坚定支持“七十七国集团＋中国”，共同维护广大发展中国家的集体利益。

① Karl Hallding, Marie Olsson, Aaron Atteridge, et al. “Together Alone: BASIC countries and the climate change conundrum.” Nordic Council of Ministers, Copenhagen, 2011, pp: 92.

二、基础四国在全球气候治理中面临的挑战

进入 21 世纪，《京都议定书》之后每一轮谈判，发达国家的主要目标都是打破目前《公约》对发达国家和发展中国家的一般性区分，以进一步转嫁减排责任；发展中国家则尽力维护以“共同但有区别的责任”原则为基础的“两分法”。此次 2015 协议谈判也不例外。各方最核心的分歧也正是集中体现在如何体现各国的区分以及如何理解“共同但有区别的责任”上。客观反映发展中国家责任和能力的变化以及各国之间的区别是解决 2015 协议所有问题的关键。占据全球人口和排放量三分之一的“基础四国”成为矛盾的焦点。因此，“基础四国”的一举一动都将成为 2015 协议谈判的重要决定因素。

目前，各方对 2015 协议的结构和内容均有不同诉求。欧盟强调“自上而下”的减排机制，各国承担相应减排义务，并设定定期审评机制以确保全球减排努力能够符合将全球温升控制在 2℃以内的全球目标。美国认为应由各国“自下而上”自主提出减排贡献，不应设立任何强制性减排目标，并牢牢紧盯中、印等排放大国。美欧诉求虽有不同，但都旨在消除发达国家和发展中国家的区别。发展中国家阵营中，小岛屿国家和最不发达国家受气候变化不利影响最为严重，因此呼吁主要排放大国均要承担减排义务。在此情况下，“基础四国”作为发展中国家代表和“七十七国集团 + 中国”内部的核心力量，必须团结广大发展中国家，表达共同的利益诉求，避免 2015 协议由发达国家主导并对发展中国家未来排放空间形成过度的限制，才能更好地维护自身利益。因此，“基础四国”在扮演重要角色的同时，也面临着巨大的挑战。

一是要提出能够与欧美相抗衡的具体解决思路。无论是欧盟提出的“承诺加审评”模式还是美国以“国家自主贡献”为主的 2015 协议框架都有其合理性，并逐渐被各方所接受，成为 2015 协议的主流共识。对于“基础四国”而言，在坚持《公约》框架，维持发达国家和发展中国家基本区分的前提下，应该看到发展中国家内部差异的存在，以

及以“基础四国”为代表的新兴经济体的能力提高，不能一直停留在坚持原则和重申立场上，需要更加理性地提出建设性的、相对具体的解决方案。

二是要妥善面对和处理发展中国家内部的分化。“七十七国集团+中国”内部的分化在哥本哈根会议前开始表面化。“基础四国”与其他发展中国家之间的分歧开始显现。2009年哥本哈根会议时，由于“基础四国”在最后关头曾与美国联手促成《哥本哈根协议》，不少发展中国家将此结果看成是大国意志的强加。诸多发展中小国在对《公约》下谈判公平性和程序性问题产生微词的同时，对“基础四国”的不信任也趋于表面化。如何在维护自身利益的同时不失去广大发展中国家的战略依托，将是“基础四国”在未来谈判中所面临的一大挑战。

三是如何看待和处理四国内部的立场差异。“基础四国”机制本身具有一定松散性，且在各国利益诉求方面存在差异。对于“基础四国”谈判集团的性质，国际社会一直存在争议。一种观点认为，“基础四国”是一个新型联盟，是“七十七国集团+中国”内部强有力的子集团，四国对《京都议定书》第二承诺期发达国家义务的明确要求及其历史责任的认定是四国立场一致的基础。[①] 而更多的看法则是，“基础四国”仅是一个松散的“政治结社”(political association)[②]，而非一个“紧密的谈判集团”(tight negotiating bloc)[③]。尽管四国部长每年定期举办会议并发表联合声明，但声明内容多以重复各项原则为主，鲜有实质性合作成果，侧面反映出四国在统一立场上所面临的困难。在2015协议形

① Sander Happaerts and Hans Bruyninckx. “Rising Powers in Global Climate Governance. Negotiating in the New World Order.” Leuven Centre for Global Governance Studies. Working Paper No. 124 – October 2013.

② Lesley Masters. “What future for BASIC? The emerging powers dimension in the international politics of climate change negotiations.” Institute for Global Dialogue. Issue 95/March 2012.

③ Karl Hallding, Marie Olsson, Aaron Atteridge,et al. “Together Alone: BASIC countries and the climate change conundrum.” Nordic Council of Ministers, Copenhagen, 2011, pp: 95.

式、各国义务等诸多谈判的关键问题上，四国各自立场开始出现明显不同，出现“松散化”的发展趋势。

在“基础四国”中，中国和印度两个排放大国立场较为一致，而巴西和南非在谈判模式、2020 年后减排目标等关键问题上与中、印两国存在分歧。巴、南两国并不坚持在谈判中严格区分发达国家和发展中国家，认为可以达成对所有缔约方具有法律约束力的 2020 年后减排目标。这种立场是由巴、南两国在气候变化问题上的特殊国情所导致的。巴西面临的减排压力相对较小，在气候变化国际谈判中，主要诉求是通过维护发展中国家整体利益，增强本国在发展中国家和全球的话语权。巴西对发达国家和发展中国家的区分更多是道义上的支持，而非其实质利益上的核心关切。南非人口、经济、能耗和排放规模相对较小，根据《公约》官方网站的数据，不考虑土地部门，南非的温室气体排放量约为巴西的一半，印度的四分之一，不到中国排放量的十分之一。因此，在国际层面，南非承担减排义务的压力更小。南非参加气候变化国际谈判与其以“非洲复兴”为指导思想的整体外交战略密切相关。南非谋求在国际事务中的话语权和大国地位，既立足于非洲这个大后方，也特别积极地寻求发达国家的支持。这也就导致南非在气候变化国际谈判中，不仅时刻以非洲国家集团的利益为最主要考虑，也与以欧盟为代表的发达国家立场相近，即希望在 2020 年后建立一个“自上而下”的减排机制，且每个国家（发达国家和发展中国家）都承担有法律约束力的减排义务。随着 2015 协议谈判的具体化，这些分歧不断表面化。如何在内部求同存异是“基础四国”面临的最为根本和严峻的挑战。

三、“基础四国”气候变化合作前景

虽然“基础四国”机制面临着诸多挑战，但基于国情、利益诉求的相似性和共同面对的压力，“基础四国”机制仍将继续下去。为了使该机制能够更好地发展下去并发挥更大的作用，“基础四国”还需要在实践过程中不断完善和加强。

未来，四国应进一步拓宽在气候变化领域的合作，从谈判立场的协调逐步扩展到务实合作。尽管“基础四国”在谈判进程中不断要求发达国家率先行动，但实际上四国都已经采取了大量务实的应对气候变化的行动。目前，四国中已有中国、印度和巴西提交了国家自主贡献。中国贡献包括二氧化碳排放2030年左右达到峰值并争取尽早达峰；单位国内生产总值二氧化碳排放比2005年下降60%—65%、非化石能源占一次能源消费比重达到20%左右、森林蓄积量比2005年增加45亿立方米左右；印度提出到2030年单位GDP排放强度在2005年的基础上降低33%—35%，通过加强造林力度增加25亿—30亿吨的碳汇，并在国际支持下，到2030年将非化石燃料在其能源结构中所占比重从目前的30%增加到40%左右，由此在2022年增加1.75吉瓦的可再生能源生产能力；巴西提出到2025年减少37%的温室气体排放，到2030年减少43%的温室气体排放，并使可再生能源在能源结构中占比达到45%。这些目标的提出，将引领四国加速低碳发展转型，也为四国在相关领域开展务实合作奠定了基础。相似的国情和发展阶段意味着“基础四国”在应对气候变化行动上面临同样的困难，但也有大量的经验可以分享。因此，“基础四国”机制在协调立场的同时应该进一步加强务实合作，这也有助于各国增进相互了解，促进在谈判中寻找和建立更多共同的立场。

同时，“基础四国”在气候变化国际谈判中应坚持“以我为主”的战略，提出新的应对气候变化的理念，引导全球应对气候变化的进程。“基础四国”在已经采取的大量应对气候变化的行动的基础上，应进一步将国内的务实行动带到国际舞台上。“基础四国”应更广泛传播自身应对气候变化的理念和主张，更加积极地推动应对气候变化的全球合作。

【参考文献】

[1] 李俊峰，张晓华，陈怡，等．美《总统气候行动计划》解读与分析 [Z].2013-7-6.
[2] 美国环保协会．美国《清洁电力计划》解读 [Z/OL]，http://www.cet.net.cn/web/getdetailnews.action?id=2437, 2015-10-05.
[3] 张晓华，傅莎，祁悦，等．中美开展气候和能源合作的现状和建议 [Z].2015-5-13.
[4] 张晓华，徐丽丽，胡晓．美国能源部《多元化能源战略——通向可持续经济增长之路》评述 [Z/OL]，2014-6-4.
[5] 张晓华，祁悦，刘林蔚．中美欧 2020 后减缓气候变化行动目标的比较分析 [J]. 中国能源，2015，37(7):11 － 16.
[6] 张晓华，祁悦，刘林蔚．美国"国家自主贡献"内容分析及对我相关工作的建议 [Z].2015-4-2 .
[7] 邹骥，张晓华，傅莎，等．对中美气候变化联合声明的几点评论 [Z/OL], http://www.ncsc.org.cn/article/yxcg/yjgd/201411/20141100001254.shtml, 2015-10-05.

第八章

中国在全球气候治理中的定位

Chapter Eight

China's Role in Global Climate Governance

摘要

当前全球温室气体排放格局已发生显著变化，中国碳排放总量和人均碳排放量的快速增长在全球范围内日益突显，经济体量大、能源消费多、碳排放总量高的特征明显，但其历史累积排放相比美、欧仍有较大差距。随着经济发展和中国工业化、城市化进程的进一步推进，中国 CO_2 排放的增长趋势不可避免，且建筑和交通部门的排放量和在总排放中的比重都将会大幅增加，进一步加大减排压力。

尽管中国已成为全球第一大年度排放国和第二大经济体，但其主要发展指标和长期存在的显著“二元结构”特征均显示了其发展中国家的定位，和发达国家在所处的发展阶段、面临的发展需求、承担的历史责任和具备的综合能力等方面仍存在巨大差别。此外，中国还面临能源资源禀赋以煤为主，地域辽阔，人均资源存量、质量、有效人均活动空间均十分紧缺等特殊国情。中国在参与全球气候治理时需充分考虑其发展阶段、国情和能力，努力探索一条优于发达国家传统发展道路的创新低碳发展路径。

与此同时，中国作为最大的发展中国家，应立足于本国国情和客观定位，在可持续发展框架和《公约》原则下，运用大国思维以积极务实姿态参与国际气候进程，实现从参与者到建设者的角色转变，推动促成于我有利的“公平、有效、共赢”的全球气候体制的达成，发挥负责任大国的建设性作用，主动承担与自身发展阶段、能力、定位和国情相符的应对气候变化全球责任。

中国提出 2020 年后的国家自主贡献目标是中国承担应对气候变化全球责任的核心体现，体现了中国积极应对气候变化，努力控制温室气体排放，提高适应气候变化的能力，以及深度参与全球治理，承担合理国际责任的姿态和决心。

第一节　累积排放与瞬时排放变化趋势

当前全球温室气体排放格局已发生显著变化，中国碳排放总量和人均碳排放量的快速增长在全球范围内日益突显。经济体量大、能源消费多、碳排放总量高的特征明显，已经成为仅次于美国的第二大经济体，能源活动和水泥生产过程的二氧化碳排放总量接近于欧盟、美国和日本碳排放量的总和。

中国能源活动相关 CO_2 排放量从 1990 年的 23.4 亿吨上升至 2012 年的 89 亿吨，占全球比重从 11.2% 上升至 27.5%，超过欧盟和美国排放量的总和。2005—2012 年中国能源相关 CO_2 排放量增长 28 亿吨，占全球 CO_2 排放增量的 67%。[①] 与此同时，中国人均 CO_2 排放也有了快速增长，继 2007 年超过世界平均水平后，2013 年我国（能源活动和水泥过程）人均 CO_2 排放为 7.4 吨，同期世界人均排放为 4.9 吨 / 人，美国、加拿大、澳大利亚的人均 CO_2 排放均超过 15 吨 / 人，分别达到 16.6、15.7、16.9 吨 / 人。欧盟各国和日本的排放水平虽然远低于美、加、澳三国，但同期欧盟 28 国、德国、英国、日本等国的人均 CO_2 排放也分别达到 7.4、10.2、7.5、10.7 吨 / 人。[②] 中国人均 CO_2 排放已赶上欧盟平均水平，但距离发达国家平均水平和美、澳、加、德、日等发达国家仍有一定差距。

从历史累积排放的角度 (见表 8-1)，1751—2012 年，全球累积能源相关加水泥工艺过程 CO_2 排放（以下简称 CO_2 排放）达到 14080.3 亿吨 CO_2，其中，附件一国家共排放了 10215.2 亿吨 CO_2，占全球的

---

①数据来源：中国 1990 年碳排放由《中国能源统计年鉴 2013》中 1990 年的能源消费数据计算得到，2012 年碳排放由《中国统计摘要 2015》中 2012 年的能源消费数据计算得到。全球除中国数据参考 IEA CO_2 Emissions From Fuel Combustion Highlights 2014 报告计算得到。

② PBL/JRC report，"Trends in global CO_2 emissions; 2014 report", December 2014，Available at http://www.pbl.nl/en/publications/trends-in-global-co2-emissions-2014-report.

表 8-1 主要国家能源相关 CO_2 加水泥历史累积排放（1751—2012）

国家	历史累积排放（亿吨 CO_2）	占全球比重（%）	人均历史累积排放（吨 CO_2/ 人）
英国	753.4	5.4%	1191.5
美国	3658.1	26.0%	1165.3
德国	852.1	6.1%	1040.5
俄罗斯	1107.6	7.9%	771.7
法国	354.4	2.5%	539.5
日本	560.6	4.0%	439.5
南非	175.0	1.2%	342.0
墨西哥	153.7	1.1%	127.2
中国	1512.8	10.7%	112.0
巴西	119.2	0.8%	60.0
印度	423.6	3.0%	34.3
EU28	3347.4	23.8%	657.6
OECD	8336.2	59.2%	663.4
附件一国家	9587.8	68.1%	736.0
世界	14080.3	100.0%	199.8

数据来源：作者根据美国橡树岭实验室 CDIAC(CO_2 信息分析中心）数据计算得到，人口数据来自世界银行数据库。欧盟为 27 国数据，未包含克罗地亚。

72.5%。OECD 国家共排放了 8336.2 亿吨 CO_2，占全球的 59.2%。美国占全球累积 CO_2 排放的比重为 26%，欧盟占 23.8%，而同期中国只占 10.7%。从人均历史累积排放看，1751—2012 年全球人均累积 CO_2 排放为 199.8 吨 CO_2/ 人，附件一国家为 736 吨 CO_2/ 人，为世界平均水平的 4 倍。中国同期人均历史累积 CO_2 排放仅为 112 吨 CO_2/ 人，不到世界同期水平的 60%，和发达国家相比还有较大差距。在同期，OECD 国家平均人均历史累积 CO_2 排放达到了 663.4 吨 CO_2/ 人，而美国、欧盟、英国、德国、日本的人均历史累积 CO_2 排放更是分别达到了 1165.3、657.6、1191.5 和 1040.5 吨 CO_2/ 人，分别是中国的 10.4、5.9、10.6 和 9.3 倍。尽管中国已经是世界第一温室气体排放大国，历史累积排放第二，但其 1751—2012 年的历史累积 CO_2 排放仍仅分别为美、欧的 41% 和

45%，人均历史累积 CO_2 排放仅分别为美、欧的 9.6% 和 17%。

"十一五"以来，中国采取了大量的措施如"十大重点节能工程"，"千家企业"，"淘汰落后产能，上大压小"等降低单位 GDP 能源强度。截止到 2014 年，中国单位 GDP 的能源强度相比 2005 年已下降了 30%。同期，由于核能、水电、风电等非化石能源的发展，能源结构在一定程度上实现低碳化，单位 GDP 的 CO_2 排放强度下降了 33.8%。但即便如此，由于中国仍处于以重化工为主的工业化进程中且受以煤为主的能源资源禀赋的影响，中国单位 GDP 的能源强度相对其他国家仍然较高。2013 年中国单位 GDP CO_2 排放强度达到 1.8 $kgCO_2$/ 美元（2005 年价），分别是美国、日本、欧盟的 4.8、7.2、6.6 倍。①

从部门排放看（见图 8-1），工业是中国能源相关 CO_2 排放的第一大部门，尽管"十一五"以来，工业在总排放中的比重略有下降，从 2005 年的 42.2% 下降到了 2010 年的 40.8%。电力是第二大排放部门，而且比重自 2005 年以来不断上升，从 2005 年的 35.6% 上升到了 2010 年的 38.6%。交通是第三大排放部门，比重从 2005 年的 8.3% 上升到了 9.0%。建筑部门作为第四大排放部门，由于其电气化程度的提高和集中供热比重的提高，排放比重略有下降，从 2005 年的 7.3% 下降到了 2010 年的 6%。与发达国家相比，中国工业排放的比重明显偏高，而交通和建筑的排放比例明显较低。②

---

① P. Friedlingstein, et.al, Persistent growth of CO_2 emissions and implications for reaching climate targets, Nature geoscience, 2014.

② 中国现有统计体系部门划分方法与国际通行方法不同，无法充分反映能源消费特性，也不便于制定针对性措施。例如统计年鉴中的"交通运输、仓储和邮政业"仅包括对从事运营的交通运输企业的统计，而不包括非运营企事业单位和私人交通工具能源消费，不利于未来对交通用能和排放的检测和管理。因此，本文采用国际通用口径对温室气体排放部门重新划分。其中交通部门采用"大交通"概念，包括所有交通运输工具，即从事运营的交通运输工具、非运营企事业单位和私人拥有的交通工具。建筑部门主要涵盖建筑运行能耗和排放，即建筑照明、采暖、空调、电器和其他各类建筑内用能设备运行过程中的能耗和排放。电力、热力的生产和供应业作为能源加工和转换部门，从现有的工业部门中提取出来，形成单独的电力生产部门和热力生产部门。综上所述，现有统计口径下的温室气体排放部门经过重新调整，划分为电力生产、热力生产、农业、工业、交通和建筑六大部门。

图 8-1 在计算 CO_2 排放时将电力、热力相关的 CO_2 排放分别计入了电力、热力部门。如果将电力、热力相关的间接排放计入终端消费部门，中国能源相关 CO_2 的排放结构将发生变化。因为工业部门消耗了绝大多数的电力和热力，因此在分摊以后工业部门排放大幅度上升，2010 年占总排放的比重达到了 73%。建筑部门的比重也有大幅上升，2010 年占总排放的比重从 6% 上升到了 15.5%。交通部门由于主要消耗油品，因此其比例基本不变，2010 年为 9.5%（图 8-2）。可以看到，尽管考虑了电力、热力的间接排放，但是交通和建筑部门的比重与发达国家两者各占 1/3 左右的比重相比，还有很大差距。

图 8-1 2005—2010 年中国分行业的 CO_2 排放变化

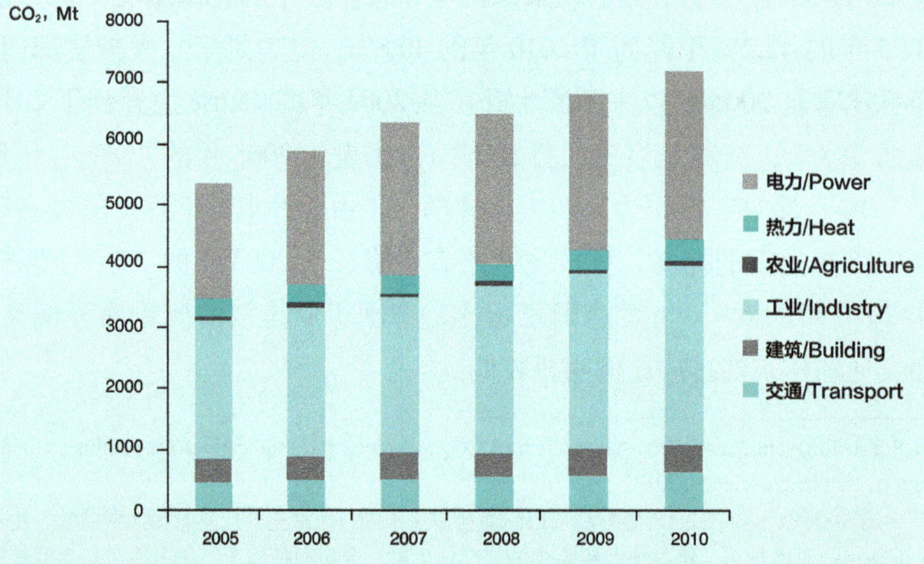

数据来源：根据《中国能源平衡表 2005—2010》计算得到，基础数据来自《中国能源统计年鉴 2006—2011》。

在工业内部，钢铁是最大的排放源，2010 年占工业总排放的比重接近 38%。除钢铁外，石油加工、化工、非金属制品和有色金属等行

图 8-2 中国分行业的 CO_2 排放（将电力、热力对应排放分摊到终端部门）

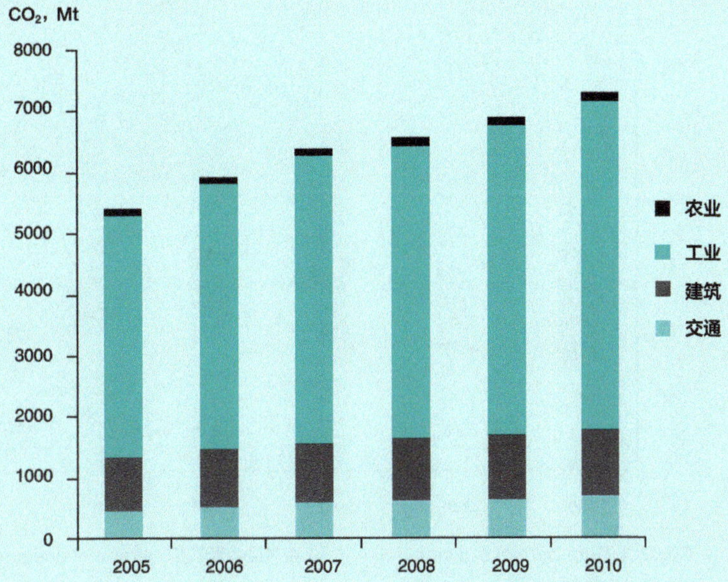

数据来源：根据《中国能源平衡表 2005—2010》计算得到，基础数据来自《中国能源统计年鉴 2006—2011》。

业排放比例也较高。这五大行业 2010 年占工业排放的比重超过了 71%。

如考虑电力、热力的间接排放，建筑部门排放从 2005 年的 8.5 亿吨上升到了 2010 年的 11.2 亿吨，年均增长 5.8%（如图 8-3）。在建筑部门内部，公共建筑排放增长速度要高于居民建筑的排放增长。2010 年公共建筑的 CO_2 排放在 2005 年基础上增长 39%，达到 4.1 亿吨 CO_2，占建筑部门总排放的 36.8%。总体而言，目前建筑部门排放占总排放比重与发达国家相比仍然较低，随着未来旅游、会展等第三产业进一步发展和人均住宅面积的提高以及居民生活水平的改善，对建筑部门排放的控制将越来越重要。

从长期经济发展趋势上看，中国目前处于快速工业化和城市化进程中，基础设施建设仍未完成。随着经济发展，人民群众的住房、交通等消费水平将迅速上升，这会进一步促进能源需求以及 CO_2 排放的

图 8-3 中国 2005—2010 年建筑部门排放[①]

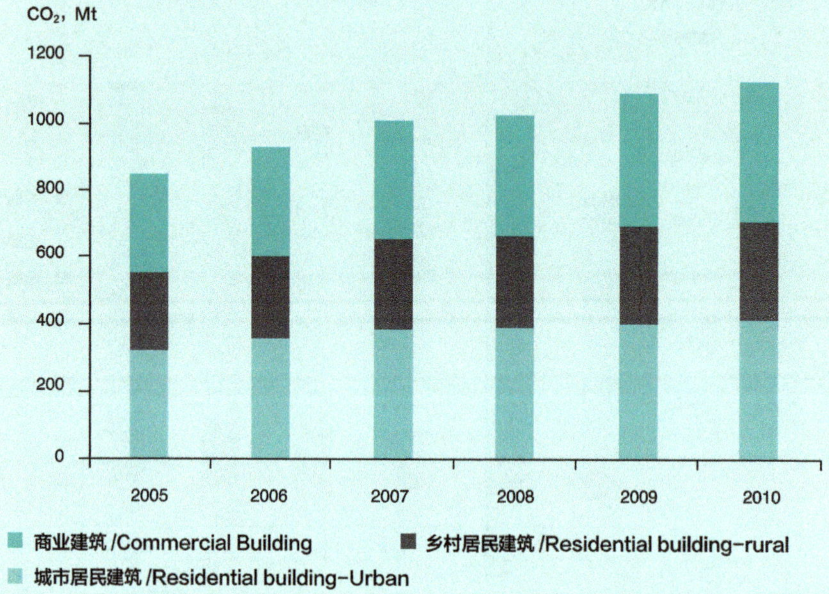

数据来源：根据《中国能源平衡表 2005—2010》计算得到，基础数据来自《中国能源统计年鉴 2006—2011》。

较快增长。因此在未来相当长的时期内，中国 CO_2 排放的增长趋势不可避免。而随着中国工业化和城市化进程的进一步推进，中国建筑和交通等部门的排放量和在总排放中的比重都将会大幅增加，进一步加大减排压力。未来，中国的碳排放水平对于全球的排放路径至关重要，作为全球第二经济大国和第一温室气体排放大国，中国面临的压力将不断增大，在谈判中日益成为各方关注的焦点。

① 中华人民共和国商务部网站 . 世界各国国内生产总值情况排行榜 , http://www.mofcom.gov.cn/article/i/dxfw/jlyd/201309/20130900321950.shtml, 2013-09-25.

第二节　中国的发展进程、发展阶段、国情和能力

整体来看，尽管中国已成为全球第一大排放国和第二大经济体，但其主要发展指标和长期存在的显著“二元结构”特征均显示了其发展中国家的定位，和发达国家在所处的发展阶段、面临的发展需求、承担的历史责任和具备的综合能力等方面都存在巨大差别，以减贫、提高收入、提高社会保障程度、提高基础设施等公共服务覆盖面、提高生活质量为主要内容的发展，依然是压倒一切的战略重点。在看到“中国不是乍得”[①]的同时，还必须看到“中国也不是北美、欧洲、日本和澳大利亚”。

从经济发展阶段来看，根据国际普遍对经济发展阶段的划分，中国总体上处于工业化中期向后期过渡的阶段，以重工业和加工制造业为核心的工业化仍需持续一段时期。2014 年服务业增加值比重仅为 48%，而发达国家达到人均排放峰值时的服务业比重普遍达到 60% 左右。2012 年，中国以占全球 11.5%[②]的国内生产总值，消耗了全球 45.7% 的钢铁、57.8% 的水泥、21.9% 的能源和 50.2% 的煤炭；中国工业部门能源消费总量占全国能源消费总量的比重达到 62.8%[③]，尤其是采掘、钢铁、建材水泥、造纸、化工、电力等高耗能行业比重较高，且高耗能行业单位 GDP 能耗和主要产品单位能耗均高于主要能源消费国家平均水平，加重了我国经济发展的高碳特征，大部分产品单位能耗水平与国际先进水平相比仍有约 25%—40% 的节能空间。

① 美国前国务卿希拉里·克林顿语。

② 中华人民共和国商务部网站．世界各国国内生产总值情况排行榜，http://www.mofcom.gov.cn/article/i/dxfw/jlyd/201309/20130900321950.shtml, 2013-09-25.

③ 中华人民共和国国家统计局数据，http://data.stats.gov.cn/easyquery.htm?cn=C01.

从城市化进程和区域建设发展模式来看，中国还没有完成城市化进程，仍处于城市化加速发展阶段。2014 年城市化率为 54.8%，发达国家达到人均排放峰值时的城市化率水平普遍在 70% 以上。在未来 25 年内，中国预计将有 3.5 亿人从农村进入城市，超过目前美国人口总数。城市化率每提高一个百分点，就意味着约有 1400 万左右的人口从农村涌向城市，同时将伴随着城市住房、道路、管网和卫生设施等基础设施的大规模建设和钢铁、水泥和化工等高载能产品的大量消耗[①]。随着中国城市化及城镇化过程中人口、产业的高度集聚和快速扩张，资源消耗和污染排放在城市也呈现快速聚集增加态势，城市成为解决中国资源、环境、应对气候变化等最复杂、最紧迫问题的核心战场。中国急需加快各类大中小城市集约、高端、智慧、低碳为导向的综合变革，推动市民生产生活环境向更加生态、休闲、便捷的方向发展，全面提升城市的综合发展质量和可持续发展能力。

从能力角度看，中国和发达国家在资金、技术能力和软实力上的能力差距仍然显著存在。一个经济体应对气候变化的能力应从国家、企业、居民等不同利益相关者去考察。自改革开放以来，中国经济有了巨大的发展。1978 年至今的近 40 年来，中国 GDP 以年均 9.4% 的速度增长，2010 年已超越日本成为世界第二大经济体。2014 年人均 GDP 超过 7000 美元，已迈入上中等收入国家的行列，解决了 13 亿人民的温饱问题，初步达到小康水平。但中国仍然是一个发展中国家。作为大国，中国具有举国效应，可以集中力量办大事，但同样是大国，中国面对众多尚不富裕的人口和数以亿计的贫困人口，在提供社会保障、脱贫、基础设施等公共服务方面的责任和负担也是巨大的，其掌握的财力仍然有较大缺口。中国 2013 年人均 GDP 全球排名仅为第 81 名，人类发展指数全球排名仅为 91 名，距离发达国家达到人均峰值时的普遍人均 GDP 水平（2 万—2.5 万美元，2010 年价）仍有较大差

① 邹骥、傅莎等，中国人类发展报告 2007/08，对外翻译出版社，2010.

距，还处于库兹涅茨曲线左侧的“爬坡期”。中国仍存在大量的贫困人口。若按人均 1 天消费 1.25 美元的贫困标准（以 PPP 计算），中国 2008 年仍有 13.1% 的人口，即 1.73 亿人口生活在这一标准以下。而若将贫困标准设为人均每天消费 2 美元（以 PPP 计算），则中国 2008 年有 29.8% 的人口，即 3.95 亿人口生活在这一标准下[①]。近年来中国低碳技术发展虽取得长足进步，但仍明显落后于全球低碳技术革命的步伐，大量核心技术仍掌握在发达国家跨国公司和国家实验室手中。中国仍处于全球产业链和价值链的中低端，在全球利益分配中处于劣势。此外，中国在法制和监管、利益相关者协调的日常社会治理能力、意识等方面还比较薄弱，缺少优质的人才，软实力是短板。

除了发展进程、发展阶段和能力外，中国未来低碳发展战略还应充分考虑其特殊国情。总体上看，中国是一个人均资源存量、质量、有效人均活动空间都十分紧缺的国家，必须走高效发展道路。

从能源资源禀赋上看，中国是世界上少数几个以煤为主的国家，且以煤为主的能源结构在未来较长一段时间仍将持续。以煤为主的能源结构使得中国单位能源碳强度比美国、英国、日本等发达国家高 50% 左右，更是法国、巴西等国的 3 倍左右[②]，这将给中国控制温室气体排放带来特殊挑战。此外，中国能源资源分布的不均匀，可再生能源和其他清洁能源资源中心与用能需求中心不匹配也进一步加大了中国调整能源结构的难度。

从地域幅度和人口、资源、空间分布看，中国作为一个大国且人口、资源空间分布和发展不均衡，面临着人口、货物、资源、能源大范围输送流动需求，这也会进一步增加温室气体排放控制的压力。

从发展要求看，中国未来发展面临多重挑战，如经济发展、缩小贫富差距和区域差距、减少环境污染、保障能源安全等。如何保证经

① 数据来源：世界银行数据库。

② 数据来源：IEA, CO_2 Emissions from Fuel Combustion (2014 Edition), IEA, Paris.

济增长、应对气候变化与其他发展目标协同并进是中国实现可持续发展的重要任务，也是中国设置未来低碳发展目标时的考虑重点。

综上所述，中国在参与全球气候治理时需充分考虑上述发展特征、国情和能力。一方面，中国受其所处的发展阶段、能力和国情的影响，CO_2排放总量和人均排放在今后一定时期内仍将持续增长。但与此同时，中国传统的长期依靠要素投入增加驱动高速增长的发展模式难以为继，面临着国际国内形式的不断变化和如下一系列风险：高度依赖投资和出口的粗放型增长方式后续动力不足，面临着陷入很多国家曾经经历的“中等收入陷阱”的风险；经济增长和环境、能源、资源间的矛盾日益凸显，面临着更为严峻的能源、资源和环境约束；贫富差距和地区间发展的不平衡加剧；国际贸易争端频发；落后的社会保障导致的一系列民生问题，等等。这些国内外形势的变化说明中国已不再具备沿袭发达国家以高能源/资源消耗为支撑的现代化道路的国内外环境，中国的可持续发展不仅受到国内资源和环境的严重约束，而且面临国际环境容量空间限制。

低碳发展作为一种高效能的集约式、内涵式经济模式，通过应用与新的资源环境约束条件相匹配的技术和制度，催生新的经济增长点和推动传统行业新的增长模式，从而促进经济发展质量和效益持续提升，将成为倒逼推进中国经济转型发展的强大动力。对中国来说，其首要任务是努力探索一条优于发达国家传统发展道路的低碳发展路径，以比发达国家更低的人均 GDP 水平达到更低的人均 CO_2 排放峰值，按照党的十八大提出的大力推进生态文明建设的要求，坚定不移地走可持续发展道路，从我国基本国情和发展的阶段性特征出发，采取有力的政策措施，积极应对气候变化，着力推进低碳发展，为实现两个“一百年目标”保驾护航。

第三节　中国参与全球气候治理的责任、权利和国家目标

气候变化已成为事关国际道义、国际责任、国家形象、国家发展权利的重要议题。如前文所述，中国作为最大的发展中国家，已成长为全球气候变化问题中的“大户”，位置更加关键，处境也更加敏感。中国在气候变化问题上的积极作为，一方面对于全球应对气候变化进程有重要意义，另一方面也为中国提供了一个提高对自身大国地位驾驭能力的舞台，对中国提高国际影响力和提升国家形象都具有重要的现实意义。习近平主席在2013年APEC峰会上作出了“为应对全球气候变化作出新的贡献”的庄严承诺。关于气候变化，他强调“不是别人要我们做，而是我们自己要做，采取了许多措施，今后我们还会这样做”。中国应立足于本国国情和客观定位，在可持续发展框架和《公约》原则下，运用大国思维以积极务实姿态参与国际气候进程，实现从参与者到建设者的角色转变，推动促成于我有利的“公平、有效、共赢”的全球气候体制的达成，发挥负责任大国的建设性作用，主动承担与自身发展阶段、能力、定位和国情相符的应对气候变化全球责任。

具体而言，中国可从以下方面承担应对气候变化的全球责任：

一、积极推动“公平、有效、共赢”的国际气候体制建设

作为一个排放超级大国和即将完成工业化进程的新兴发展中大国，随着中国经济的发展，以购买力平价计算，2020年中国的经济总量有望达到美国当前的水平，有能力在众多国际事务中发挥更为积极的作用。中国应逐步由国际应对气候变化进程的积极参与者转变为“公平、有效、共赢”的国际气候制度的制定者和主导者，实现从“顺势而为”

到“主动出击”的战略转身，充分展现负责任大国的积极姿态，促进国际气候体制建设。

二、通过管控温室气体和尽早达到峰值为实现全球控制温室气体目标作出贡献

作为一个后发的发展中排放大国，不论是从全球应对气候变化需求还是国内资源环境约束看，中国都有必要改变现有发展模式，在发达国家适当支持下，通过发展模式的跃迁，开创一条比美欧等发达国家更为低碳的、更早达到更低峰值的创新发展道路。而且，中国创新发展路径的努力和实践对于其他面临同样发展需求的发展中国家而言，也有着很重要的借鉴意义。中国应在充分认识气候变化问题的严重性和紧迫性的基础上，本着对人类长远发展高度负责的精神，坚定不移地走创新的可持续发展道路，采取积极措施管控峰值。以转型为特征的峰值贡献目标与中国创新发展路径的需求和意图非常契合，可考虑作为中国 2020 年后贡献的重要组成部分。

三、为全球应对气候变化科技创新作出贡献

在未来应对气候变化的工作中，中国可推动加强对气候变化领域科学问题的基础研究和国际合作，推动对减缓、适应气候变化关键技术的研发，强化对气候变化各领域专家力量的培育、组织和引导，积极参与、强化和利用国际应对气候变化的科研平台如政府间气候变化专门委员会，建立国际应对气候变化的智库合作平台。

四、为帮助发展中国家应对气候变化作出贡献

在南南合作框架下，在“两廊一圈”、“一带一路”等区域合作战

略目标的指引下，从较为单一的产品赠送和能力建设转向更为广泛的包含技术援助、政策实践交流和战略规划在内的多层次综合南南合作体系。还可创新南北合作模式，为发展中国家第三方提供质量更高的援助。与相关国际机构开展合作，利用其援助资金、先进技术和理念以及成熟的外援网络和渠道，向发展中国家输出中国技术，分享中国经验。

与此同时，中国在参与全球气候进程时仍要清楚认识自身作为发展中国家的定位。中国应做发展中国家的旗帜，在国际舞台上充分强调发展中国家的阶段特点，领导发展中国家适当承担道义责任，坚决捍卫发展中国家的发展权，联合要求发达国家切实履行减排责任，努力争取发达国家对发展中国家的技术、资金补偿。

五、中国参与全球气候治理的国家目标

总体而言，中国参与全球气候治理时应统筹考虑如下目标：第一，为完成中国工业化、城镇化和发展路径转型的平稳过渡，避免硬着陆，争取峰值到来之前合理的必要排放空间并留有余地；第二，推动形成“公平、有效、共赢”的国际气候体制，支持应对气候变化，最大限度降低气候变化带来的不利影响；第三，争取实现更低更早峰值的发展机会和发展权，最大限度降低国民经济对温室气体排放和能耗的依赖程度，形成绿色低碳发展的长期竞争力；第四，从全球气候体制的参与者逐渐转变成为建设者和主导者，以此为平台之一促进提高中国在全球治理规则制定与实施中的地位、主导权和话语权；第五，树立负责任、公平正义的良好形象。

第四节　对中国国家自主贡献目标的分析[①]

《联合国气候变化框架公约》第 19 次缔约方大会（华沙会议）决议中首次使用了“预期的国家自主决定贡献 (Intended Nationally Determined Contribution , INDC)”这一概念。根据华沙会议和利马会议决定，各主要缔约方需在 2015 年年底在巴黎举办的第 21 次缔约方大会前（准备好的国家可在第一季度）提交其预期的国家自主决定贡献。

需要注意的是，“预期的国家自主决定贡献”（以下简称 INDC）概念虽直接源自华沙决定（1/CP.19）第 2 段，但它是气候谈判妥协的产物，其授权应综合考虑《公约》第 2、4 条及德班平台决议（1/CP.17）等。《公约》、德班、华沙决议和利马决议共同为预期的国家自主决定的贡献奠定了框架，应从如下三个基本角度去理解 INDC 这一在气候谈判中形成的概念：

一是要用“在《公约》下”、“以《公约》的原则为指导”去理解 INDC。即 INDC 是各方为应对气候变化、实现《公约》最终目标提出的进一步强化“承诺”或“行动”，应充分考虑各国，特别是发达国家和发展中国家不同的历史责任、能力、发展阶段和国情。

二是 INDC 是德班平台谈判的一个工具或过程，通过 INPC，各国在国内自主研究、磋商、制定各自应对气候变化的近、中、长期目标，推进谈判进程，为 2015 年在巴黎达成新一轮联合国气候变化协议就各缔约方可能作出的承诺或宣布的新目标在国内增强共识、加强准备。

三是 INDC 是在 20 多年来的《公约》进程中首次明确就“国家自主决定”其“贡献”目标达成共识，标志着国际气候治理结构向更加“自

① 相关结论也可参见傅莎、邹骥、刘林蔚的“解读中国国家自主贡献（INDC）目标”，国家气候战略中心网站，http://www.ncsc.org.cn/article/yxcg/zlyj/201506/20150600001484.shtml.

下而上”的目标模式迈进一步。

根据要求，2015 年 6 月 30 日，中国政府提交了其国家自主贡献（INDC），对其 2020 年后的强化减缓和适应行动作出了安排[①]，具体目标如下：

——二氧化碳排放 2030 年左右达到峰值并争取尽早达峰；

——2030 年单位国内生产总值 CO_2 排放比 2005 年下降 60%—65%；

——2030 年非化石能源占一次能源消费比重达到 20% 左右；

——2030 年森林蓄积量比 2005 年增加 45 亿立方米左右；

——继续主动适应气候变化，在农业、林业、水资源等重点领域和城市、沿海、生态脆弱地区形成有效抵御气候变化风险的机制和能力，逐步完善预测预警和防灾减灾体系。

中国 2020 年后的国家自主贡献是中国承担应对气候变化全球责任的核心体现。本节基于创新发展路径的视角，从中国提出 INDC 的出发点和考虑、中国 INDC 对本国低碳转型和全球应对气候进程的含义、中国实现 INDC 面临的困难和挑战等角度对中国提出的国家自主贡献开展评估。

一、基于“创新发展路径”视角分析中国提出 INDC 目标的基本考虑

如上文所述，尽管中国已成为全球第一大排放国和第二大经济体，但其主要发展指标和长期存在的显著“二元结构”特征均显示了其发展中国家的定位，以减贫、提高收入、提高社会保障程度、提高基础设施等为主要内容的发展，依然是压倒一切的战略重点。与此同时，中国传统的长期依靠增加要素投入驱动高速增长的发展模式难以为继，

①中国政府 . 强化应对气候变化行动——中国国家自主贡献 . 联合国气候变化框架公约网站，2015.

面临陷入以资源、环境为主要制约的“中等收入陷阱”的风险，必须创新发展路径，促成经济增长模式转向“新常态”，使经济增长的驱动力从主要依靠要素投入增长转变为更加依靠要素生产率的提高，减少经济对能源、资源和环境要素的依赖，培育新的经济增长点和竞争力，走上一条高要素效率的可持续低碳发展路径。

中国的国家自主贡献选取了以发展路径转型创新为主题的一揽子指标体系，包括意向性峰值时间目标和以提高碳要素效率为核心的量化转型过程指标（包括碳强度、非化石能源比重、碳汇等指标），致力于实现发展路径变迁，努力探索一条不同于美欧传统路径的创新型可持续低碳发展路径，并注重加强实现路径转型所必需的技术、资金和其他必要条件及适宜环境的建设。

为满足有关国家自主贡献应涵盖所有要素（减缓、适应、资金、技术、能力建设和透明度）的授权并有效应对已经发生的气候风险，中国的国家自主贡献还包含了中国适应气候变化的众多举措和政策。同时，中国还在其贡献中包含了进一步加强气候变化南南合作，建立南南合作基金，全面拓展气候变化南南合作规模和领域的目标。

二、“创新发展路径”视角下的贡献评价方法和指标

如何认识贡献的力度和公平性是目标谈判中的热点，也是各方评估的重点。目前国际上已有多家机构和智库——如国际能源署（IEA）、联合国环境规划署（UNEP）、世界银行、荷兰环境评估署（PBL）、德国波兹坦研究所（PIK）、气候行动跟踪（CAT）、德国新气候机构、世界资源研究所（WRI）、法国巴黎可持续发展与国际关系研究所（IDDRI）等——在筹划开展对各国 INDC 的个体力度、公平性以及全球整体力度的评估。

从创新发展路径的角度，本节采用了如下指标框架评估中国 INDC 对本国低碳转型和全球应对气候进程的含义：

评估中国 INDC 力度的基本出发点是其是否开创了一条不同于美欧的创新发展路径，是否能成功实现更早达到更低峰值的目标。除此之外，报告还从以下角度评估中国 INDC 力度：

（1）和中国自身比较。一是评估贡献目标的递进性，即纵向比较 2020 年后贡献与 2020 年前行动目标，力度是否加强，是否存在倒退，是否呈现加速增长态势；二是评估贡献目标的额外性，即评估贡献目标在现有政策框架（或基准情景）之外所需付出的额外努力。

（2）与美欧等发达国家比较。即从实现峰值的时间、条件和峰值水平、累积排放、新增非化石能源装机和发电量等指标出发横向比较中国 INDC 力度。

（3）和全球 2℃目标比较。即评估中国 INDC 是否与 2℃目标要求的长期减排路径相一致。

评估中国 INDC 的公平性主要看中国 INDC 是否符合《公约》要求，是否符合中国发展中国家定位，是否符合中国的责任、能力、发展阶段和国情。

三、中国 INDC 对本国低碳转型和全球应对气候进程的含义评估

基于评估结果，整体来看，中国 INDC 体现了中国积极应对气候变化，努力控制温室气体排放，提高适应气候变化的能力，以及深度参与全球治理，承担合理国际责任的姿态和决心。

（一）从创新发展路径的角度评估中国 INDC 力度

作为一个后发的发展中大国，中国有望成功实现转型，开创一条比美欧等发达国家更为低碳的发展路径，以更低的收入水平达到更低的峰值。根据上文分析，主要发达国家人均排放达到峰值对应的人均 GDP 水平，大体在 2 万—2.5 万美元（2010 年价格）左右，人均峰值水平约为 10—22 吨 CO_2。而按照中国的 INDC 目标，中国可望在人均 GDP 达到 1.4 万美元时就达到人均排放峰值，且人均峰值可望保持在 8.3

吨左右 CO_2 的较低水平上（见图 8-4）。这是中国作为后发发展中大国创新发展路径的重要标志。若考虑中国世界工厂的定位及以煤为主的资源禀赋，中国人均消费端排放峰值和能源消费峰值与发达国家相比将更低。

（二）从目标连续性、递进性和额外性角度评估中国 INDC 力度

1．从目标连续性看，通过实施国家自主贡献，中国经济将进一步实现和碳排放脱钩

从表 8-2 和图 8-5、8-6 可以看出，经济增长是中国 CO_2 排放的主要增排因素。2020、2030 和 2050 年，中国人均 GDP 分别约是 2005 年水平的 3.2 倍、5.2 倍和 11 倍。而通过实施 INDC，同期中国能源相关 CO_2 排放只分别相当于 2005 年水平的约 1.8 倍、2 倍和 0.8 倍，中国经济与碳排放实现进一步脱钩，中国经济发展对碳排放的依赖日益减轻，

图 8-4 主要经济体人均GDP和人均CO_2排放轨迹

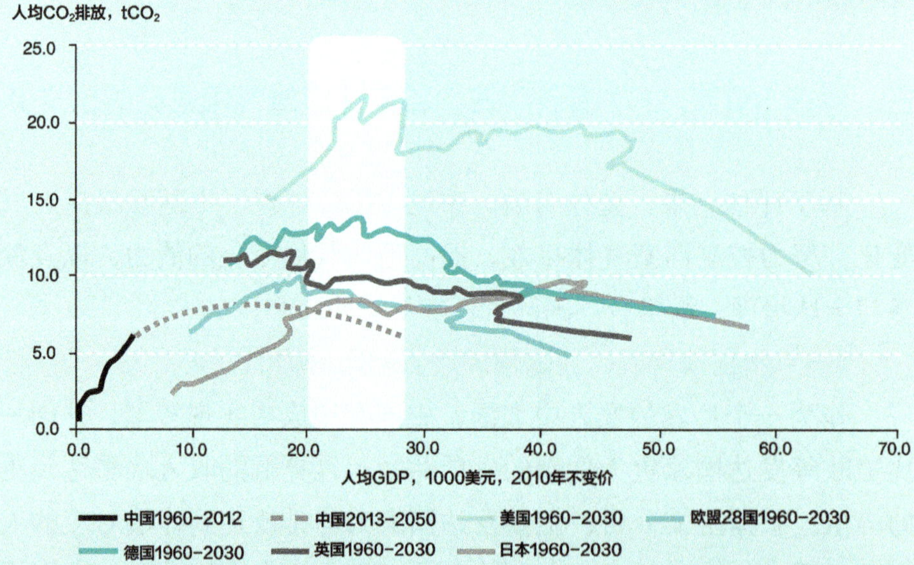

数据来源：仅包含能源相关CO_2，1960—2012年的历史CO_2排放数据来自美国橡树岭实验室二氧化碳分析中心（CDIAC）；人口和GDP数据来自世界银行。2012年后数据由作者根据各国国家自主贡献目标测算得到。

表 8-2 中国国家自主贡献下未来主要驱动因子变化，2005=100

	2005	2010	2015	2020	2030	2040	2050
人口	100	103	105	108	112	112	111
人均 GDP	100	166	235	321	517	783	1103
单位 GDP 能源强度	100	81	68	59	43	29	18
单位能源碳强度	100	98	94	89	80	63	39
能源相关 CO_2 排放	100	135	158	182	201	158	84

数据来源：2005、2010 年数据来自《中国统计年鉴》、《能源统计年鉴》和中国官方公布的目标完成情况。2015 年之后的数据基于国家气候战略中心和中国人民大学 PECE 模型核算的 INDC 情景研究结果。

图 8-5 中国国家自主贡献下的未来主要驱动因子变化，2005=100

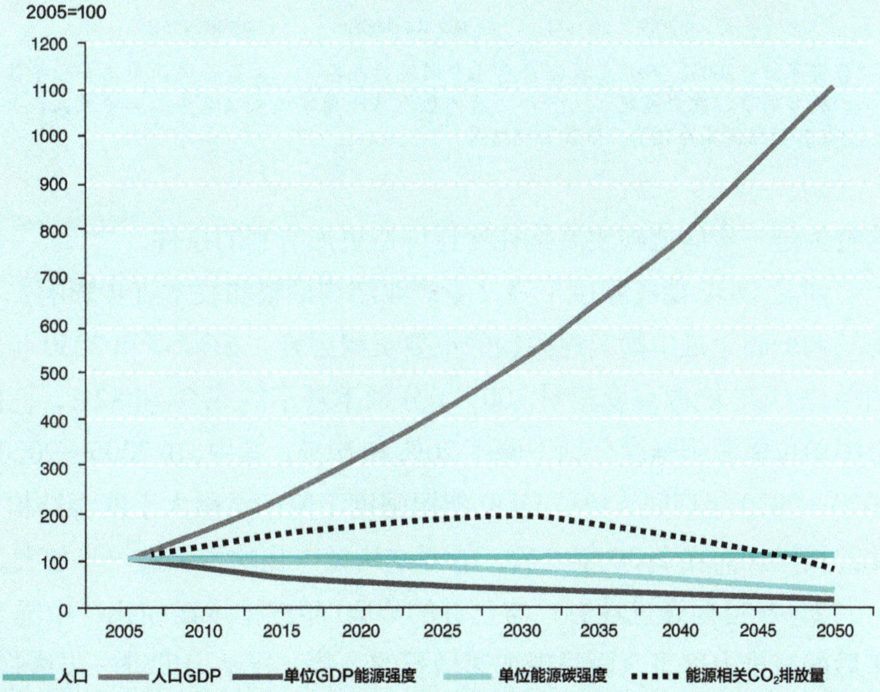

数据来源：2005、2010 年数据来自《中国统计年鉴》、《能源统计年鉴》和中国官方公布的目标完成情况。2015 年之后的数据基于国家气候战略中心和中国人民大学 PECE 模型核算的 INDC 情景研究结果。

图 8-6 中国能源相关CO_2排放影响因素对排放变化的贡献率

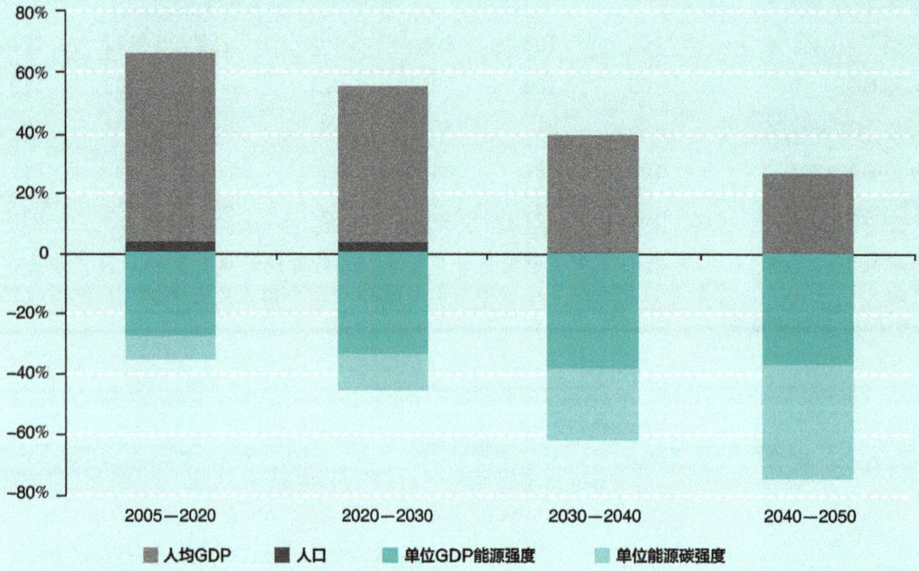

数据来源：2005、2010 年数据来自《中国统计年鉴》、《能源统计年鉴》和中国官方公布的目标完成情况。2015 年之后的数据基于国家气候战略中心和中国人民大学 PECE 模型核算的 INDC 情景研究结果。

这将为进一步向低碳发展路径变迁创造更加有利的条件。

单位 GDP 能源强度下降（受产业结构调整和技术进步影响）和能源结构低碳化是中国实现脱钩的主要贡献因素。2030 年和 2050 年的中国单位 GDP 能源强度相对 2005 年分别下降了约 57% 和 82%，而同期中国单位能源碳强度分别下降了 20% 和 61%。其中，在 2005—2020 年、2020—2030 年期间，单位 GDP 能源强度下降的贡献大于能源结构低碳化的贡献。而在 2030 年之后，能源结构低碳化的贡献进一步加大。

2．从目标递进性看，与已有的 2020 年减排承诺相比，中国 2020 年后的减排力度将全面呈现加速增长的态势，行动力度进一步增强。

从表 8-3 可以看出，2020—2030 年中国单位 GDP 的 CO_2 排放强度下降速率将进一步提高，从 2005—2020 年的平均 3.9% 上升到 4.4%，并有望进一步提高到 2030—2040 年的 6.3% 和 2040—2050 年的 9.2%。

表 8-3 中国在降低碳强度和发展非化石能源方面的行动力度

	2005—2020	2020—2030	2030—2040	2040—2050
单位 GDP 的 CO_2 强度年均下降率	3.9%	4.4%	6.3%	9.2%
年均新增非化石能源装机 (GW)，其中	41.5	62.8	79.6	90.1
年均新增风能装机 (GW)	13.9	23.0	31.0	35.0
年均新增太阳能装机 (GW)	7.0	24.5	33.0	40.8
年均新增核能装机 (GW)	3.4	9.0	9.3	10.5

数据来源：基于国家气候战略中心和中国人民大学PECE模型核算的INDC情景结果计算得到。其中2020年的碳强度和非化石能源目标分别根据45%和15%计算，2030年的碳强度和非化石能源目标根据65%和20%计算。

与此同时，中国将加快非化石能源项目的部署和建设，推动提高非化石能源装机的增速。年均非化石能源装机需从2005—2020年的4150万千瓦上升到2020—2030年的6280万千瓦，并进一步上升到2040—2050年的9000万千瓦左右。

3．从目标额外性看，与已有政策情景相比，中国实现INDC目标需要付出大量额外努力

为实现中国INDC目标，中国需要进一步采取多种致力创新发展路径，降低经济增长对能源排放要素投入的依赖，推动产业结构转型和能源结构低碳化的政策，包括“十三五”开始逐步实施化石能源消费总量和碳总量管理，加大对非化石能源发展的政策支持力度，启动全国碳市场，合理控制建筑行业发展速度，优化交通运输模式，等等。与现有政策情景相比，在INDC情景下，中国能源相关CO_2排放量将有大幅下降，2030年能源相关CO_2排放量将比现有政策情景下减少30亿吨CO_2，2030年单位GDP二氧化碳强度相对2005的下降幅度也将进一步由现有政策情景下的57%上升到60%—65%。中国需要为此付出大量额外努力。

（三）从国际比较的角度看中国INDC的力度

1．从需新增非化石能源装机看，中国远高于美欧

2005—2030年，中国非化石能源比重将从7.4%提高到20%。据

测算，如果中国能实现2030年非化石能源比重达到20%的目标，届时可再生能源和核能的装机规模将达13亿千瓦左右，相当于美国当前发电装机总量。其中水电装机将约达4亿千瓦左右，风电、太阳能发电的装机各达4.4亿、3.5亿千瓦左右，核电装机1.5亿千瓦左右。2014—2030年，中国非化石能源供应量仍需以年均7%左右的速度增长，2030年中国非化石能源电力装机预期需在2014年的基础上增加9亿千瓦左右，与2014年全国火电总装机基本相当，相当于每年需新投产风电机组和太阳能发电装机各约2000万千瓦左右，每年新增核电装机约800万千瓦，远高于美欧同期（见图8-7）。

2．从历史累积排放看，通过落实INDC，到2030年中国CO_2历史累积排放将始终保持低于美欧的水平

IPCC第五次评估报告肯定了全球温升和累积排放之间的近似线性

图8-7 中、美、欧2030年相对2014年非化石能源装机变化

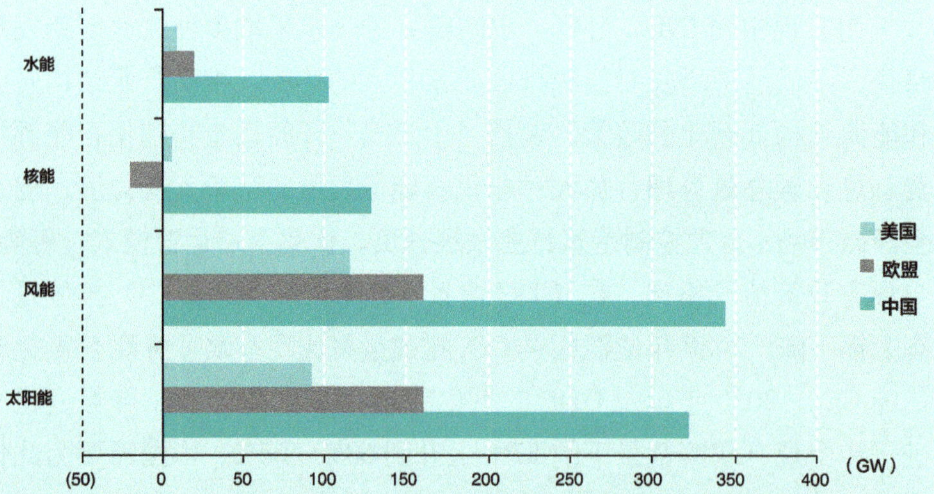

数据来源：中国数据来自国家气候战略中心和中国人民大学PECE模型核算的INDC情景结果。美、欧数据来自IEA《2015年全球能源展望特别报告》中的INDC情景结果，为2030年相对2013年的增量。

关系。从图8-8中可以看出，发达国家温室气体排放的高峰期是19世纪中叶至20世纪90年代的先行工业化阶段，以高度依赖化石能源，伴随大量温室气体排放为特征完成其工业化进程。而中国能源相关CO_2排放快速增加始于20世纪中叶，集中于改革开放以来30多年的快速工业化阶段。在国家自主贡献目标下，中国将确保通过创新发展路径，使其到2030年的能源相关CO_2累积排放水平仍低于美欧同期水平（届时美、欧和中国在1751—2030年间的历史累积排放的相对比例为1.4 ：1.2 ：1.0），甚至低于美欧当前的（1751—2012年）的累积排放水平。如果从人均历史CO_2累积排放（图8-9）看，中国同期（1751—2030年）的人均历史累积排放将控制在美国的18%和欧盟的30%。

（四）从全球2℃目标要求角度看中国INDC的力度

根据最新发布的IPCC第

图8-8　中、美、欧CO_2历史累积排放比较

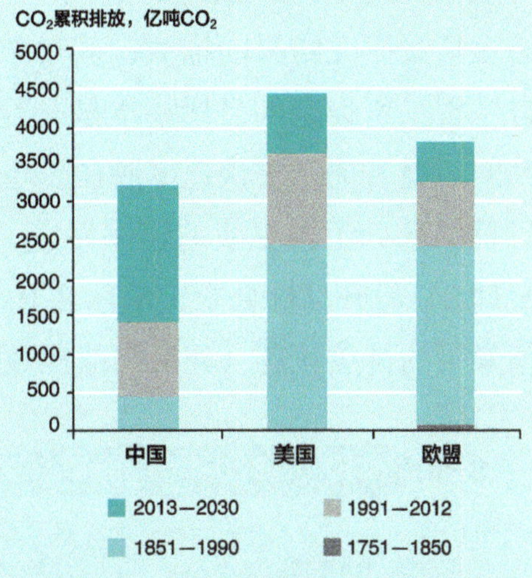

数据来源：仅包含能源相关CO_2, 1751—2012年数据来自美国橡树岭国家实验室二氧化碳信息分析中心（CDIAC），2013—2030年数据由国家气候战略中心基于各国国家自主贡献测算。

图8-9　中、美、欧人均CO_2历史累积排放比较

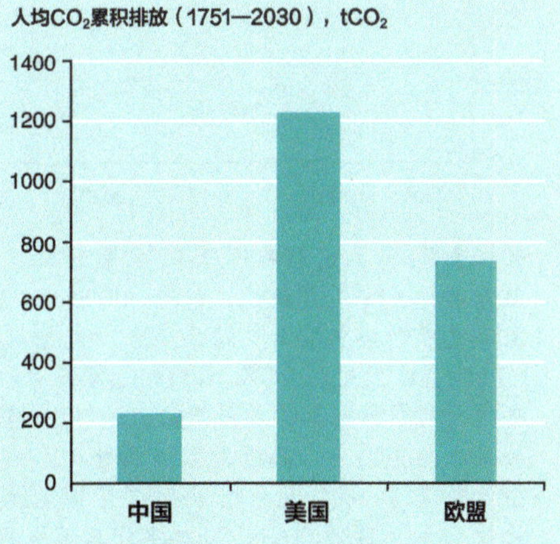

五次评估报告中较为符合中国实际的、有可能实现 2℃温升目标（50%可能性实现 2℃，包含 450 情景和 500 情景，未考虑限制技术情景）的诸多情景研究结果的推演，其对中国 2030 年前的减排要求和中国国家自主贡献所设定的目标基本是一致的，核心关注在于中国 2030 年达到峰值后是否能以更快速率脱碳。也就是说，如果中国能在实现 2030 年目标的基础上以更快的速率脱碳，那么中国长期发展路径将有望满足全球 2℃温升目标要求。从图 8-10 中可以看出，虚线代表的中国 INDC 情景下的排放轨迹（分别为能源数据调整前后的）正好落入 IPCC 第五

图 8-10 中国INDC目标情景和全球2℃温控情景的关系

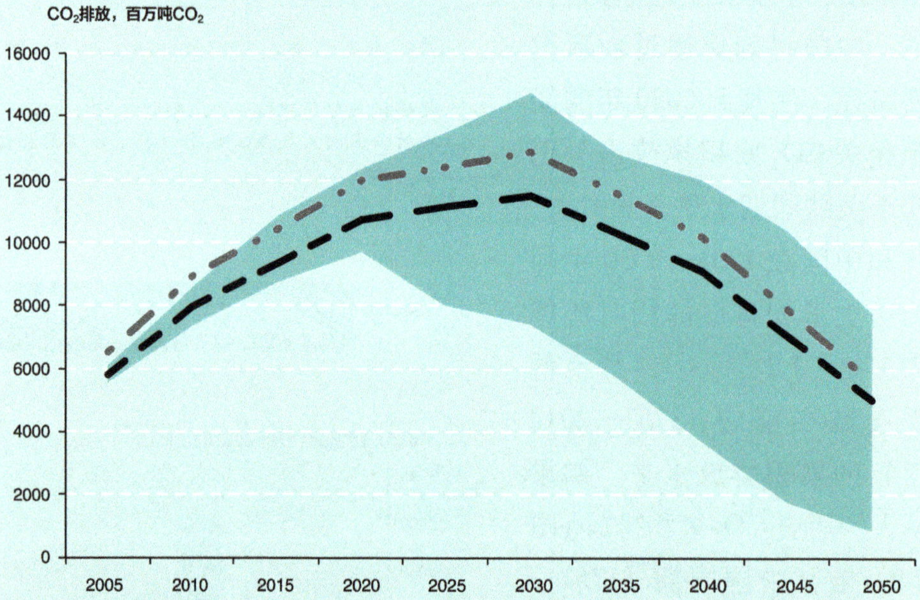

数据来源：蓝色阴影部分为 IPCC 第五次评估报告中较为符合中国实际的有可能实现 2℃温升目标（50% 可能性，包含 500 和 450 浓度情景）的诸多情景研究（30 余个情景，未考虑限制技术情景）结果集合，包含 GCAM, IMAGE, MERGE, MESSAGE, POLES, REMIND, TIAM-ECN, WITCH 等多个全球著名模型组的结果，其中 2010 年和 2015 年排放与中国实际排放差距较大的 30 余个情景被剔除。黑线是中国 INDC 目标情景下的排放趋势（未考虑能源数据调整），由国家气候战略中心和中国人民大学 PECE 模型计算得到。灰线是考虑 2014 年中国能源数据调整后的粗略趋势。为和全球数据可比，此处 CO_2 为能源相关 CO_2 加水泥工艺过程 CO_2 排放。

次评估报告中满足 2℃温升要求且符合中国实际的情景范围之内（图中蓝色阴影）。考虑到中国的国家自主贡献目标包含了大量以创新发展路径为导向的技术、资金、政策、机制和能力建设方面的务实安排，且致力于通过科学规划（城市空间布局）和行政命令管控（限制新增高耗能产业产能等）等手段规避城市建设、基础设施、高碳产业的锁定效应，因此，通过实施中国国家自主贡献，中国有望在发展理念、民意基础、政策与制度储备、技术系统储备、创新能力、资本存量与流量、综合与专项能力等方面形成更加厚实的积累，为 2030 年后的加速转型奠定坚实的基础。

（五）中国 INDC 的公平性评估

1．中国 INDC 符合《公约》要求和中国责任、能力、发展阶段和国情

如上文所述，中国的主要发展指标和长期存在的显著“二元结构”特征均显示了其发展中国家的定位，和发达国家在所处的发展阶段、面临的发展需求、承担的历史责任和具备的综合能力等方面都存在巨大差别。中国仍是一个拥有超过 13 亿人口的发展中国家，人均 GDP 刚超过 7000 美元，尚有 1.2 亿人口生活在贫困线以下，目前正处在工业化、城镇化的发展阶段，面临着发展经济、改善民生、消除贫困、保护环境的多重挑战，以减贫、提高收入、提高社会保障程度、提高基础设施等公共服务覆盖面、提高生活质量为主要内容的发展，依然是压倒一切的战略重点，即便到 2030 年，中国成为第一经济大国，其人均 GDP 也仅为欧美的 1/3—1/4（2005 年不变价美元）。中国的主要任务是改变现有发展模型，通过发展路径的跃迁，开创一条比美欧等发达国家传统发展路径更为低碳的，更早达到更低 CO_2 排放峰值的创新发展路径。由此可以看出，中国 INDC 选取了以发展路径转型创新为主题的一揽子指标体系是符合发展中国家定位，符合中国国情、发展阶段和能力，符合《公约》的规定和要求的，是实现全球应对气候变化目标努力的重要组成部分。

2．中国 INDC 可为发展中国家后续发展提供示范和借鉴，并通过

南南合作等方式传播经验、提供支持

作为最大的发展中国家，中国被很多发展中国家视为发展样板。中国可通过创新发展，开创一条低碳可持续发展路径，为后发的发展中国家提供可资借鉴的示范模型和经验，支持这些发展中国家规避传统高碳路径依赖和锁定效应，尽早走上“低污染低排放”的高效创新发展道路。同时，中国也能通过自身转型促进全球转型，为重塑全球发展路径作出贡献。

四、中国实现国家自主贡献面临的困难和挑战

低碳发展虽已是中国未来必然的发展方向，但是，中国的社会经济发展阶段、经济结构、能源资源禀赋、能源效率、技术能力储备、体制政策基础以及所面临的国际政治经济格局等也共同决定了中国想要实现 INDC 目标仍将面临以下种种巨大挑战：

第一，资源的可得性将增加实现 INDC 目标的风险。例如，要实现国家自主贡献，需要获取大量的天然气等清洁能源以替代部分煤炭，如何在控制成本的情况下确保资源可靠的供给、保障能源安全是中国实现目标的巨大挑战。

第二，中国经济目前仍处于工业化和城镇化进程中，以机械制造、钢铁、建材、化工等为代表的具有重化工业特征的行业在经济中仍占据较大比重，同时，城镇化正进入高速发展阶段，大规模的基础设施建设不断推行。这些都必然带来能源消费和碳排放的持续增长。尽管中国政府已开始着力于调整经济结构、转变增长方式，但这种努力不可能是一蹴而就的，且可能随着经济下行压力的增大而出现反复，需要克服巨大的困难。

第三，随着中产人口占比的逐年提升，交通、建筑等方面的消费排放压力会上升，甚至在中长期会超过制造业的排放压力。如何塑造年轻一代消费者的低碳消费模式和生活方式，对中国实现国家自主贡

献目标至关重要。

第四，技术本身的可靠性和不确定性是约束中国贡献目标实现的另外一个重大挑战。例如如何确保可再生能源的上网和电网的稳定供应，如何解决发展水电、核电时的额外影响如环境影响、核安全问题等，如何解决碳捕集与封存技术（CCS）本身的不确定性等。

第五，作为发展中国家，中国的整体科技水平仍比较落后，技术研发能力有限，尤其是在关键的低碳技术和适应技术领域仍存在欠缺。如何通过有效的国际技术合作机制确保技术的可得性是中国落实国家自主贡献目标时需要重点解决的问题。

第六，落实减缓和适应贡献目标面临巨额投资需求和“结构化”成本。实现中国贡献目标需要大量增量投资，这些投资有一部分能带来经济回报，但也有很大部分会带来巨大的经济成本，诸如碳捕集与封存（CCS）等技术的应用。除了经济成本，中国实施贡献目标短期内还可能面临其他社会成本，例如淘汰落后产能，会引发结构性失业和其他社会问题，需要采取措施以缓解和消除这些社会成本。

第七，社会主要利益相关者的基本能力普遍不足，意识、体制、社会机制的改善任务十分繁重。中国应对气候变化的宏观管理能力尚有不足，法律法规、政策体系还非常不健全，许多领域仍然处于空白状态。中国目前还缺少健全的温室气体核算体系，对温室气体排放的监测监控核算能力和执法能力还比较薄弱。这些都将对中国落实国家自主贡献目标形成挑战。

附　录

1. PECE 模型简介[①] 和情景说明

本文采用了由国家气候战略中心和中国人民大学共同开发的PECE 模型开展情景研究。PECE 模型是一个以自下而上的单国能源技术选择模型（PECE-ES）为核心的耦合了社会经济模块和能源服务需求模块的简单混合模型。它包含自下向上能源供给和终端消费技术模型（PECE-ES）、基于生产函数的社会经济模型（PECE-SE）、基于计量分析的能源需求服务模型（PECE-ESD）三个子模型。PECE 能源技术模型（PECE-ES）是 PECE 模型的主要组成部分，采用最小成本法进行优化模拟。

模型模拟了经济中的价值流、能源流和物质流全过程，从经济产出到能源供应，从一次能源和原材料供给，通过加工转换到二次能源和材料供给，再到满足终端用户需求。模型通过对这一过程中典型技术和经济活动的描述全面展示了这一过程。在 PECE 能源技术模型中，每项技术可以进一步分解为不同设备的组合。模型含有丰富的技术信息。技术的选择是基于使得总成本最小的非线性优化框架完成的，同时，这一技术选择过程还受到多种约束的限制，包括对最终能源服务需求的满足，能源和材料的供给情况，以及其他系统约束等。系统成本包括技术成本（包括固定投资和运行成本）、能源成本和政策成本如环境、资源、能源、技术税费、补贴等。模型适用于对不同政策情景进行分析。

①傅莎．基于能源系统模型的中国低碳发展技术战略和政策选择研究 [D]. 中国人民大学博士论文，2012.

附图 PECE模型的基本结构

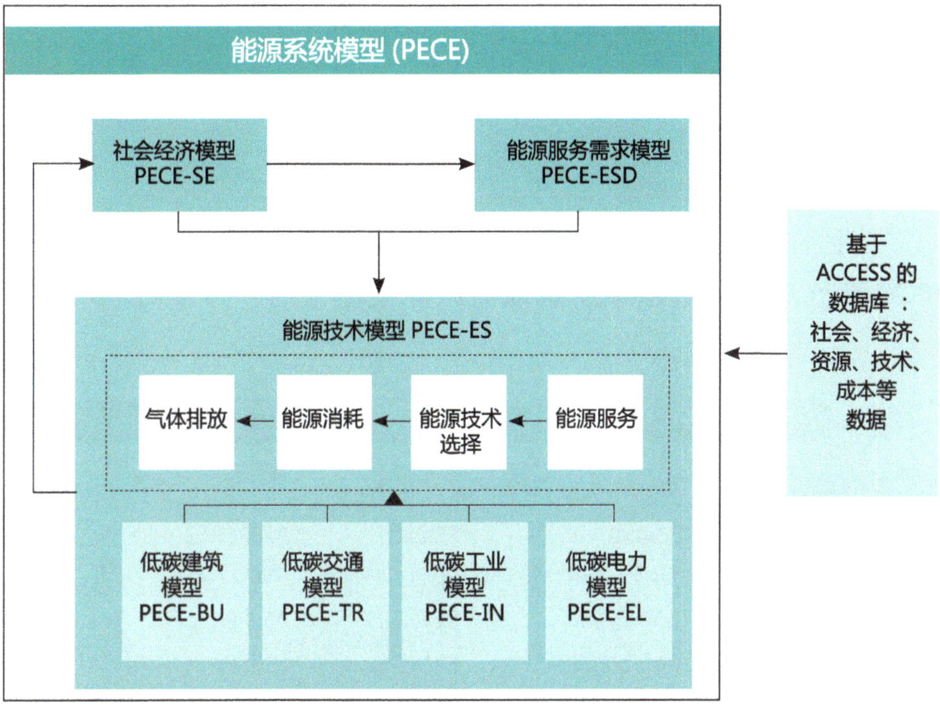

模型的基本结构如附图所示。

PECE 模型描述和求解的是大型复杂的动态非线性优化问题，是基于通用数学建模系统 GAMS（General Algebraic Modeling System）平台开发的，模型求解一般选用 CONOPT 求解器。

本文主要采用 PECE 模型模拟了中国的 INDC 情景。本文中的 INDC 情景以中国 INDC 文件中承诺的碳排放达峰、非化石能源比例和强度目标作为约束条件，计算满足既定能源服务需求的成本最优或者减排潜力最大的减排方案，描述了中国为对全球应对气候变化行动作出更大贡献而付出的努力，如进一步采取多种致力创新发展路径，降低经济增长对能源排放要素投入的依赖，实行产业结构转型和能源结构低碳化政策（如“十三五”开始逐步实施化石能源消费总量和碳总量管理，加大对非化石能源发展的政策支持力度，启动全国碳市场等政策后的能源消耗和排放趋势），等等。

研究以 2005 年为基年，以 2005—2050 年为分析区间，重点分析 2020 年、2030 年和 2050 年这三个关键时间点。

2. 情景数据库及情景筛选依据说明

为评估中国 INDC 目标与全球 2℃温控目标的联系，本文选取了 IPCC 第五次评估进程中所选取的国际模型论坛（如 AME, LIMITS,

附表 文章选取的2℃情景清单

模型	情景	模型	情景
GCAM	AMPERE2-450	POLES	AMPERE2-450
GCAM	LIMITS-StrPol-450	POLES	EMF27-450
GCAM	LIMITS-StrPol-500	REMIND	AMPERE2-450
GCAM	LIMITS-RefPol2030-500	REMIND	LIMITS-500
GCAM	LIMITS-RefPol-450	REMIND	LIMITS-StrPol-450
GCAM	LIMITS-RefPol-500	REMIND	LIMITS-StrPol-500
IMAGE	AMPERE2-450	REMIND	LIMITS-RefPol2030-500
IMAGE	LIMITS-450	REMIND	LIMITS-RefPol-450
IMAGE	LIMITS-500	TIAM-ECN	LIMITS-StrPol-500
IMAGE	LIMITS-StrPol-450	WITCH	AMPERE2-450
IMAGE	LIMITS-StrPol-500	WITCH	EMF27-450
IMAGE	LIMITS-RefPol2030-500	WITCH	LIMITS-500
IMAGE	LIMITS-RefPol-450	WITCH	LIMITS-StrPol-450
IMAGE	LIMITS-RefPol-500	WITCH	LIMITS-StrPol-500
MERGE	EMF27-450	WITCH	LIMITS-RefPol2030-500
MESSAGE	AMPERE2-450	WITCH	LIMITS-RefPol-450

数据来源：LIMITS Scenario Database，available: http://www.iiasa.ac.at/web/home/research/researchPrograms/Energy/LIMITs_Scenario_Database.html. AMPERE Scenario Database, available: http://www.iiasa.ac.at/web/home/research/researchPrograms/Energy/AMPERE_Scenario_database.html. EMF 27 database, available: http://www.iiasa.ac.at/web/home/research/researchPrograms/Energy/EMF27DB.html.

AMPERE, ROSE, EMF 等）情景数据库中包含详细中国信息、较为符合中国实际的有可能实现 2℃温升目标（50% 可能性，包含 500 和 450 浓度情景）的 30 余个情景结果，包含 GCAM, IMAGE, MERGE, MESSAGE, POLES, REMIND, TIAM-ECN, WITCH 等多个全球著名模型组的结果。其中 2010 年和 2015 年排放与中国实际排放差距较大的 30 余个情景被剔除。同时，本文选取的所有情景均假设技术是完全可获取的，限制技术情景并未纳入本文分析范围。此外需要说明的是，所有 2℃情景均未考虑中国能源数据调整的影响。具体选取的情景目录见附表。

【参考文献】

[1] 傅莎．基于能源系统模型的中国低碳发展技术战略和政策选择研究 [D]. 中国人民大学博士论文，2012.

[2] 傅莎，邹骥，刘林蔚．解读中国国家自主贡献（INDC）目标．国家气候战略中心网站 http://www.ncsc.org.cn/article/yxcg/zlyj/201506/20150600001484.shtml.

[3] 国家统计局．中国统计摘要 2015[M]. 北京：中国统计出版社，2015.

[4] 国家统计局．中国能源统计年鉴 2006—2013[M]. 北京：中国统计出版社，2007—2014.

[5] 邹骥，傅莎，等．中国人类发展报告 2007/08[M]. 北京：对外翻译出版社，2010.

[6] 中国政府．强化应对气候变化行动——中国国家自主贡献，联合国气候变化框架公约网站，2015.

[7] AMPERE Scenario Database, available: http://www.iiasa.ac.at/web/home/research/researchPrograms/Energy/AMPERE_Scenario_database.html.

[8] EMF 27 Scenario Database, available: http://www.iiasa.ac.at/web/home/research/researchPrograms/Energy/EMF27DB.html.

[9] EU.Intended Nationally Determined Contribution of the EU and its Member States, UNFCCC website, 2015.

[10] IEA. CO_2 Emissions from Fuel Combustion (2014 Edition), IEA, Paris.

[11] LIMITS Scenario Database，available: http://www.iiasa.ac.at/web/home/research/researchPrograms/Energy/LIMITs_Scenario_Database.html.

[12] PBL/JRC. Trends in global CO_2 emissions; 2014 report", December 2014，Available at：http://www.pbl.nl/en/publications/trends-in-global-co2-emissions-2014-report.

[13] P.Friedlingstein, et.al, Persistent growth of CO_2 emissions and implications for reaching climate targets, Nature geoscience, 2014.

[14] US, Intended Nationally Determined Contribution of US, UNFCCC website, 2015.
[15] WORLD BANK. 2015. World Bank Global Development Indicators Database [Online]. Available: http://databank.worldbank.org/data/reports aspx?source=worlddevelopment-indicators 2015.

第九章 结论和建议

Chapter Nine Conclusion and Recommendation

一是全球气候变化是不争的科学事实，自工业革命以来累积排放的温室气体导致大气中温室气体浓度升高进而导致地球表面平均温度上升，造成一系列全球范围的负面影响和风险。考虑到控制温升的措施要到本世纪末才能发挥显著效果，气候变化已成为 21 世纪全人类面临的重大挑战，没有一个国家可以独善其身、置身其外，需要全人类集体行动，共同应对，需要建立全球气候治理体系。

二是气候变化问题具最大时空尺度的外部性。温室气体排放成因及其减缓效果涉及跨几十年甚至上百年的时间尺度，往往超越常规决策者的思考视野和一般决策及投资周期尺度。对大时空尺度的认知和把握往往充满着不确定性，这在确定长期愿景和目标、界定并认识历史责任方面引起了极大的争议。同时，需要特别应对超主权范围外部性决策问题。在以民族国家主权为基础的当代人类国际政治文明发展水平下，全球治理并不存在一个道义、政治和法律意义上超主权实体实施治理管制的基础，只能靠主权国家之间通过多边和双边谈判而形成共同认可的国际法律条约来对各个主权国家的行为进行规范。一方面，这样的谈判会因为不同国家价值观、利益和政治结构的区别而变得十分缓慢艰难，另外一方面谈判作出的妥协安排往往十分脆弱，在实施上大打折扣。但一旦达成协议，作出的安排往往也具有较强的稳定性。

三是反思财富、幸福的观念和人类发展的前途，科学技术发展的历史，应当建立世界转型和人类发展路径创新的中长期愿景和视野。对可持续发展的方向及低碳转型大势作出一个基本的判断和政治认可。未来的世界必将面临人类发展路径的创新，需要使发展对资源的依赖程度下降，效率的提高将是长期发展日益重要的源泉。技术将持续更新，能源结构会发生重大的低碳变化，生产方式、生活方式都将发生新的变革，产业结构和业态也将发生变化，世界经济的面貌会向着低碳方向发生深刻的演变。竞争力评价的内容也将有所转变。这些都是人类发展路径创新的相关内容，将引致世界各主要经济体去开拓与发达国家在历史上创立的传统高碳发展路径完全不同的崭新低碳发展路径。面向新的发展前景和科技、能源与产业革命的浪潮以及低碳转型的发展趋势，为了实现人类发展路径的创新，需要在全球范围内建立气候治理体系，支撑和加速全球低碳转型。

四是全球气候治理旨在为各国提供制度上的持久约束和动力，调整国家战略行为，为各国实现绿色低碳转型提供激励；也是要在无超主权实体力量的情况下，基于互信，基于谈判的协议，推进各主权国家采取行动，承担并履行责任。这个治理要面对世界的多样性，特别是发达国家与发展中国家在历史责任、发展阶段和优先发展议程、能力、国情等方面差别的基本事实。要为发达国家与发展中国家量身定制绿色低碳转型、发展路径创新的制度和机制安排。因此，“共区原则”和公平是建立全球气候治理体系必须遵循的基本原则。这样的全球治理体系业已形成并在进一步发展。

五是通过对世界近现代主要经济体工业化历史及科学技术史、能源发展史和二氧化碳排放历史实证数据的考察，发现达到相似的发展水平，可以有不同的发展路径和排放水平。投入的生产要素组合（如能源结构）、技术水平、经济结构、收入水平、人口总量素质与分布，都会影响到以排放路径为表征的发展路径。主要经济体都经历了排放在经济发展进入工业化和城镇化的早期阶段而上升，而后随着收入水平的提高而达到峰值并在进入后工业社会发展阶段后呈现稳定下降趋势的“库兹涅茨曲线”特征。

六是全球气候治理与全球治理具有广泛的关联，是全球治理的重要有机组成部分，事关打造人类命运共同体和管理经营全球公共财富，对全球政治、经济、安全治理具有显著的外溢效应。在政治上涉及全球事务的主导权和话语权、影响力；在经济上，涉及未来世界经济的总体趋势和面貌、贸易投资与金融的标准和竞争力；在安全上由于气候负面影响的巨大损失以至于成为非传统安全的重要内容。气候变化是横跨全球治理各主要领域的综合性议题。

七是《联合国气候变化框架公约》多边进程是实施全球气候治理的主渠道，为全球气候治理奠定了道义的、政治的和法律的基础。《公约》所确定的公平原则、"共区"原则和各自能力原则以及其他原则与条款，依然适用于指导今天及今后一个很长时期全球气候治理规则的制定。尽管在《公约》签署 20 多年以来世界形势与战略格局发生了重大的变化，发展中国家在全球人口、GDP、能耗和排放等的占比持续上升，但基于对累积排放、发展阶段和发展水平、能力等指标的实证检验表明，《公约》上述原则所依据的科学事实并未改变，发达国家与发展中国家在历史责任、发展阶段、能力和国情等方面的差别依然显著甚至有加大的趋势。《公约》原则和条款依然是指导"巴黎协议"谈判的指针。

八是正确理解和设置全球气候治理的目标。基于对全球气候变化科学原理、历史由来与未来趋势的科学评估和判断，需要在科学认知上、技术和产业发展方向上、政治意愿上确认全球和国别社会经济发展的低碳方向与低碳转型的紧迫性。同时，又要充分考虑科学认识水平的历史局限性和主要依据模型模拟技术而得出的数量分析结果的不确定性，以及社会经济系统决策的复杂性和进一步叠加的不确定性。不能简单机械地将具有很大不确定性的科学评估数据结果直接作为长期目标谈判的数量依据以致陷入"数据游戏"和"谴责博弈"的僵局，反而影响了达成共识的进程。可行的做法是在长期目标、愿景的性质和方向上达成共识，避开难以达成共识的数量争论，给各个缔约方以自主决定目标数量程度的自主权，促其尽快启动行动，助推并加速技术和市场的变动，从而在一段时间内形成更新和建立新的目标的现实基础。全球气候治理的目标必然是一个不断实施、科学与技术和市场信息不断反馈并得到评估、动态滚动更新的实施与确立的交互过程。此外，目标应当是一个一揽子目标，包括最终的减缓与适应目标及针对影响温室气体排放等驱动因素的、可直接指导行动的过程目标。

九是在“自上而下”与“自下而上”管理模式上寻找恰当的平衡和适用内容。经过《京都议定书》的实践并考虑美国国内政治因素对国际多边气候进程的影响，对各缔约方减排和资金支持目标承诺的“自上而下”的治理模式，在没有超主权国际强制力的情况下陷入困境。在新一轮的多边气候协议谈判过程中，作为“自下而上”管理模式标志的“国家自主贡献”目标在2013年华沙决定中应运而生，且为大多数缔约方所接受。但围绕着旨在对“国家自主贡献”目标进行周期性全球盘点、审评乃至影响进一步更新的“自上而下”的安排，依然是一个争论的焦点有待在“巴黎协议”中得到安排。透明度问题也被视为适用于“自上而下”管理模式的领域。此外，法律强制约束力适用的内容、法律形式等，也是落实全球气候治理中需要解决的问题。在兼顾全球治理的有效性和可行性的情况下，一种可行的治理模式是：在缔约方减排目标与责任的承诺问题上可更多采取“自下而上”的模式，而在涉及目标更新的盘点、审评及核算规则与透明度及议事规则和决策程序等相关问题上，在《公约》原则与条款的基础上，更多采取“自上而下”的模式，以此形成全球气候治理的“混合模式”。

十是平衡处理好大国影响与广泛参与之间的辩证关系。全球气候治理的国际政治运作基础，在当前只能是战后建立的“联合国”这一处理多边国际事务的基本国际政治秩序，《公约》决策规则的核心在于“协商一致”原则。它要求各个缔约国在决策中具有平等的地位，要求充分考虑包括公民社会在内的各方诉求，要求广泛参与的国际民主和旨在加强政治互信的透明。大国需要发挥带头作用和示范作用，将其所具有的更大实际影响力作为对全球气候治理的推动力量。应平衡处理好大国基于带头示范作用的较大影响力和广大中小缔约方广泛参与之间的辩证关系，以保证多边国际气候进程得以公正有效地推进。要切实加强发展中国家参与全球气候治理的能力，在议题设置、协议内容要素安排等问题上满足在资金、技术和能力建设方面对发展中国家提供援助的要求。

十一是中国在全球气候治理中要坚持发展中国家定位，但同时意识到其大国和转型的特殊性。尽管新中国成立以来特别是改革开放以来中国的社会经济发展取得了惊人的成就，步入中上等收入国家行列，正在迈向全面建成小康社会的发展阶段，但从中国社会经济依然存在的“二元结构”特征、按照国际标准定义的中等收入人群占比和贫困人口占比以及社会保障和公共服务的覆盖质量，加之中国在全球产业链中的中低端定位和尚十分薄弱的自主创新能力与其他软实力水平看，中国依然是一个发展中国家，在全球治理中的责任依然由其作为一个发展中国家的历史责任、发展阶段、能力和国情所决定。但同时也应当意识到，中国与一般的多数发展中国家相比又有两个基本不同的特征，一个是大国地位，人口和年度排放居世界第一，经济总量居世界第二，其政策和行动具有“举国效应”；另外一个是在发展中国家阵营内部具有较高的发展水平，正在进入向发达经济体转型的发展阶段。这个转型表现在价值观念、体制、技术、市场、生产与消费方式等诸多方面。这样的特征也必然要反映到中国在全球气候治理中的立场主张的发展之中。

十二是中国的特殊定位决定了中国要以负责任的发展中大国的身份建设性地参与并逐步寻求主导全球气候治理，促进形成公正、合理、有效、共赢的全球气候治理体系。中国一方面要积极主动地承担与自身发展阶段、能力和国情相适应的国际责任，最大限度地实现自身经济发展“转方式、调结构”、“提质升级”、克服中等收入陷阱的战略诉求与全球气候保护要求的一致性；另一方面要主持正义公正，捍卫发展中国家的利益，坚持与发展中国家形成战略同盟及南南合作；同时在合作共赢的基础上积极主动地加强与发达国家的务实合作，促进南北合作。在推进全球气候治理的过程中，中国应当坚持原则，坚持要求发达国家率先减排和援助发展中国家的要求，坚持《公约》的原则和条款，并采取务实的态度，寻求更加广泛的共识和平衡的争议解决方案。

作者简介

Biography

邹骥 ZOU Ji

邹骥教授分别于 1984 年和 1990 年在清华大学获得环境工程学士学位和技术经济硕士学位；1997 年在中国人民大学获得人口、资源与环境经济学博士学位。2001 年至 2012 年任中国人民大学环境学院副院长，教授和博士生导师。2009 年至 2012 年期间，还曾任世界资源研究所中国区首席代表。他曾任政府间气候变化专门委员会（IPCC）第三工作组第四次评估报告的主要作者，IPCC 第三工作组第五次评估报告的主要协调作者（CLA）。2013—2014 年他出任联合国政府间可持续发展融资专家委员会委员。2000—2009 年和 2012 年至今，他作为中国政府代表团成员参加联合国气候谈判，并在多边谈判和对话中担任主持人或联合主持人。2012 年以来，他出任中国国家发展和改革委员会国家应对气候变化战略研究和国际合作中心副主任，研究员。

邹骥教授长期从事能源与气候变化政策经济分析领域研究工作，主攻技术变动和国际气候体制经济分析、可持续城市规划等研究方向。他先后主持完成了中国低碳发展宏观战略国际环境研究项目和技术战略研究项目，财政部“金砖银行气候变化战略研究”，中国清洁发展机制基金赠款项目“IPCC 第五次评估报告第三工作组支撑研究”和“关于德班平台重大战略问题和总体战略研究”等研究项目。主要著作和文章有《论全球气候治理——构建人类发展路径创新的国际体制》，《2009/2010 中国人类发展报告：迈向低碳经济和社会的可持续未来》，解读“中国国家自主贡献”（INDC）目标等。

Professor ZOU Ji was awarded a B.S. in Environmental Engineering (1984), a M.S. in Engineering Economics (1990) at Tsinghua University, and a Ph.D. in Population, Resources and Environmental economics (1997) at Renmin University of China (RUC). He served as Deputy Dean of School of Environment and Natural Resource at RUC during 2001-2012. He worked for World Resources Institute (WRI) as China Country Director from 2009 to 2012. He was a Lead Author of Working Group III for the Fourth Assessment Report of

Intergovernmental Penal on Climate Change (IPCC), and he served as a Coordinating Lead Author (CLA) for the Fifth Assessment Report of IPCC Working Group III. He had been a member of the UN Intergovernmental Committee of Experts on Sustainable Development Financing from 2013-2014. He was appointed as a delegate of China for UN Climate Talks during 2000 - 2009 and continued to be since 2012, and worked as facilitator or co-facilitator for several issues in multilateral negotiation and dialogue. Since 2012, he has served as the Deputy Director General of the National Center for Climate Change Strategy and International Cooperation (NCSC) under China National Development and Reform Commission.

Professor ZOU has been working in areas of economic analysis of energy and climate policies with focuses on technology changes, international climate architecture and sustainable city planning, etc. He is the team leader for projects including "Study on International Environment" , "Technology Strategy Analysis" sub-projects under "China Macro Strategy for Low Carbon Development Program" ; "Study on Climate Strategy for New Development Bank (BRICS)" with Department of Finance; "Study on Major Strategic Issues and Overall Strategy of Durban Platform" , "Technical Supporting Project for the Fifth Assessment Report of IPCC Working Group Ⅲ" . His major publications includes "Global Governance for Climate: Shaping an International Architecture to Boost Innovation of Human Development Paths" , "China Human Development Report 2009/10: Towards a Low Carbon Economy and Society" , "An Analysis of China' s INDC" , etc.

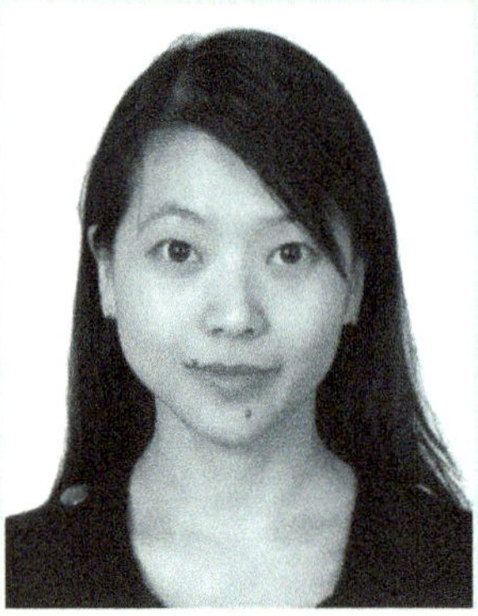

傅莎 FU Sha

傅莎，博士，助理研究员。毕业于中国人民大学人口、资源和环境经济学专业，2005年、2007年和2012年分别获得管理学学士、经济学硕士和经济学博士学位。2012年进入国家应对气候变化战略研究和国际合作中心国际部工作。长期从事能源与气候变化领域的研究工作，涉及气候变化国际谈判、国际气候体制研究、能源-环境-经济系统模型构建和分析，能源和GHG减排技术经济评价、国家和地方层面的能源和低碳发展战略研究、能源与气候政策的经济影响分析等领域。她是中国谈判代表团成员，重点跟踪长期目标和峰值、审评、德班平台的谈判。她还是联合国气候变化框架公约下的技术执行委员会委员和联合国环境规划署（UNEP）2015年减排差距报告的主要作者。近年来作为核心成员参与了多项能源和气候变化研究领域的国家重点课题和国际合作课题。

FU Sha，Ph.D, Assistant Professor. She is graduated from Renmin University and major in Human, Natural resources and Environmental economics. She has got her B.A, M.Sc and Ph.D. degree in 2005, 2007 and 2011 respectively. She joined National Center for Climate Change Strategy and International Cooperation (NCSC) in year 2012, and now a member of Department of International Cooperation. With long experience as a researcher and policy analyst focusing on energy and climate change issue in China, her area of specification is as follows: (1) Energy-Economic-Environment System Modeling and Analysis; (2) National and local Strategy Design for Energy and GHG Emission Mitigation; (3) International Climate architecture study, especially Post 2020 International Climate Regime; (4) Assessment of social economic impact of Climate polices; (5) Technology Roadmap Development and Needs Assessment. She is a member of Chinese delegation to UNFCCC Climate negotiation since 2012. She also serves as member of Technology Executive Committee under the UNFCCC and core lead author of UNEP GAP report 2015. In recent years, as a key team member, She has participated and coordinated several national and international studies in the field of energy and climate change.

陈济 CHEN Ji

陈济于2009年毕业于英国邓迪大学，获得能源与环境经济博士学位。在获得博士学位前，他分别在英国约克大学和华中科技大学获得环境经济学和建筑环境与设备专业的硕士和学士学位。在英国留学期间，他主要从事电力市场改革、排污权交易等领域的经济和政策研究。回国后，开始从事工业节能和能效政策的研究以及节能技术应用分析和推广工作。2010年至今，他主要关注国际气候变化政策和体制，并作为中国代表团成员参与气候变化国际谈判。近年来，他主要关注气候友好技术的开发与转让，从事气候技术转让、低碳产业战略与政策的研究。他目前担任世行气候技术需求评估项目项目办主任，是联合国应对气候变化框架公约下气候中心与网络（CTCN）的中方联络人。

Dr. CHEN Ji is an energy and environmental policy analyst. He holds PhD in Energy and Environment Policy and Economics from Dundee University (UK), MSc in Environmental Economics from York University (UK) and BEng in Building Service Engineering from Huazhong Science and Technology University (China). He had diversified work experience in energy and climate change field, including research institution, consulting firm and government agency. He is now an assisting professor of the National Centre for Climate Change Strategy and International Cooperation (NCSC), focusing mainly on climate-related technologies and policies. While conducting researches, he has also been serving as a negotiator representing the Chinese government to participate in the UNFCCC process since 2010. He is the director of the Project Management Office for China's Technology Needs Assessment project funded by Global Environment Facility and China's contact person of the Climate Technology Centre and Network (CTCN).

祁悦 QI Yue

祁悦，2007 年毕业于北京大学环境学院，获得理学学士学位，2013 年毕业于中国科学院地理科学与资源研究所，获得理学博士学位，2010 年至 2012 年参加德国马普学会 - 中国科学院联合培养博士生项目，在德国马普化学所完成博士论文的主要工作。攻读学位期间主要的工作包括：陆地生态系统碳汇和碳蓄积服务及其价值评估、碳排放空间分配、碳足迹测算、地基和卫星光谱观测大气成分、干旱区土壤氮氧化物排放特征等。2013 年 7 月加入国家应对气候变化战略研究和国际合作中心国际合作部，主要关注应对气候变化国际机制、国际气候谈判战略、城市适应气候变化、低碳政策国际比较、气候变化南南合作等领域的研究，并作为中国代表团成员参与气候变化国际谈判，也曾作为技术专家参与发达国家缔约方国家信息通报的技术审评。

Dr. QI Yue graduated from the College of Environmental Sciences of Peking University in 2007. From 2007 to 2013, she studied in the Institute of Geographic Sciences and Natural Resources Research and participated in the Joint Doctoral Training Programme between the Chinese Academy of Sciences and Max-Planck Institute. Her research mainly focused on the evaluation of ecosystem services, particularly carbon sink service, assessment of carbon footprint, and ground-based and satellite remote sensing of atmospheric composition. Since 2013, she joined the national center for climate change strategy and international cooperation. Currently, her work mainly focus on the UNFCCC negotiation process, global governance on climate change, south-south cooperation on climate change, urban adaptation, and comparison of low-carbon policies, etc. She is also a member of Chinese delegation to the UNFCCC negotiation, and has served as a member of technical expert team for the technical review of national communications.

王克 WANG Ke

王克博士，任教于中国人民大学环境学院，是中国人民大学能源与气候经济学项目执行主任，中国人民大学国家发展与战略研究院研究员。曾就读于中国人民大学和清华大学。主要研究和工作领域为能源环境经济模型构建与政策模拟、国际气候制度的经济分析、低碳技术需求评估与低碳技术发展战略、低碳城市规划等。曾受聘担任联合国气候变化框架公约（UNFCCC）秘书处、联合国气候变化框架公约技术转让专家小组（EGTT）、世界银行、亚洲开发银行、联合国开发计划署、瑞士发展援助署等国际机构的咨询专家。担任亚洲开发银行技术援助项目“青岛低碳发展战略与规划研究（TA 7219-PRC）”的专家组副组长，支持青岛、贵阳等国家低碳试点城市编制低碳发展规划。他还是国家发展改革委领导的由世行执行的全球环境基金中国技术需求评估项目的减缓技术需求评估专家。作为课题和子课题负责人多次参与气候变化领域的重大课题，包括国家973计划、国家科技支撑计划、教育部人文社会科学重大攻关项目、中国低碳宏观战略项目等。他是联合国开发计划署和中国人民大学联合编写的《中国人类发展报告2009/10：迈向低碳经济和社会的可持续未来》的主要作者和项目协调人。

Dr. WANG Ke is an assistant professor of Environmental Economics at School of Environment and Natural Resources, Renmin University of China (RUC), He is also the executive director of Programme of Energy and Climate Economics, research fellow of National Academy of Development and Strategy. He got his bachelor degrees both from Renmin University and Tsinghua University, and doctor’s degree from Tsinghua University.

His research areas focus on energy-environment-economic modeling and policy simulation, economics analysis of international climate change regime, low carbon technology innovation strategy and promotion policy, low carbon municipal planning, etc.

He provided technical support for Chinese UNFCCC delegation focusing on technology transfer issues. He also provided consultancy services to UNFCCC secretariat, UNFCCC Expert Group on Technology Transfer (EGTT), World Bank, Asian Development Bank, UNDP and Swiss Agency for Development and Cooperation (SDC) etc. He was the leader author and coordinator of " China Human Development Report 2009/10: China and a Sustainable Future Towards a Low Carbon Economy and Society" jointly published by UNDP China and Renmin University of China. He served as the deputy team leader for ADB Technical Assistance project to support Qingdao to develop low carbon strategy and blueprint and supported Qingdao to develop "Qingdao's low carbon development plan (2014-2020)" . He is also the mitigation expert for GEF China Technology Needs Assessment project lead by NDRC and implemented by World Bank. As the project and branch project leader, he attended many key projects on climate change area, such as Key Program on China's low carbon macro strategy, National 973 Plan, National Key Technology Support Program, The Major Research Project on Humanities and Social Science under Ministry of Education.

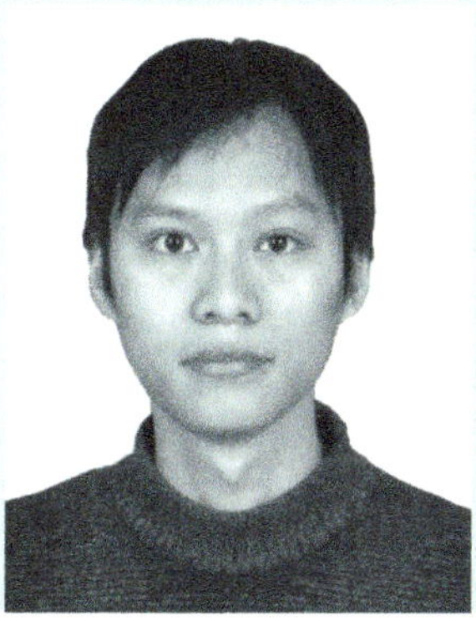

崔学勤 CUI Xueqin

崔学勤，中国人民大学环境学院博士研究生。在攻读博士学位之前，分别于2009年和2011年毕业于中国人民大学环境学院，获得管理学学士和经济学硕士学位。2014年9月至2015年1月在意大利 Fondazione Eni Enrico Mattei (FEEM) 研究所访问学习，开展关于全球碳预算公平分配的联合研究。在气候变化和低碳发展领域的主要工作包括：(1) 全球碳预算空间公平分配研究；(2) 低碳技术发展战略；(3) 温室气体核算与驱动因素分析；(4) 低碳城市规划等。在硕博期间，参与了"中国人类发展报告2009/10：迈向低碳经济和社会的可持续未来"、中国低碳发展宏观战略、世界银行"支持制定中长期减排目标政策分析"项目、973项目"气候变化谈判中的技术转让机制研究"、青岛市低碳发展规划等研究课题。

CUI Xueqin is a Ph.D candidate at School of Environment & Natural Resources in Renmin University of China. Prior to starting his Ph.D. education, Xueqin received his Master Degree of Economics and Bachelor Degree of Management from Renmin University. Cui Xueqin worked and studied at Fondazione Eni Enrico Mattei (FEEM) in Milan, Italy from September 2014 to January 2015 as visiting student, and conducted a joint research on equitable share of future carbon budget among regions and countries. Xueqin has been focusing on energy and climate change economics, Including fair share of global carbon budget, GHG accounting and driving force analysis, low-carbon technology development strategy, low-carbon city development planning. During his Ph.D period, Xueqin has been involving in many research projects, including China Human Development Report 2009/10, China Macro Low-Carbon Development Strategy, World Bank Mid and Long Term Mitigation Target Setting, City Level GHG Inventory Methodologies in China, Technology Transfer Mechanism in Climate Change Negotiation (973 Project), and Low Carbon Blueprint in Qingdao and Guiyang, etc.

刘俊伶 LIU Junling

刘俊伶，中国人民大学环境学院硕博连读研究生，2015 年 1 月— 8 月参加“国家建设高水平大学公派研究生项目”，在法国巴黎的国际能源署进行博士生联合培养与联合研究，参与完成“世界能源展望 2015”报告以及“气候变化特别报告”。主要研究方向包括：(1) 能源环境经济模型构建与政策模拟；(2) 能源与气候变化战略与政策研究；(3) 多边投入产出分析及全球贸易隐含碳研究；(4) 低碳城市规划。在硕博期间，参与了中国低碳发展宏观战略国际环境、中国低碳发展宏观战略技术战略等国家低碳宏观战略研究课题、973“气候变化谈判中的技术转让机制研究”等国家重点基础研究课题，以及亚行、英国使馆、瑞士使馆等基金资助的青岛、贵阳等多个城市低碳发展规划研究课题。

LIU Junling is a Ph.D candidate at School of Environment & Natural Resources in Renmin University of China. She studied at IEA (International Energy Agency) in a State-Sponsored Graduate Scholarship Program for Building High-level Universities in 2015 for 7 months and participated in the research work of “World Energy Outlook 2015” and “WEO 2015 Special Report on Energy and Climate Change” . LIU Junling has focused her research on: Energy-Environment-Economy modeling and policy simulation; Energy and Climate Strategy and policy study; multi-region input-output analysis and trade embodied carbon emissions calculation; low-carbon city development planning. During her Ph.D period, she has been involved in National Macro Low-Carbon Strategic Studies Programs including “Study on International Environment for China Macro Low-Carbon Development Strategy” and “Technology Strategy Study for China Macro Low-Carbon Development Strategy” . She also participated in some National Key Basic Research Programs (973 Program) such as “Study on Technology Transfer Mechanism for Climate Change Negotiation” , and many projects on city-level low carbon development planning sponsored by various funds from Asian Development Bank, British Embassy, Swiss Embassy, etc.

张小丽 ZHANG Xiaoli

张小丽，中国人民大学环境学院硕士研究生。中国人民大学能源与气候经济学项目组成员之一。2014 年毕业于辽宁大学经济学院，获得经济学学士学位。主要研究方向包括：能源环境经济模型构建与政策模拟，能源与气候变化战略与政策研究，低碳城市规划。硕士期间，参与世界银行“支持制定中长期减排目标政策的分析”项目，城市低碳发展规划等课题的研究工作。

ZHANG Xiaoli is a post-graduate candidate at School of Environmental & Natural Resources in Renmin University of China. She is also one of the members of Programme of Energy and Climate Economics. She received her Bachelor Degree of Economics from the School of Economics, Liaoning University in 2014. ZHANG Xiaoli has focused her research on: Energy-Environment-Economy modeling and policy simulation; Energy and Climate Strategy and policy study; low-carbon city development planning. During her post-graduate period, she has been involved in World Bank “Mid and Long Term Mitigation Target Setting”, and projects on city-level low carbon development planning, etc.